TOPOLOGY AND DYNAMICS OF CHAOS
In Celebration of Robert Gilmore's 70th Birthday

WORLD SCIENTIFIC SERIES ON NONLINEAR SCIENCE

Editor: Leon O. Chua
University of California, Berkeley

Series A.   MONOGRAPHS AND TREATISES*

Volume 68: A Nonlinear Dynamics Perspective of Wolfram's New Kind of Science
(Volume III)
L. O. Chua

Volume 69: Modeling by Nonlinear Differential Equations
P. E. Phillipson & P. Schuster

Volume 70: Bifurcations in Piecewise-Smooth Continuous Systems
D. J. Warwick Simpson

Volume 71: A Practical Guide for Studying Chua's Circuits
R. Kiliç

Volume 72: Fractional Order Systems: Modeling and Control Applications
R. Caponetto, G. Dongola, L. Fortuna & I. Petráš

Volume 73: 2-D Quadratic Maps and 3-D ODE Systems: A Rigorous Approach
E. Zeraoulia & J. C. Sprott

Volume 74: Physarum Machines: Computers from Slime Mould
A. Adamatzky

Volume 75: Discrete Systems with Memory
R. Alonso-Sanz

Volume 76: A Nonlinear Dynamics Perspective of Wolfram's New Kind of Science
(Volume IV)
L. O. Chua

Volume 77: Mathematical Mechanics: From Particle to Muscle
E. D. Cooper

Volume 78: Qualitative and Asymptotic Analysis of Differential Equations
with Random Perturbations
A. M. Samoilenko & O. Stanzhytskyi

Volume 79: Robust Chaos and Its Applications
Z. Elhadj & J. C. Sprott

Volume 80: A Nonlinear Dynamics Perspective of Wolfram's New Kind of Science
(Volume V)
L. O. Chua

Volume 81: Chaos in Nature
C. Letellier

Volume 82: Development of Memristor Based Circuits
H. H.-C. Iu & A. L. Fitch

Volume 83: Advances in Wave Turbulence
V. Shrira & S. Nazarenko

Volume 84: Topology and Dynamics of Chaos:
In Celebration of Robert Gilmore's 70th Birthday
C. Letellier & R. Gilmore

*To view the complete list of the published volumes in the series, please visit:
http://www.worldscientific.com/series/wssnsa

WORLD SCIENTIFIC SERIES ON NONLINEAR SCIENCE

Series Editor: Leon O. Chua

Series A   Vol. 84

TOPOLOGY AND DYNAMICS OF CHAOS

In Celebration of Robert Gilmore's 70th Birthday

edited by

Christophe Letellier
CORIA — University of Rouen, France

Robert Gilmore
Drexel University, USA

NEW JERSEY · LONDON · SINGAPORE · BEIJING · SHANGHAI · HONG KONG · TAIPEI · CHENNAI

Published by

World Scientific Publishing Co. Pte. Ltd.
5 Toh Tuck Link, Singapore 596224
USA office: 27 Warren Street, Suite 401-402, Hackensack, NJ 07601
UK office: 57 Shelton Street, Covent Garden, London WC2H 9HE

British Library Cataloguing-in-Publication Data
A catalogue record for this book is available from the British Library.

World Scientific Series on Nonliear Science, Series A — Vol. 84
TOPOLOGY AND DYNAMICS OF CHAOS
In Celebration of Robert Gilmore's 70th Birthday

Copyright © 2013 by World Scientific Publishing Co. Pte. Ltd.

All rights reserved. This book, or parts thereof, may not be reproduced in any form or by any means, electronic or mechanical, including photocopying, recording or any information storage and retrieval system now known or to be invented, without written permission from the Publisher.

For photocopying of material in this volume, please pay a copying fee through the Copyright Clearance Center, Inc., 222 Rosewood Drive, Danvers, MA 01923, USA. In this case permission to photocopy is not required from the publisher.

ISBN 978-981-4434-85-0

Printed in Singapore.

Preface

I met Robert Gilmore by chance while I was visiting Pierre Glorieux's group in Lille. It was around 1996. That was also the first meeting with Marc Lefranc. All these names and many of the contributors were already familiar to me from their papers I read during my Ph.D thesis which was devoted to the topology of chaotic attractors. Very quickly I started to interact with Robert Gilmore about chaotic attractors with symmetries. Today, Robert and Claire Gilmore are spending a month in Rouen every year. This is always a great opportunity to have stimulative discussions, not only about the topology and the symmetry of chaos, but also about problems related to fundamental physics. This is also an opportunity to learn about classical music, Bach especially, and the métro of Paris, one of Bob's main interests.

In June 2011, it was a wonderful opportunity to thank Bob for all these warm interactions and to have a workshop mostly devoted to the topological analysis of chaos. It was also an opportunity to have a "family meeting", most of the contributors having had the joy to learn with Bob. We were also pleased to have with us Joan Birman, Otto E. Rössler, Christian Mira and René Lozi. This book provides an insight into the science which was discussed. The rest is in our memory and cannot be rended in a book!

Christophe Letellier
Rouen, October 2012

Participants to the International Workshop "from Laser dynamics to topology of chaos". June 28, 2011.

From left to right: Christophe Letellier, Ricardo Lopez, Mario Natiello, Hernan Solari, Emad Yacoub, Jorge Tredicce, Ludomir Oteski, Luc Pastur, Juan Carlos Martin, Keith Gilmore, Claire Gilmore, Robert Gilmore, Jacob Katriel, François Lusseyran, Joan Birman, Valérie Messager, Daniel Cross, Martin Rosalie, Joe Birman, Sifeu Takougang Kingni, Adrien Kerfourn, Christian Mira, Elise Roulin, Amélie Danlos, Roomila Naeck, Dounia Bounoiare, Claudia Lainscsek, Cristina Sadi Rochas da Coelho, Marc Lefranc, Xinzhi Liu, Sylvain Mangiarotti, Aziz Alaoui, Otto E. Rössler, René Lozi, Ubiratan Freitas.

Contents

Preface	v
1. Introduction to topological analysis *Christophe Letellier & Robert Gilmore*	1
Emergence of a Chaos Theory	**21**
2. The peregrinations of Poincaré *R. Abraham*	23
3. A Toulouse research group in the "prehistoric" times of chaotic dynamics *Christian Mira*	39
4. Can we trust in numerical computations of chaotic solutions of dynamical systems ? *René Lozi*	63
5. Chaos hierarchy — A review, thirty years later *Otto E. Rössler & Christophe Letellier*	99
Development of the Topology of Chaos	**125**
6. The mathematics of Lorenz knots *Joan S. Birman*	127

7. A braided view of a knotty story — 149
   Mario Natiello & Hernán Solari

8. How topology came to chaos — 169
   Robert Gilmore

9. Reflections from the fourth dimension — 205
   Marc Lefranc

10. The symmetry of chaos — 227
    Christophe Letellier

Applications of Chaos Theory — 249

11. The shape of ocean color — 251
    Nicholas Tufillaro

12. Low dimensional dynamics in biological motor patterns — 269
    Gabriel B. Mindlin

13. Minimal smooth chaotic flows — 281
    Jean-Marc Malasoma

14. The chaotic marriage of physics and financial economics — 303
    Claire Gilmore

15. Introduction of the sphere map with application to spin-torque nano-oscillators — 317
    Keith Gilmore & Robert Gilmore

16. Robert Gilmore, a portrait — 331
    Hernán G Solari

Author Index — 341

Subject Index — 345

Chapter 1

Introduction to topological analysis

Christophe Letellier

CORIA UMR 6614 - Rouen University
F-76801 Saint-Etienne du Rouvray cedex
France

Robert Gilmore

Physics Department
Drexel University
Philadelphia, PA 19104, USA

The topological analysis of chaotic attractors is briefly reviewed. The main concept on which this book is based is thus introduced with a simple case (the attractor solution to the Sprott D system) explicitly treated. Then some perspectives to extend the topological analysis are discussed.

Contents

1. Strange versus Chaotic 1
2. Topological Analysis: the Program 4
3. Topological Analysis: an Explicit Case 8
4. Classification 12
5. Hopes for the Future 14
References 16

1. Strange versus Chaotic

In the mid of 1970's, the word chaos designated aperiodic solutions that could not be described more accurately. In its etymologic sense, chaos corresponds to the indescribable state of the World before God organized it. "Chaos" was thus used to express that there are solutions which were not possible to describe. If this choice was justified in the beginning (i.e. in the 1970's), this is no longer the case now. The first two parts of this book are devoted to the topological analysis whose aim is to provide a description of the object in the neighborbood of which a "chaotic" solution is organized. In fact, in the 1970's, three different terms were commonly used to designate what is now called "chaos". "Turbulence" was mostly used by physicists, following David Ruelle and Floris Takens' paper.[1] Although being a mathematician,

Ruelle spent much effort to promote the idea that some "complicated" behaviors could be produced by rather simple systems: he thus triggered the contribution of Harry Swinney's group to experimental evidences of chaotic behaviors in fluid mechanics and chemical reactions.[2,3] Mathematicians felt more comfortable with the term "aperiodic" as used by Edward Lorenz,[4] Mary-Lucy Cartwright & John Littlewood,[5] Norman Levinson,[6] etc. The term "chaotic" was used by a young generation of scientists with hybrid background and quite acquainted with numerical simulations: let us mention Otto E. Rössler, the Santa Fe group (Doyne Farmer, James Crutchfield, Robert Shaw, Norman Packard), among others. At that time, chaotic was often confused with "strange" inherited from Ruelle and Takens' paper.[1]

The terms "strange" and "chaotic" were used equivalently at least up 1984, that is, up to the article "strange attractors that are not chaotic" by Celso Grebogi, Edward Ott, Steve Pelikan and James Yorke.[7] These scientists choose to use these two words as follows:

> *Chaotic* refers to the dynamics on the attractor, while *strange* refers to the geometrical structure of the attractor. [...]
>
> *Definition.* A chaotic attractor is one for which typical orbits on the attractor have a positive Lyapunov exponent [a].

Chaotic means "sensitive to initial conditions" and strange is associated with the property announced by Ruelle and Takens,[1] that is, an "attractor which is locally the product of a Cantor set and a piece of two-dimensional manifold". Thus, although Guckenheimer and coworkers used strange, their formal definition can be taken for chaotic. Replacing "strange" with "chaotic" in the original text,[8] we get

> By a "chaotic attractor", for a map $f(\cdot)$ we mean an infinite set Λ with the following properties:
>
> (1) Λ is invariant under $f(\cdot)$, i.e. $f(\Lambda) = \Lambda$.
> (2) Λ has an orbit which is dense in Λ.
> (3) Λ has a neighborhood a consisting of points whose orbits tend asymptotically to Λ : $\lim_{t \to \infty} f^{(t)}(a) \subset \Lambda$.
>
> The requirement that Λ be infinite guarantees that Λ consists of more than a single periodic orbit.
>
> The central — and startling — fact about chaotic attractors is that orbits on or near them may behave in an essentially [...] unpredictable fashion. Thus, despite the fact that the model is completely deterministic, the dynamical behavior of trajectories can only be predicted statistically!

Items (1) and (3) correspond to the definition of an attractor. Item (2) implies that the attractor is not trivial, that is, a singular (or fixed) point or a limit cycle. To distinguish an arbitrarily long periodic orbit from an aperiodic one, Guckenheimer and

---

[a]From this single definition, many wrong conclusions were published about "chaotic" experimental time series since this positive Lyapunov exponent is not sufficient to ensure the chaoticity of the dynamics underlying a given data set.

co-workers added the requirement for Λ infinite. As René Lozi discusses in chapter 4, distinguishing an aperiodic orbit from an arbitrarily long periodic orbit seems to be impossible when numerical simulations are considered.[9] Moreover, when experimental data are investigated, the unavoidable noise contamination clearly rends impossible such a distinction. Some indirect necessary conditions were introduced: the asymptotic trajectory must have at least one positive Lyapunov exponent or must be characterized by a fractal dimension; the first-return map to a Poincaré section must satisfy the Li-Yorke theorem[10] as used by Rössler (see chapter 5). Unfortunately, none of these conditions can lead to a conclusive statement. The two former are only necessary but not sufficient conditions: the Li-Yorke theorem only implies that the existence of a period-3 orbit forces the existence of at least one solution of any given period. In particular, it does not forbid the existence of a limit cycle among a collection of unstable periodic orbits, with at least one of any given period.

Showing that a first-return map of an interval contains a period-3 orbit guarantees that any period is realized in such a map. But there is no rigorous proof that the solution has not an arbitrarily long period (see Lozi's chapter, 4). The proof can be easily provided at a certain error level ϵ. Physicists are happy when ϵ is less than the noise-to-signal ratio but, most of the time, mathematicians remain frustated. How can we proceed in practice? The easier situation to manage is when a well-known route to chaos can be identified in the system. For instance, if your system — experimental or numerical — presents a period-doubling cascade and a smooth unimodal map (like the logistic map), then the existence of a chaotic behavior is guaranteed with its related underlying determinism. When such a route to chaos is not identified, the problem is more difficult. As we already mentioned, a positive Lyapunov dimension and a fractal dimension are only necessary conditions and, worse, their computations are quite sensitive to noise contamination. There are the "well-known enemies" to use a terminology introduced by Hernan Solari, Mario Natiello and Gabriel Mindlin.[11] Consequently, as pointed out by Leon Glass,[12] before any statement about the chaotic nature of a behavior, some proof for the existence of a determinism must be provided. The deterministic nature is obvious when the studied dynamics is produced by numerical simulation of differential equations or discrete maps, but this is a rather difficult task when experimental data are investigated. From our point of view, global modeling, that is, getting a set of differential, difference or discrete equations which reproduces the observed attractor, is the best technique to obtain such a proof (see[13] for a review). Once there is a proof for an underlying determinism, a topological analysis can be attempted when the measured data are long enough (typically one hundred of cycles with at least 20 data points per oscillation). The requirement for having at least one hundred intersections with a Poincaré section is induced by the fact that a topological analysis is based on the population of (unstable) periodic orbits embedded within the attractor. As pointed out by Henri Poincaré,[14] one could always choose a periodic solution as a first approximation of an aperiodic orbit. Much later, came the idea that these periodic orbits are the skeleton around which a chaotic orbit is organized.[15-17] Then as performed for the first time by Gabriel Mindlin and Robert Gilmore,[18] the relative organization of the unstable periodic orbits is investigated in terms of linking numbers, relative rotation rates, etc. and then synthesized under the form of a branched manifold, also called a knot-holder or a template (see Joan Birman's chapter).

2. Topological Analysis: the Program

The driving force behind the development of the topological analysis program was the analysis and understanding of data from a chaotic dynamical system.[19] Once we understood that periodic orbits could be considered as the skeleton of a chaotic attractor,[15-17] our direction was set. First we had to convince ourselves that whatever mechanism created the attractor simultaneously created and organized all the unstable period orbits in it. This was easy. Then we had to decide how it could be possible to describe the organization of the unstable periodic orbits. After several false starts we found two successful routes.[20-23] The first, and most general, involved computing sets of linking numbers for pairs of unstable periodic orbits in the attractor. This was done using the Gauss linking integral introduced by Carl Friedrich Gauss in a brief note on a page of his personal diary dated January 22, 1833

> Of the *geometria situs*, which was forseen by Leibnitz, and into which only a pair of geometers (Euler and Vandermonde) were granted a bare glimpse, we know and have, after a century and a half, little more than nothing.
>
> A principal problem at the interface of *geometria situs* and *geometria magnitudinis* will be to count the intertwinings of two closed or endless curves.
>
> Let x, y, z be the coordinates of an undetermined point on the first curve; x', y' z' those of a point on the second and let
>
> $$\int\int \frac{(x'-x)(dydz'-dzdy')(y'-y)(dzdx'-dxdz')(z'-z)(dxdy'-dydx')}{[(x'-x)^2+(y'-y)^2+(z'-z)^2]^{\frac{3}{2}}} = V$$
>
> then this integral taken along both curves is $= 4m\pi$, m being the number of intertwinings.
>
> The value is reciprocal, that is, it remains the same if the curves are interchanged.

This result was not published elsewhere than in the fifth volume of Gauss' *Werke* (1867). That year, James Clerk Maxwell wrote a letter to Peter Tait dated December 4, 1867 in wich he rediscovered the Gauss integral, apparently unaware of Gauss' note.[25] In Section 420 of his treatise on electromagnetism,[26] he published his derivation of Gauss' integral. In the next section, he added

If [two curves] are intertwined n times in the same direction, the value of the integral is $4\pi n$. It is possible, however, for two curves to be intertwined alternately in opposite directions, so that they are inseparatly linked together though the integral is zero (Fig. 1).

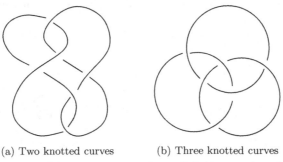

(a) Two knotted curves (b) Three knotted curves

Fig. 1. Two (a) and three (b) linked curves with a null linking number. Examples drawn by Maxwell in his letter to Tait (1867). Only case (a) was inserted in his book (1873).

There is an equivalent way for estimating the linking number, that is, by counting signed crossings in projections.[24] A second method, less general but more powerful when it could be used, was to compute relative rotation rates.[20] This could only be done when the attractor "had a hole in the middle". Since all the attractors we studied early on had a hole in the middle, both methods could be used.

To use either method, though, we first needed to have a set of unstable periodic orbits in hand. This problem was resolved by searching for close returns.[27] We found the close returns search extremely effective in most instances. Further, this method could be used either on flow data or on map data: returns of the flow to a Poincaré section. But before we could use the close returns method we had to have an embedding of the data. For this part of the program there already existed a standard procedure: the time-delay embedding.[28,29] But for technical reasons (laziness) we preferred another type of embedding: the differential embedding in which each successive independent variable is defined as the first time derivative of the previous:

$$\begin{cases} \dot{x}_1 = x_2 \\ \dot{x}_2 = x_3 \\ \vdots \\ \dot{x}_n = f(x_1, x_2, \ldots, x_n). \end{cases} \quad (1)$$

The reason is that in three dimensions, the projection of the embedding onto the x_1-x_2 plane immediately provides linking number information.[18,19,30,31]

Once a reasonable number of unstable periodic orbits had been extracted from a data set and their pairwise linking numbers computed, half the problem was solved.

The remaining half involved identifying a branched manifold[52,53] (see Rössler and Letellier, chapter 5; Birman, chapter 6; Gilmore, chapter 8). The first branched manifold for describing a chaotic attractor was drawn in 1963 by Edward Lorenz using "isopleths" (Fig. 2a). Then Otto E. Rössler drew in 1976 a "blender" (Fig. 2b) to sketch his first chaotic attractor.[32] The same year, Robert Williams investigated the "Lorenz mask" which was topologically equivalent to the branched manifold proposed by Lorenz.[33]

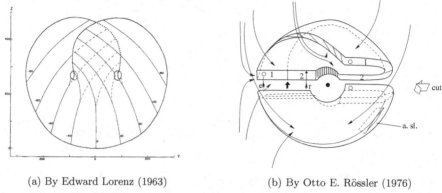

(a) By Edward Lorenz (1963)   (b) By Otto E. Rössler (1976)

Fig. 2. The first branched manifolds drawn to described chaotic attractors.

To do this we had to guess a branched manifold; construct periodic orbits on this manifold that could be compared with those extracted from experimental data; identify 1:1 experimental orbits with orbits on a branched manifold; compute linking numbers of the orbits on the branched manifold; compare the two tables of linking numbers.[19] This whole program started at a haphazard level (... guess a branched manifold ...) and became only mildly more robust with each successive step. Unexpectedly, the most difficult step was the step that seems like it should be the easiest: compute linking numbers of orbits on a branched manifold. The unexpected problem here is that we had trouble counting. In order to facilitate this step we decided to automate it. This means writing a computer code to do the counting for us. The output would be a table of linking numbers. One of the sets of input data would be the set of orbits for which the linking numbers would be computed, expressed in terms of their symbol set. The other input would have to be an algebraic description of the branched manifold. A suitable algebraic description was given in terms of a pair of matrices: a square $n \times n$ matrix, where n is the number of distinct symbols in the alphabet needed to describe all trajectories; and an $n \times 1$ array[19,30,31] (c.f., Gilmore, chapter 8), or using the linking matrix built according to the standard insertion.[34]

After some time this numerical algorithm worked for any kind of branched manifold corresponding to a chaotic attractor with a hole in the middle. We could create a return map for an experimental attractor (with a hole in the middle), use it to identify the symbol set in each of the periodic orbits extracted using the method of close returns, and use this orbit name to associate each experimental periodic orbit with a mathematical orbit on a (hypothetical) branched manifold. If the two tables agreed then we had identified, via the branched manifold, the mechanism that simultaneously created the chaotic attractor and organized all the unstable periodic orbits in it.

Little by little, the words began to roll around in our heads: algebraic description ... branched manifold. The exigencies of writing a computer code had forced us to develop an algebraic way to characterize branched manifolds. We could use topological pictures or algebraic structures, whichever is easier. In the United States with its computers everywhere, algebra was easier. In France, we think as a reaction to Bourbaki, pictures were preferred as used by Poincaré. Either way we had this gorgeous insight. Only a few orbits needed to be used to identify the branched manifold. When two symbols were required, three orbits sufficed.[18,19,31] On the basis of a small number of orbits the algebraic description of the template could be determined. This description could then be used to compute all other linking numbers. If those computed (on the branched manifold) were the same as those "measured" (using experimental orbits) then we had "nailed" the template. If *any single integer* in the two tables disagreed we had to *reject* the hypothesis that the description of the branched manifold and orbit set was correct. Here for the first time in the quantitative analysis of chaotic dynamical systems was a solid hypothesis test. There are no error bars. There is no underlying statistical theory that must be invented or fibbed about. Either the result is right or it is wrong. (Statisticians would say more correctly: the result is wrong or it is not wrong!)

In the first topological analysis,[19] we found nine orbits of period up to eight in a sample of data from the Belousov-Zhabotinskii chemical reaction.[35,36] The labeling on one orbit, of period six, was uncertain because one point in the orbit occurred at the critical point of the first return map. The 9×9 table of linking numbers had 36 off-diagonal linking number entries. We used three of the nine orbits, with three independent linking numbers, to identify the underlying template. This template was used to compute the remaining 33 linking numbers in the table. One of the two choices for the uncertain period-six orbit gave incomplete agreement with the table of experimental linking numbers. The alternative choice for the period-six orbit gave perfect agreement with the table of experimental linking numbers. In short, we used this topological analysis method to simultaneously identify the mechanism that created this experimental data set and also to refine the identification of one of the unstable periodic orbits extracted from the data. The topological analysis was also used to validate a global model in the context of a copper electrodissolution,[37] a string experiment[38] and a Belousov-Zhabotinskii reaction.[39]

3. Topological Analysis: an Explicit Case

A topological analysis can start by building a torus surrounding the attractor.[40] A bounding torus is a semi-permeable surface enclosing the attractor but excluding the fixed points around which the attractor is structured. The genus g of such a torus provides the number of components required to compute a Poincaré section. When $g \geq 3$, $g - 1$ components are required to compute a Poincaré section. There is no genus-2 bounding torus surrounding an attractor,[41] and a single component is required for an attractor bounded by a genus one attractor. This last case is illustrated using the chaotic attractor solution to the Sprott D system[42]

$$\begin{cases} \dot{x} = -y \\ \dot{y} = x + z \\ \dot{z} = xz + ay^2, \end{cases} \quad (2)$$

which has a single fixed point located at the origin of the phase space $\mathbb{R}^3(x, y, z)$. For $a = 2.3$, the Sprott D system produces a chaotic attractor which is bounded by a genus-one bounding torus (Fig. 3a). This toroidal surface can be viewed as a semi-permeable surface: the flow can only cross it from outside to inside. Once the trajectory is inside, it cannot return outside. If rigorously proving the existence of such a surface is not necessarily an easy task, a numerical proof is quite manageable.

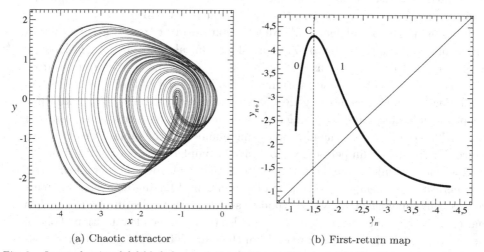

(a) Chaotic attractor (b) First-return map

Fig. 3. Inverted unimodal folded chaos solution to Sprott D system (2). The Poincaré section is conveniently defined by $y_n = 0$ and $\dot{y}_n < 0$ (thick horizontal line). Parameter value: $a = 2.3$.

An attractor surrounded by a genus-one bounding torus must be investigated with a single component Poincaré section as shown in Fig. 3a. In the present case, the Poincaré section can be chosen as

$$\mathcal{P} = \{(x_n, z_n) \in \mathbb{R}^2 \mid y_n = 0, \dot{y}_n < 0\} . \tag{3}$$

A first-return map to the Poincaré section (Fig. 3b) appears as a smooth-unimodal map made of an increasing monotonic branch and a decreasing branch. Both are separated by a critical point C which defines a partition of the attractor. Such a partition allows to distinguish different domains of the attractor: these domains are topologically inequivalent as shown for the first time by Otto Rössler (see Chapter 5). Most often, these domains differ by a local torsion since all of them are stretched under the action of the flow. The attractor solution to the Sprott D system is thus constituted by two topologically non-equivalent domains: the domain associated with the increasing branch — encoded by "0" — has an even local torsion. This means that the tangent space to a trajectory in this domain of the attractor presents an even number of π-twists. The second domain, associated with the decreasing branch — encoded by "1" — has an odd local torsion (an odd number of π-twists). Details can be found in.[30,43,44]

The skeleton of the attractor is made of the population of unstable periodic orbits embedded within it. They can be extracted numerically using a close returns technique.[18] They are then encoded according to the partition provided by the first-return map to the Poincaré section $\mathcal{P}$ (Fig. 3b). Periodic orbits embedded within the attractor and whose period is less or equal to 10 are reported in Tab. 1. These periodic orbits are ordered according to the kneading theory (see[45–47] and Hao Bai-Lin's Book[48] for a review).

Table 1. Population of unstable periodic orbits embedded within the chaotic attractor solution to the Sprott D system (2). The orbits are ranked according to the unimodal order (see[44] for details).

1	1011101011	10111110	101111110
10	101110	10111111	1011111
1011	101111	1011111110	1011110
10111010	1011111011	1011111111	101111010
1011101010	1011111010	101111111	101111011

Topological properties can then be investigated by computing the linking numbers that can be obtained by counting oriented crossings in a regular plane projection — there are no more than two segments crossing at the same point in that projection — of a couple of periodic orbits.[43] An example with the knot made of the period-2 orbit (10) and the period-4 orbit (1011) is shown in Fig. 4 with the oriented crossings (determined taking into account the third coordinate at each intersection as explained in[43,44]). The linking number is the half-sum of the oriented crossings, hence (Fig. 4)

$$\mathrm{lk}(10, 1011) = \frac{1}{2}[-8 + 2] = -3 . \tag{4}$$

This means that orbit (10) turns three times in the negative direction around orbit (1011). Linking numbers are topological invariants: they remain unchanged under an isotopy (a continuous deformation without any cutting).

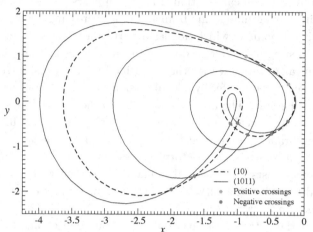

Fig. 4. A knot between two unstable periodic orbits embedded within the chaotic attractor solution to the Sprott D system (2). Case of the period-2 orbit (10) and the period-4 orbit (1011). The linking number lk(10,1011) is equal to -3.

The attractor can thus be schematically represented by a branched manifold with two stripes, one with an even local torsion and one with an odd local torsion. Continuity conditions commonly require that the local torsion of the two branches differs by $\pm\pi$, altough some counter-example where a tearing mechanism split the attractor in two bands, one visiting a region of the phase space quite far from the other; both bands are then squeezed into a single one. One of these counter-examples was provided by Otto E. Rössler[49] in 1976 and is shown in Fig. 5.

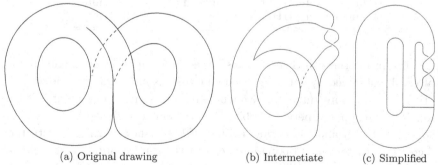

(a) Original drawing (b) Intermetiate (c) Simplified

Fig. 5. Three topologically equivalent branched manifolds of the case investigated by Rössler where the local torsions in two successive branches of the attractor differ by 2π.

The upper left part of the attractor solution to Sprott D system presents a "global torsion" that corresponds to π-twists applied to the whole attractor, that is, to the two stripes of the branched manifold as shown in Fig. 6a.

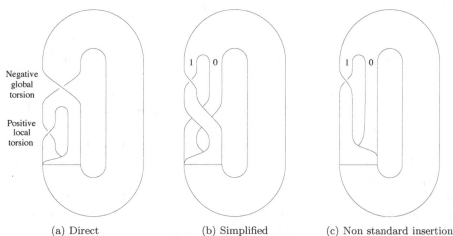

(a) Direct (b) Simplified (c) Non standard insertion

Fig. 6. Branched manifold describing the topology of the chaotic attractor solution to the Sprott D system. The direct branched manifold — with the global torsion (a) — can be transformed into a simpler template (b). The same branched manifold without the standard insertion (c).

From the knot shown in Fig. 4, it clearly occurs that the global torsion is negative and that there is a positive local torsion located at the lower left part of the attractor. A branched manifold can thus be represented as shown in Fig. 6a. This provides the actual topology of the attractor, taking into account all dynamical features observed. Nevertheless, an isotopy can be applied to simplify the branched manifold to obtain the template shown in Fig. 6b. Both representations are topologically equivalent and the simpler one (Fig. 6b) is preferred for computations. However the direct template is the one to be used in deciding if the attractor is new or not.[50]

The simplest template can be described by what is called a linking matrix M_{ij}[43,44] where M_{ii} defines the local torsion in the ith branch (or stripe) and M_{ij} defines the permutations between the ith and the jth branch. Therefore for the template in Fig. 6b the linking matrix is

$$M_{ij} = \begin{bmatrix} -1 & -1 \\ -1 & 0 \end{bmatrix}. \quad (5)$$

Gilmore and co-workers are using a slightly different way to algebraically describe a template. They do not use the standard insertion convention. They thus have to add an array to the linking matrix: the latter is different from the previous one since permutation between the branches to obtain the standard insertion is not required.

They would use Fig. 6c to describe the attractor and the two matrices

$$M_{ij} = \begin{bmatrix} -1 & 0 \\ 0 & 0 \end{bmatrix} \quad (6)$$

$$\begin{bmatrix} 1 & 0 \end{bmatrix}.$$

The second matrix describes the order according which the branches are squeezed together, from the top to the bottom. With the linking matrix, the topological characterization is complete.

The attractor solution to the Sprott D system is thus characterized by a simplified template that is topologically equivalent to the template associated with the Rössler system.[44] Nevertheless, when plotting the knot (10,1011) embedded within the Rössler attractor, only six negative crossings are counted and not eight negative and two positive crossings as counted in Fig. 4. This results from the negative global π-twist imposed to the attractor combined with the positive folding (characterized by a positive local torsion. In the Rössler attractor, there is just a negative local torsion). The Sprott D attractor corresponds to what is called a reverse horseshoe, that is, a folding combined with a global torsion by an odd number of π-twists.

4. Classification

The use of matrices to characterize branched manifolds lead to the idea that we could possibly classify branched manifolds algebraically. Of course, a branched manifold exists to characterize the organization of all the periodic orbits that can exist on it. In fact, "different" branched manifolds can predict identical spectra of linking number pairs. So one is faced with a "representation theory" for branched manifolds. When are two apparently different pictorial representations of a strange attractor equivalent? When are two aspparently different algebraic representations of a strange attractor equivalent? These are interesting questions that need to be worked on.

Other problems arose. Once a branched manifold has been proposed, it is typical to find many more orbits on it that can be found in a chaotic data set. The branched manifolds describe "full shift" dynamics characteristic of a hyperbolic dynamical system and experimental systems are not hyperbolic. Hyperbolicity is an assumption that facilitates mathematical discussion: Nature abhors hyperbolicity more than a vacuum. We have never seen a hyperbolic system in nature. So another question arose: How to restrict the set of orbits on a branched manifold so that it is a good aproximation to the spectrum of unstable periodic orbits that can exist in a chaotic attractor? Various approaches have been taken. The most important of these are reviewed and summarized by Natiello and Solari (chapter 7). In short, it seems possible to find a short list of orbits whose presence forces the presence of all orbits, up to some fixed period, that exist in an experimental attractor. There is an algorithm for constructing these *basis sets of orbits*, but it was an *ad hoc* construction[30,31] forced by an endeavor to explain many sets of laser data[51] (c.f., Gilmore, chapter 8). We are sure there will be many improvements in years to come (c.f., Natiello and Solari, chapter 7; Lefranc, chapter 9).

Our methods worked very well for attractors with a "hole in the middle", and even though all the data sets that we studied early on had "a hole in the middle", there was one notable exception: the Lorenz attractor. One could argue that this was just a mathematical game, but then experiments were done that generated data sets like that of the Lorenz attractor.[4] With T. D. Tsankov, we investigated more complicated attractors: those with more than one hole.[40,41] This lead to the realization that in three dimensions we were lucky a second time (the first time for the Gauss linking number). Every chaotic attractor could be embedded in a three-dimensional manifold whose boundary was a two-dimensional surface. Mathematically, such surfaces are classifiable. They are simple! They are tori, characterized by one number, an integer g, called the genus.[24] An attractor with a hole in the middle lives in a torus of genus one: the "one" is the hole. The Lorenz attractor lives in a torus of genus $g = 3$ (the third hole is along the z axis). The figure eight branched manifold[52] lives in a torus of genus $g = 9$. In fact, the two-dimensional surface of the bounding torus in not a plain vanilla surface: it is "dressed" with additional information. This is the restriction to the surface of the flow generating the chaotic attractor. Thus two indices are required to describe attractors with "many holes". One is an integer, the genus $g \geq 3$. The second is a group theoretical index related to the permutation group P_{g-1}. Why $g-1$ for $g \geq 3$? It is because we can show that the global Poincaré section for an attractor that can be embedded in a genus-g bounding torus consists of the union of $g-1$ disks[30,40,41] (c.f., Gilmore, chapter 8).

The algebraic description of branched manifolds in a genus-g bounding torus is within reach. The computation of linking numbers on such branched manifolds is also within reach. The construction of forcing relations (basis sets of orbits) is for now ridiculously out of reach. The importance of genus-g attractors is growing. Genus $g > 1$ is a signature that tearing[54,55] occurs as part of the stretching and squeezing process that, repeated infinitely often, builds up the foliated structure characteristic of a strange attractor. Any dynamical system with a finite symmetry group[56–59] can be expected to undergo tearing for some control parameter values, and therefore lives within a genus-g bounding torus. A number of experiments have been carried out on physical systems that exhibit tearing phenomena as a path to chaos.[60–66]

Even before a topological analysis can be carried out, data from an experiment must be visualized. That means an embedding has to be constructed. An embedding is 1:1 differentiable map — a diffeomorphism — from an unseen attractor to one that can be treated mathematically. Embeddings often are carried out on scalar time series. Questions naturally arise: Is your embedding equivalent to mine (equivalence is by isotopy)? Will the results of my topological analysis be the same as yours?

The answer is: not necessarily.[67] Here is an example to show why not. Imagine a nonlinear oscillator that is periodically driven with period T, and data $x(t)$ recorded. From $x(t)$ we can construct the time derivative $\dot{x}(t)$. Assume the data satisfy $x^2 + \dot{x}^2 < R^2$ during the experiment. Then we can create a simple three-dimensional embedding by constructing a simple closed curve

$$(X(\theta), Y(\theta), Z(\theta)) = R(\cos\theta, \sin\theta, 0),$$

synchronizing the physical time parameter t and the mathematical curve parameter θ by $t/T = \theta/2\pi$, and multiplying the normal to the curve at θ by $x(t)$ and the binormal by $\dot{x}(t)$. This maping will create an embedding.[67] If, instead of using the unknot, Gilmore uses some other inequivalent knot, he constructs an inequivalent embedding. Each inequivalent knot provides an inequivalent embedding. This argument extends to chaotic attractors that are embeddable in tori of genus g. Even less is known about inequivalent embeddings of genus-g tori into $\mathbb{R}^3$ than is known about embedding of genus-1 tori (knots) into $\mathbb{R}^3$.[68,69] These considerations demand the construction of a representation theory for strange attractors. For three dimensional dynamical systems a large part of this program has been completed.[70-74]

5. Hopes for the Future

There is still more to learn about low-dimensional (three-dimensional) dynamical systems that exhibit chaotic behavior. We have learned a great deal about such systems when the phase space is flat: when the attractor is contained in $\mathbb{R}^3$. We know much less when the ambient manifold containing the attractor is three-dimensional but not $\mathbb{R}^3$. In fact, we don't even have a good mechanism for determining when a chaotic attractor can be squeezed into a three-dimensional manifold. The Lyapunov dimension criterion $\lambda_1 + \lambda_3 < 0$ when $\lambda_2 = 0$ (listing exponents in descending order) doesn't seem appropriate as the exponents are averaged quantities and this statistic seems unsuited for guaranteeing that the attractor's space is less than three *everywhere* on the attractor. It would be extremely useful to have an inertial manifold theorem for low-dimensional sets of ordinary differential equations. Even when this is done we still have the problem of learning how to generalize integer invariants (Gauss linking numbers) and topological structures (branched manifolds) to three-dimensional manifolds that are not flat and could very well be multiply connected. This program presents exciting challenges.

The heart of the topological analysis program is the determination of the organization of the unstable periodic orbits that coexist in abundance with chaotic attractors. These have been used as a tool to identify an underlying branched manifold. The branched manifold has then been used to identify a mechanism giving rise to chaotic behavior.

In dimensions greater than three (even flat spaces $\mathbb{R}^n$, $n > 3$) this program comes to a screeching halt because links "fall apart" in four and more dimensions. Is it possible to find useful topological structures to replace the knots that have been fundamental for the development of the topological analysis program for chaotic attractors in $\mathbb{R}^3$? So far: No! Some of Gilmore's mathematician friends tell him to give up. Perhaps... , but Bob be such an optimist it's hard. Perhaps a more useful line of approach would be to skip the first, knotty step, even the second template step, and jump right to the end of this chain: find some way to make statements directly about the mechanism that creates chaotic behavior. How to do that ??

If and when we begin to understand how to describe chaotic attractors in manifolds of higher dimension, what tools will play an important role? It would be nice to have a theorem that plays the role of the Morse theorem for mappings $V : \mathbb{R}^n \to \mathbb{R}^1$ and that forms the basis for Catastrophe Theory.[75–78] This allows to isolate the small number of variables in phase space that are important for describing the main features of the potential function V. Right now the *germ* x^2 of the simplest potential $V(x) = \frac{1}{3}x^3 - ax$ describes the fold singularity underlying the Smale horseshoe.[79] There is a hope that the theory of singularities and unfoldings can be transferred to the study of relatively low dimensional dynamical systems.[30]

Group theory, too, should play an important role in any future developments of topological analyses in higher dimensions. Right now the permutation group plays an important role in classifying bounding tori of genus g that are dressed with a flow.[40,41] The spirit behind representation theory has breathed life into the theory of representations of dynamical systems.[70,71,74] We know already that inequivalent embeddings (diffeomorphisms) can be constructed for low dimensional chaotic attractors. We also know that as the dimension of the embedding space increases, obstructions to isotopy drop away, and in sufficiently high dimensions all embeddings of any one particular attractor become equivalent. For attractors that can be embedded in $\mathbb{R}^3$ the magic dimension is $\mathbb{R}^5$. Below dimension 5, some information gleaned about a low dimensional attractor by topological methods depends on the attractor and some depends on the embedding. For $n \geq 5$ there is only one universal embedding, so information gleaned from an embedding into $\mathbb{R}^n, n \geq 5$ depends only on the attractor. Unfortunately, we don't yet know a good topological signature in $\mathbb{R}^5$ for describing three-dimensional attractors.

To summarize, if $\mathcal{A}$ is a chaotic attractor and $\mathcal{M}^n$ is an n-dimensional manifold,

$$\begin{array}{ccc} \mathbb{R}^3 & \to & \mathcal{M}^3 \\ \downarrow & & \downarrow \\ \mathbb{R}^n & \to & \mathcal{M}^n \end{array}$$

we can carry out a topological analysis only in the upper left hand corner of this diagram ($n > 3$).

References

1. D. Ruelle & F. Takens, On the nature of turbulence, *Communications in Mathematical Physics*, **20**, 167-192 (1971).
2. H. L. Swinney & J. Gollub, Onset of turbulence in a rotating fluid, *Physical Review Letters*, **35**, 927-930 (1975).
3. J. Maselko & H. L. Swinney, Complex periodic oscillations and Farey arithmetic in the BelousovZhabotinskii reaction, *Journal of Chemical Physics*, **85**, 6430-6441 (1986).
4. E. N. Lorenz, Deterministic nonperiodic flow, *Journal of the Atmospheric Sciences*, **20**, 130-141 (1963).
5. M. L. Cartwright & J. E. Littlewood, On non-linear differential equations of the second-order, *Annals of Mathematics*, **48** (2), 472-494 (1947).
6. N. Levinson, Second-order differential equation with singular solutions, *The Annals of Mathematics*, II, **50** (1) 127-153 (1949).
7. C. Grebogi, E. Ott, S. Pelikan & J. A. Yorke, Strange attractors that are not chaotic, *Physica D*, **13**, 261-268 (1984).
8. J. Guckenheimer, G. F. Oster & A. Ipaktchi, Periodic solutions of a logistic difference equation, *Journal of Mathematical Biology*, **4**, 101-147 (1976).
9. R. Lozi, Giga-periodic orbits for weakly coupled tent and logistic discretized maps, In: *Modern Mathematical Models, Methods and Algorithms for Real World Systems*, A. H. Siddiqi, I. S. Duff & O.Christensen (Eds), Anamaya Publishers, New Delhi, pp. 80-14 (2006).
10. T. Y. Li & J. A. Yorke, A period-three implies chaos, *American Mathematics Monthly*, **82**, 985-992 (1975).
11. H. G. Solari, M. A. Natiello & G. B. Mindlin, *Nonlinear Dynamics*, Institute of Physics Publishing (1996).
12. L. Glass, Chaos and heart rate variability, *Journal of Cardiovascular Electrophysiology*, **10**, 1358-1360 (1999).
13. L. A. Aguirre & C. Letellier, Modeling Nonlinear Dynamics and Chaos: A Review, *Mathematical Problems in Engineering*, **2009**, 238960 (2009).
14. H. Poincaré, *Les Méthodes Nouvelles de la mécanique céleste*, Tome I, Gauthier-Vilars, Paris, pp. 81-82, (1892).
15. D. Auerbach, P. Cvitanovich, J.-P. Eckmann, G. Gunaratne, & I. Procaccia, Exploring chaotic motion through periodic orbits, *Physical Review Letters*, **58**, 2387-2389 (1987).
16. C. Grebogi, E. Ott, & J. A. Yorke, Unstable periodic orbits and the dimension of chaotic attractors, *Physical Review A*, **36**, 3522-3524 (1987).
17. D. P. Lathrop and E. J. Kostelich, Characterization of an experimental strange attractor by periodic orbits, *Physical Review A*, **40**, 4028-4031 (1989).
18. G. B. Mindlin & R. Gilmore, Topological analysis and synthesis of chaotic time series, *Physica D*, **58**, 229-242 (1992).
19. G. B. Mindlin, H. G. Solari, M. A. Natiello, R. Gilmore, & X.-J. Hou, Topological analysis of chaotic time series data from the Belousov-Zhabotinskii reaction, *Journal of Nonlinear Science*, **1**, 147-173 (1991).
20. H. G. Solari & R. Gilmore, Relative rotation rates for driven dynamical systems, *Physical Review A*, **37**, 3096-3109 (1988).
21. E. Eschenazi, H. G. Solari & R. Gilmore, Basins of attraction in driven dynamical systems, *Physical Review A*, **39**, 2609-2627 (1989).
22. N. B. Tufillaro, H. G. Solari, & R. Gilmore, Relative rotation rates: Fingerprints for strange attractors, *Physical Review A*, **41**, 5717-5720 (1990).
23. G. B. Mindlin, X.-J. Hou, H. G. Solari, R. Gilmore, & N. B. Tufillaro, Classification of strange attractors by integers, *Physical Review Letters*, **64**, 2350-2353 (1990).

24. D. Rolfsen, *Knots and Links*, Berkeley, CA: Publish and Perish Press (1976).
25. R. L. Ricca & B. Nipoli, Gauss' linking number revisited, *Journal of Knot Theory and its ramification*, **20** (10), 1325-1343 (2011).
26. J. C. Maxwell, *A Treatise on Electricity and Magnetism*, Clarendon Press, Oxford (1873), reprinted by Dover (1951).
27. J. P. Eckmann, S. O. Kamphorst, & D. Ruelle, Recurrence plots of dynamical systems, *Europhysics Letters*, **5**, 973-977 (1987).
28. F. Takens, Detecting strange attractors in turbulence, *Lecture Notes in Mathematics*, **898**, 366-381 (1981).
29. N. H. Packard, J. P. Crutchfield, J. D. Farmer, & R. S. Shaw, Geometry from a time series, *Physical Review Letters*, **45**, 712-715 (1981).
30. R. Gilmore & M. Lefranc, *The topology of chaos*, Wiley (2002); 2nd Edition (2011).
31. R. Gilmore, Topological analysis of chaotic dynamical systems, *Reviews of Modern Physics*, **70**, 1455-1530 (1998).
32. O. E. Rössler, Chaotic behavior in simple reaction system, *Zeitschrift fr Naturforschung A*, **31**, 259-264, 1976.
33. R. Williams, The Lorenz Attractor, *Lecture Notes in Mathematics*, **615**, 94-112 (1977).
34. L. Le Sceller, C. Letellier, & G. Gouesbet, Algebraic evaluation of linking numbers of unstable periodic orbits in chaotic attractors, *Physical Review E*, **49** (5), 4693-4695, 1994.
35. K. G. Coffman, W. D. McCormick, Z. Noszticzius, R. H. Simoyi, & H. L. Swinney, Universality, multiplicity, and the effect of iron impurities in the Belousov-Zhabotinskii reaction, *Journal of Chemical Physics*, **86**, 119-129 (1987).
36. E. Richetti, E. de Keeper, J. C. Roux, & H. L. Swinney, A crisis in the Belousov-Zhabotinskii reaction: experiment and simulation, *Journal of Statistical Physics*, **48**, 977-990 (1987).
37. C. Letellier, L. Le Sceller, P. Dutertre, G. Gouesbet, Z. Fei & J. L. Hudson, Topological characterization and global vector field reconstruction from an experimental electrochemical system, *Journal of Physical Chemistry*, **99**, 7016-7027 (1995).
38. N. B. Tufillaro, P. Wyckoff, R. Brown, T. Schreiber & T. Molteno, Topological time series analysis of a string experiment and its synchronized model, *Physical Review E*, **51** (1), 164-174 (1995).
39. C. Letellier, J. Maquet, H. Labro, L. Le Sceller, G. Gouesbet, F. Argoul & A. Arnéodo, Analyzing chaotic behaviour in a Belousov-Zhabotinskii reaction by using a global vector field reconstruction, *Journal of Physical Chemistry A*, **102**, 10265-10273 (1998).
40. T. D. Tsankov & R. Gilmore, Strange attractors are classified by bounding tori, *Physical Review Letters*, **91** (13), 134104 (2003).
41. T. D. Tsankov & R. Gilmore, Topological aspects of the structure of chaotic attractors in $\mathbb{R}^3$, *Physical Review E*, **69**, 056206 (2004).
42. J. C. Sprott, Some simple chaotic flows, *Physical Review E*, **50** (2), 647-650 (1994).
43. N. B. Tufillaro, T. Abbott & J. Reilly, *An Experimental Approach to Nonlinear Dynamics and Chaos*, Addison-Wesley, New York (1992).
44. C. Letellier, P. Dutertre & B. Maheu, Unstable periodic orbits and templates of the Rössler system: toward a systematic topological characterization, *Chaos*, **5** (1), 271-282 (1995).
45. N. Metropolis, M. L. Stein & P. R. Stein, On finite limit sets for transformations on the unit interval, *Journal of Combinatorial Theory A*, **15**, 25-44 (1973).
46. J. Guckenheimer, Bifurcations of maps of the interval, *Inventiones Mathematicæ*, **39**, 165-178 (1977).
47. P. Collet & J. P. Eckmann, Iterated maps on the interval as dynamical systems, *Progress in Physics*, **1**, Birkhäuser, Boston, 1980.

48. Hao Bai Lin, *Elementary Symbolic Dynamics and Chaos in Dissipative Systems*, World Scientific Publishing, Singapore (1989).
49. O. E. Rössler, Different types of chaos in two simple differential equations, *Zeitschrift fr Naturforschung A*, **31**, 1664-1670 (1976).
50. C. Letellier & L. A. Aguirre, Required criteria for recognizing new types of chaos: Application to the "cord" attractor, *Physical Review E*, **85**, 036204 (2012).
51. F. A. Papoff, A. Fioretti, E. Arimondo, G. B. Mindlin, H. G. Solari, & R. Gilmore, Structure of chaos in the laser with saturable absorber, *Physical Review Letters*, **68**, 1128-1131 (1992).
52. J. Birman & R. F. Williams, Knotted periodic orbits in dynamical systems-I: Lorenz's equations, *Topology*, **22** (1), 47-82 (1983).
53. J. Birman & R. F. Williams, Knotted periodic orbits in dynamical systems-II: Knot holders for fibred knots, *Contemporary Mathematics*, **20**, 1-60 (1983).
54. G. Byrne, R. Gilmore, & C. Letellier, Distinguishing between folding and tearing mechanisms in strange attractors, *Physical Review E*, **70**, 056214 (2004).
55. C. Letellier, T. D. Tsankov, G. Byrne, & R. Gilmore, Large scale structural reorganization of strange attractors, *Physical Review E*, **72**, 026212 (2005).
56. R. Gilmore and C. Letellier, *The Symmetry of Chaos*, Oxford University Press (2008).
57. C. Letellier & R. Gilmore, Covering dynamical systems: Two-fold covers, *Physical Review E*, **63**, 016206 (2000).
58. C. Letellier & R. Gilmore, Dressed symbolic dynamics, *Physical Review E*, **67**, 036205 (2003).
59. C. Letellier and R. Gilmore, Symmetry groups for 3D dynamical systems, *Journal of Physics A*, **40**, 5597-5620 (2007).
60. H. Haken, Analogy between higher instabilities influids and lasers, *Physics Letters A*, **53** (1), 77-78 (1975).
61. C. O. Weiss and W. Klisch, On observability of Lorenz instabilities in lasers, *Optics Communications*, **51** (1), 47-48 (1984).
62. M. Gorman, P. J. Widmann, & K. A. Robins, Chaotic flow regimes in a convection loop, *Physical Review Letters*, **52**(25), 2241-2244 (1984).
63. C. O. Weiss & J. Brock, Evidence for Lorenz-type chaos in a laser, *Physical Review Letters*, **57** (22), 2804-2808 (1986).
64. M. Gorman, P. J. Widmann, & K. A. Robins, Nonlinear dynamics of a convection loop: a quantitative comparison of experiment with theory, *Physica D*, **19** (2), 255-267 (1986).
65. L. Chua, M. Komuro, & T. Matsumoto, The double scroll family, *IEEE Transactions on Circuits and Systems*, **33**, 1073-1118 (1986).
66. J. Singer, Y.-Z. Wang, & H. Bau, Controlling a chaotic system, *Physical Review Letters*, **66** (9), 1123-1125 (1991).
67. R. Gilmore, Two-parameter families of strange attractors, *Chaos*, **17**, 013104 (2007).
68. N. Romanazzi, M. Lefranc, & R. Gilmore, Embeddings of low-dimensional strange attractors: Topological invariants and degrees of freedom, *Physical Review E*, **75**, 066214 (2007).
69. R. Gilmore, C. Letellier, & N. Romanazzi, Global topology from an embedding, itJournal of Physics A, **40** (13), 291-297 (2007).
70. D. J. Cross & R. Gilmore, Representation theory for strange attractors, *Physical Review E*, **80**(1), 056207 (2009).
71. D. J. Cross & R. Gilmore, Differential embedding of the Lorenz attractor, *Physical Review E*, **81**, 066220 (2010).
72. D. J. Cross & R. Gilmore, Equivariant differential embeddings, *Journal of Mathematical Physics*, **51**, 092706 (2010).

73. D. J. Cross & R. Gilmore, Complete set of representations for dissipative chaotic three-dimensional dynamical systems, *Physical Review E*, **82**, 056211 (2010).
74. D. Cross, *Representation theory of dynamical systems*, Ph.D. Thesis, Drexel University (2010).
75. E. C. Zeeman, *Catastrophe Theory, Selected Papers 1972-1977*, Reading, MA: Addison Wesley (1977).
76. T. Poston and I. N. Stewart, *Catastrophe Theory and Its Applications*, London: Pitman (1978).
77. R. Gilmore *Catastrophe Theory for Scientists and Engineers*, NY: Wiley (1981).
78. V. I. Arnold, *Catastrophe Theory, 2nd Ed.*, Berlin: Springer-Verlag (1986).
79. S. Smale, Differentiable dynamical systems, *Bulletin of American Mathematical Society*, **73**, 747-817 (1967).

PART 1
Emergence of a Chaos Theory

Chapter 2

The peregrinations of Poincaré

Ralph Abraham

Mathematics Department, University of California
Santa Cruz, CA 95064, USA

Dynamical Systems Theory in the spirit of Poincaré has been in vogue since the controversial award by King Oscar II of Sweden and Norway, on his 60th birthday, January 21, 1889, to Poincaré. Dynamical Systems Theory diffused Eastward (via Stockholm, Saint Petersburg, and Moscow) to Gorky in Russia, to Japan, and Westward (via Princeton, Mexico City, and Rio), to Berkeley in California. Now in the centennial year of the premature death of Poincaré, it is time for a review of these peregrinations.

Contents

1. Introduction . . . 23
2. The Origin . . . 24
    2.1. Jules Henri Poincaré . . . 24
3. Westward Journey . . . 25
    3.1. George David Birkhoff . . . 25
    3.2. Solomon Lefschetz . . . 27
    3.3. Stephen Smale . . . 28
4. Eastward Journey . . . 29
    4.1. Sophie Kovalevsky . . . 29
    4.2. Aleksandr Andronov . . . 30
    4.3. Leonid Pavlovich Shilnikov . . . 31
    4.4. Chihiro Hayashi . . . 33
    4.5. Yoshisuke Ueda . . . 33
5. Conclusion . . . 33
References . . . 36

1. Introduction

Plotting on a world map the early trajectory of dynamical systems theory from Paris in 1880, into the computational era around 1965, we find a loop, or periodic orbit, as shown in Figure 1. In this chapter we will approach the main sites on this trajectory through biographies of some of the principal agents.[a] Our main focus is on flows, and we have omitted the extensive history of iteration research.

---

[a]Some high points of this journey are recounted in.[1]

Fig. 1. The trajectory of dynamical systems theory.

East 1   Paris to Stockholm, Moscow, 1884
West 1   Paris to Harvard, 1913
East 2   Moscow to Kiev and Gorky, 1930
West 2   Kiev to Princeton, 1943
West 3   Princeton to Mexico City, Rio, and Berkeley, 1958
East 3   Berkeley to Gorky, 1963
East 4   Paris to Kyoto, 1942

2. The Origin

2.1. *Jules Henri Poincaré*

Born in Nancy, France, Poincaré earned a doctorate from the University of Paris in 1879 at age 25. He was professor at the University of Paris until his premature death at age 58. In his short career of 33 years he published 30 books and 500 papers, enriching most branches of mathematics and physics. His popular books introduced the latest ideas of math and science to a wide audience.[2–4]

Henri Poincaré
(1854-1912)

Among his accomplishments, the most important for our story is his creation of Dynamical Systems Theory, the geometric (or qualitative) theory of differential equations, in his earliest publications of 1880-1886.[5–10] For details, see Barrow-Green's book [12, p. 263]. The geometric method is introduced in the paper that begins on the very first page of his collected works.[11] Among these methods is the *phase portrait*, a visualization of the flow in the state space, or domain of the system of differential equations, originally the Euclidean plane. He introduced the fundamental idea of a limit cycle, as a closed trajectory in the state space, in 1882. [8, p. 261]

Later, in 1908, he pointed out that an electronic oscillation is represented by a periodic attractor, or stable limit cycle.[13] This idea is fundamental to the theory of nonlinear oscillations, an important part of the development of Dynamical Systems Theory prior to the advent of chaos theory in the 1960s.[14] Poincaré also pioneered analytical methods based on series expansions.

Even more to the point was his creation of the foundations of chaos theory in his paper on the three-body problem of 1890, which was awarded the prize in the 60th birthday competition of King Oskar II of Norway and Sweden. In a nutshell, the competition was conceived in 1884, announced in 1885, and closed for entries in June of 1888. Poincaré announced his intention to enter the competition to Mittag-Leffler of the University of Stockholm in 1887, explicitly mentioning the first of the four prize problems, on the stability of the three-body problem. At closing time, twelve entries had been submitted. Poincaré's entry, running to 158 pages and very difficult to read, was judged the winner. The result was announced in a local newspaper on King Oskar's 60th birthday, January 20, 1889. The winning paper was prepared for publication in October in the Swedish journal, *Acta Mathematica*, which had been founded by Mittag-Leffler with the support of King Oskar.[b]

Preparation of the paper for publication was entrusted to Lars Edvard Phragmén (1863-1937), one of the editors of the journal, who reported to Mittag-Leffler in July that there were some obscure passages in the paper. Mittag-Leffler wrote to Poincaré for clarifications, and Poincaré replied in December of 1889 that he had found a serious mistake elsewhere in the paper, and he was making major revisions. Prepublication copies of the original prize paper were quickly recalled from circulation and destroyed. Only one copy survived, which was hidden in the Institut Mittag-Leffler until recently. [12, p. 1] Amazingly, Poincaré was able to submit a revised paper by January of 1890. Fast work indeed, as the original paper purported to prove the stability of the three-body problem, while the revised paper established the opposite! All 270 pages appeared in the *Acta* later in 1890.[15]

From this hasty work around the New Year of 1890, Poincaré intuitively understood the famous *homoclinic tangle*, basic to chaos theory, that he developed in the third volume of the *New methods of celestial mechanics*.[16]

3. Westward Journey

3.1. *George David Birkhoff*

Christian Mira, who has contributed extensively to dynamical systems theory and also written much of its history, has said, "Intellectually, David Birkhoff was the greatest of Henri Poincaré's disciples." [17, p. 251]

---

[b]The full story of the prize paper is spelled out in detail in Barrow-Green's book [12, Ch. 4].

Birkhoff was born of Dutch ancestry in Overisel, Michigan. He earned the Ph.D. from the University of Chicago in 1907 at the age of twenty-three, and became a professor at Harvard University in 1911, where he stayed throughout his life. He earned many honors, and was regarded the dean of American mathematics. He contributed to many areas of mathematics, especially dynamics and ergodic theory. He became famous through his proof in 1912 of Poincaré's Last Theorem, left unproved by Poincaré at his death in 1912.[18] Thus Birkhoff continued the tradition of Poincaré in dynamics without a gap. His further work in topological dynamics was published in 1927 as *Dynamical Systems*,[19] thus giving a name to the new field we call Dynamical Systems Theory. Of special interest to our story here is his joint paper with Paul Smith of 1928, in which the *homoclinic tangle* discovered by Poincaré in 1881 was drawn and fully understood for the first time.[c] This is shown in our Figure 2. The signature of a homoclinic point was defined as well.[d]

David Birkhoff
(1884-1944)

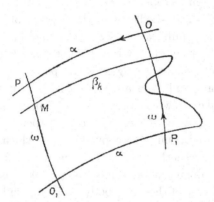

Fig. 2. The periodic point O (O_1 is its image) with its inset (stable manifold, two segments labelled ω are shown) and outset (unstable manifold, two segments labelled α are shown) intersect at the homoclinic point, P and its image P_1, and once again at M. This is Fig. 1 from Birkhoff and Smith.[20]

Serendipitously, after I had been working on the Birkhoff and Smith paper in Berkeley for two years, I moved to Columbia University in 1962, and found that Paul Smith was my chairman, having just succeeded Samuel Eilenberg.

Birkhoff's dynamical publications continued for another dozen years, until 1941. Then, for some unknown reason, the Westward trail of Poincaré ended, with an address at the University of Chicago on unsolved problems in dynamics [22, vol. 2;

---

[c] What we now call a heteroclinic point is called a doubly asymptotic point by Birkhoff and Smith.
[d] This is carefully drawn by Chris Shaw in [21, Sec. 13.5].

pp. 710-712]. Perhaps Birkhoff's anti-Semitism, as alleged by Albert Einstein and Norbert Wiener among others, played a role. The entry of the United States into World War II might also have been a factor.

3.2. *Solomon Lefschetz*

Lefschetz was born in Moscow to a Jewish family and moved to Paris at an early age. French was his first language. He studied engineering in Paris, and moved to the USA after graduation, in 1905. After losing both hands in an accident he migrated into mathematics, earning the Ph.D. in math from Clark University in 1911. He became one of the preeminent pioneers of algebraic topology. He was professor of mathematics at Princeton University from 1924 to 1953, editor of the *Annals of Mathematics* from 1928 to 1958, and president of the American Mathematical Society, 1935-1936. He had 25 excellent Ph.D. students at Princeton, Paul Smith the first of them, in 1926.

Solomon Lefschetz
(1884-1972)

In 1943 he met Nicolai Minorsky (1885-1970) who had studied in Saint Petersburg and received the Ph.D. in 1914. Minorsky emigrated to the US in 1918, eventually becoming a professor at Stanford University. His knowledge of the Russian school of dynamical systems theory, presented in his book[23] of 1947 on nonlinear oscillations, gave great impetus to the resumption of mathematical dynamics in the United States. Through Minorsky, Lefschetz discovered some of the Russian works on dynamical systems.[e]

During World War II he continued his studies on dynamics, reviving the Poincaré tradition, and founding the new field of global analysis.[25] His translation[24] of the early works of Andronov and his students into English, published in 1943, 1946, and, 1949, gave a strong boost to the reestablishment of the Poincaré tradition following its recent demise in the Birkhoff line after 1941.

He was an extraordinarily altruistic person, and decided in 1944 to further the development of mathematics in Latin America by teaching graduate courses at the Universidad Nacional Autónoma de México, UNAM, in Mexico City. It was there that I met him in 1959, while working on my Ph.D. thesis on general relativity under Nathaniel Coburn at the University of Michigan.

He attracted graduate students from all over the Americas, and in this group came the revival of the concept of structural stability of Andronov and Pontryagin,[26] and the related idea of generic properties. Peixoto improved on earlier work of H. F. de Baggis, another student of Lefschetz.

---

[e]Especially those of Andronov, Khaikin, Vitt, Krylov, and Bogolyubov. Minorsky's books cover an enormous range of topics, from European as well as Russian authors (especially, Poincaré and Van der Pol).

After retirement from Princeton in 1953, he moved to the Research Institute for Advanced Studies (RIAS) in Baltimore, which flourished from 1958 to 1964, and then created the Lefschetz Center for Dynamical Systems at Brown University.

A boom in Dynamical Systems Theory and global analysis was triggered by Mauricio Peixoto (b. 1921) of Brazil, who had studied with Lefschetz in Princeton in 1957. Lefschetz encouraged him to work on dynamical systems, and Peixoto proved an important result on structural stability in two dimensions in 1958.

3.3. *Stephen Smale*

Smale was born in Flint, Michigan, and received the Ph.D. in mathematics from the University of Michigan in 1956. That summer, at a conference in Mexico, he met René Thom, Moe Hirsch, and Elon Lima from Brazil. In the fall, all four were at the University of Chicago, along with Dick Palais and Ed Spanier, and René Thom was lecturing on transversality theory. Later that year, through Lima, Smale met Peixoto in Princeton, learned of his result, and proposed an extension to higher dimensions, using Thom's ideas on transversality in combination with his own extensive experience and ingenuity in higher dimensional differential topology.

Stephen Smale
(b. 1930)

In Mexico City in the summer of 1959, where I had met Lefschetz while working on my thesis, I had noticed that a conference was in progress. At this conference, Smale met Lefschetz and delivered a talk on Morse-Smale dynamical systems. Early in 1960 Smale visited Brazil, where he studied homoclinic points on the beaches of Rio. After arriving in Berkeley in the summer of 1960, he perfected his horseshoe example. This work completely untangled the homoclinic tangle, resolving it into symbolic dynamics. In September of 1961, Smale visited Kiev, where he announced his results. At this time, he met E. A. Leontovich, the widow of Andronov. Smale's horseshoe paper was published in 1963. In his autobiographical writings, he gives credit to the Andronov school of Gorky (see Refs. [28, p. 21] and [29, p. 73]).

All these mathematicians converged in Berkeley in the fall of 1960, just as I arrived for my first academic job following my own Ph.D. at the University of Michigan. At the daily tea in the math department in Campbell Hall, I met them all. I began working with Thom on transversality theory, and with Smale on homoclinic points and other problems of global analysis.

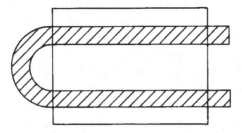

Fig. 3. The horseshoe mapping of Smale. The shaded horseshoe is the image of the plain rectangle under one iteration of the map. From Smale [30, p. 150].

4. Eastward Journey

4.1. *Sophie Kovalevsky*

The number of women in mathematics is lately on the rise, overturning a gender bias on a planetary scale of long standing. As recently as 1974, a sympathetic survey of the history of mathematics from ancient times turned up only eight significant names.[31] Kovalevsky was seventh on this list. Among her many mathematical accomplishments, her contribution to dynamical systems theory (Dynamical Systems Theory) may be the least known, and here we will tell her story.

Sophie Kovalevsky
(1850-1891)

Born Sofia Vasilyevna Korvin-Krukovskaya in Moscow in a long line of mathematicians, she grew up far away from the cultural centers of Russia, near the border with Lithuania. At seventeen, she went to Saint Petersburg to study mathematics. Her determination to continue her studies in university was frustrated as European universities were open only to men. She entered a convenience marriage in 1868 with a friend, Vladimir Kovalevsky, to obtain permission to leave Russia, and together they went to Heidelberg to continue their studies. Outside Russia, she began to use the name Sophie Kovalevsky. Although the university in Heidelburg was also closed to women, she was able to attend lectures there by Leo Königsberger (1837-1921), who had been a student of Karl Weierstrass (1815-1897) in Berlin. Weierstrass was then the most famous mathematics professor in Europe, and regularly lectured to audiences of 300 or more.

In 1870 she moved to Berlin, with a strong recommendation from Königsberger, to study with Weierstrass, but found his university also closed to women. He accepted her as a private student, and after four years as his star pupil, with already three publications, she was awarded a Ph.D. from Göttingen. She returned to Russia in 1874 with Vladimir. Here for some time, unable to find a job, she occupied herself writing theater reviews [32, p. 177]. After the birth of her daughter Fufa in 1878, she returned to Berlin to continue mathematical work. In 1882 she moved on to Paris, on the advice of Weierstrass, to interact with the mathematical community around Hermite [32, p. 206]. That year she met Poincaré, who was a prominent member of this group (see her autobiography in [33, pp. 213-229], written in 1890). She also became a friend of Poincaré sister, Aline Boutroux [35, p. 222].

Among her fellow students in Berlin were many who eventually achieved note in the history books, among them, Gösta Mittag-Leffler (1846-1927). After his Ph.D. in 1872 and a career of several years as professor of mathematics in various Scandinavian universities, he became the first professor of mathematics at Stockholm University in 1881. Kovalevsky and Mittag-Leffler became friends in Saint Petersburg in 1876. Mittag-Leffler was an activist for women's rights, and he campaigned for a position in his university for Kovalevksy. In 1883, she moved to Stockholm and began giving lectures there in German. In 1884 she was appointed Professor, the first woman in this post in any European university. In the academic year 1887-1888, she gave a course in Poincaré's work on Dynamical Systems Theory [32, p. 323]. In 1888, she was awarded the Prix Bordin by the French Academy for her paper on what is now known as Kovalevsky's top.[34]

During these years, Kovalevsky was close friends with Mittag-Leffler's sister, Anna Charlotte Leffler (1849-1892), a writer. They maintained a weekly salon, and King Oscar sometimes attended. Also, during these years, Kovalevsky made frequent trips to Saint Petersburg, Moscow, and Paris. In this way, the King became interested in the problem of the stability of the solar system, and offered the prize competition for his 60th birthday in 1889. The inspiration of this prize in 1884 was due to King Oskar himself, or Mittag-Leffler, or Kovalevsky, or all three — we do not know [12, p. 51].

The prize was offered for four problems, one of which was the stability of the three-body problem. Dirichlet (1805-1859) had announced a proof in a letter to Weierstrass in 1858, but died before publishing it. Weierstrass suggested this as a prize problem in a letter to Kovalevsky [36, p. 9].

4.2. Aleksandr Andronov

The trail of Poincaré in Russia begins with Alexsandr Mijailovich Lyapunov (1857-1918), from whom many branches diverge. He was the first and most famous of Poincaré's followers in Russia. Inspired by the works of Poincaré and Kovalevsky, he wrote his memorial paper "Problème Général de la Stabilité du Mouvement" in 1892. Here he proved a useful condition for the stability of a rest point of a dynamical system.

In the 1930s, a derivative school was established in Kiev (southwest of Moscow), known especially through the works of Krylov and his student Bogolyubov. Their book of 1935 was translated by Lefschetz from French to English in 1943.[37] Another such school was established in Gorky (also Gorkii, and now called Nizhnii Novgorod, northeast of Moscow) by

Andronov in 1932. Andronov was a student of Leonid Isaakovich Mandel'shtam (or Mandelstam, 1874-1944), a Jewish mathematical physicist in Moscow. Mandelstam had a good knowledge of the works of Poincaré as he had been a student in Strasbourg, 1899-1914 [38, p. 2].

Aleksandr Andronov
(1901-1952)

In his thesis of 1930, Andronov was the first to connect the oscillations of the Van der Pol system with the limit cycle concept of Poincaré (see [39, p. 174] and [38, p. 2]). His group included his wife, E. A. Leontovich [30, p. 147]. The group produced important books, including *The Theory of Oscillations* written in 1937, jointly authored with S. E. Khaikin and A. A. Vitt, and translated from Russian to English by Lefschetz in 1949.[24] The head of the Gorky group, for many years until his recent death, was L. P. Shilnikov [40, p. 105].

4.3. *Leonid Pavlovich Shilnikov*

Shilnikov received the Ph.D. in mathematical physics at Gorky State University in 1963. In his thesis work under the supervision of Yuri Neimark) in the late 1950s he was already interested in bifurcations of flows involving homoclinic trajectories. His first major discovery on spiral chaos in three dimensions was published in 1962 in Russian, and in English in 1970.[42] In this work he made use of Smale's work on the horseshoe map, announced in Kiev in 1961, thus completing the closed orbit shown on the world map in Figure 1 [f].

Leonid Shilnikov
(1934-2011)

The cylinder C shown in Figure 4 is the Poincaré cross-section, the domain of the first-return map, in which the horseshoe appears, as shown in that figure. This early work was greatly extended in the sequel works of Shilnikov and his students in Gorky, and those of Charles Tresser and his coworkers, Pierre Coullet and Alain Arnéodo, in Nice in the early 1980s (see Refs.[43,44] and references therein). This figure, and the simple geometric description of it given in,[45] derives from joint work of Arnéodo, Coullet, and Tresser.

We do not cover here the history of iterations of maps, although that is also part of the tradition of Poincaré. However, it is relevant that there has been a cross-over from iterations to flows. In the context of iterations, the period doubling route to chaos was discovered by Myrberg in 1963 for the quadratic map, and its topologically universal character by Metropolis, Stein and Stein in 1973 for interval

---

[f]The bibliography of Shilnikov's paper,[42] includes papers of Smale of 1961 and 1965.

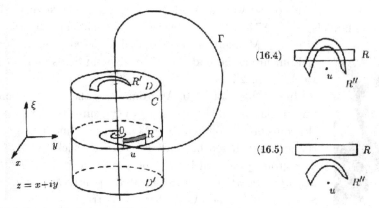

Fig. 4. Shilnikov's spiral chaos. The image under the Poincaré cross-section map of the shaded rectangle R is the horseshoe R'' shown in the upper right. From Ruelle [45, p. 117].

maps.[46] The metrically universal aspect was then discovered by Feigenbaum in 1975[47] and independently by Coullet and Tresser.[48] Coullet and Tresser discovered and explained its abundance in more general maps, and also in flows. Shilnikov also studied period doubling for flows [g].

Subsequently, Shilnikov ran the laboratory of the Department of Differential Equations headed by E. Leontovich-Andronova, and became head of that department in 1984. He wrote over 200 articles and co-authored several books. In a memorial for his 75th birthday, his students wrote [40, p. 101]:

> His works greatly influenced the overall development of the mathematical theory of dynamical systems, as well as nonlinear dynamics in general. Shilnikov's findings have became classics, and have been included in the most text and reference books which are used worldwide by mathematics students and nonlinear dynamicists to study the qualitative theory of dynamical systems and chaos. The elegance and completeness of his results let them reach to "the heart of the matter," and provide applied researchers with in-depth mathematical understanding of the outcomes of natural experiments. No doubt that this popularity is due the status of "a living classic" that Professor Shilnikov has attained over several decades for his continuous hard work on the bifurcation theory of multi-dimensional dynamical systems, mathematical chaos theory, and the theory of strange attractors.

---

[g]Many thanks to a private communication from Charles Tresser for this bit of history.

4.4. Chihiro Hayashi

Chihiro Hayashi
(1911-1987)

Hayashi began a program of experimental work as professor in the Kyoto University Department of Electrical Engineering in 1942. He published a book on this work in English in 1953,[41] and visited MIT in 1955-56. He had an excellent command of English, and had most likely learned the new results in dynamics from the books of Minorsky and Lefschetz. Minorsky,[23] Andronov and Chaikin,[24] Stoker,[49] and McLachlan (1947) are explicitly acknowledged in Hayashi's book [41, p. vii]. And in the revised and expanded 1964 edition of this work, he further credits Professors Rudenberg of Harvard and Den Hartog and Paynter of MIT.

4.5. Yoshisuke Ueda

Yoshisuke Ueda
(b. 1936)

Ueda was a student of Hayashi, earning his Ph.D. at Kyoto University in 1965. He observed chaotic behavior in computer simulation of a three-dimensional flow in 1961, and was among the first to do so. Due to Hayashi's devotion to a pre-chaotic worldview, Ueda's discovery was not published until 1970 (for the full story, see[50]). His chaotic attractor is organized around an attractive homoclinic tangle, as shown in Figures 5 and 6.

5. Conclusion

Following the famous song by Tom Lehrer: From Paris to Stockholm to Saint Petersburg to Moscow to Gorky we shall run. And then: to Princeton to Mexico City to Rio to Berkeley and back to Gorky we shall run. And lastly, back to Paris. Here are the approximate schedules for these voyages.

Westward Journey, 1880-1960

- 1880, origin at Paris (Poincaré)
- 1912, flight from Paris to Harvard (Birkhoff)
- 1928, stop at Harvard (Birkhoff and Smith)
- 1941, end of this voyage at Harvard

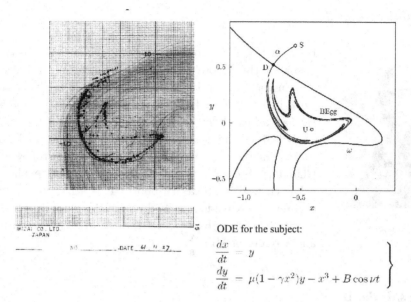

Fig. 5. Ueda's original plot of the Broken Egg (chaotic) attractor of 1961, left, and computer graphic showing the attractor as the outset of a homoclinic tangle, on the right. The figures belong to a Poincaré cross-section of a forced oscillator (three-dimensional) flow. This figure is from a lecture of Ueda, November, 2011, in Kyoto, on the 50th anniversary of his discovery of the Broken Egg. Courtesy of Professor Ueda.

- 1943, arrival of new flight from Kiev to Princeton (Lefschetz)
- 1958, stop at Princeton (Peixoto)
- 1959, arrival at Mexico City (Lefschetz, Smale)
- 1960, arrival at Rio, on to Berkeley (Smale)

Eastward Journey, 1880-1935

- 1880, origin at Paris (Poincaré)
- 1887, stop at Stockholm (Kovalevsky)
- 1892, stop at Moscow (Lyapunov)
- 1932, arrival at Gorky (Andronov)
- 1935, arrival at Kiev (Krylov)
- 1963, flight arrives from Berkeley (Smale's horseshoe)

Reunion, 1961-1970

- 1961, The arrival of Smale's horseshoe in Kiev and then Gorky connects the two journeys in a round trip
- 1970, The work of Charles Tresser and coworkers on Shilnikov's bifurcation brought the tradition back home to Paris

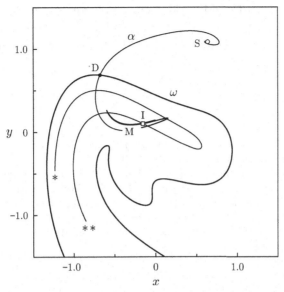

Fig. 6. Another computer graphic of the Broken Egg tangle, showing the homoclinic intersections. From (Ueda, 1992; Fig. 2, p. 102), originally published in 1973 in a paper by Ueda, Akamatsu, and Hayashi. Improved version courtesy of Professor Ueda. A similar figure (Ueda, 1982; Fig. 4, p. 86) created in 1969, is among the first such drawings to be published.

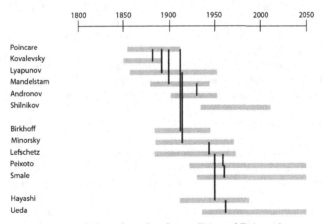

Fig. 7. A flow chart for the tradition of Poincaré.

Following this time, an explosion of mathematical work extended the Poincaré tradition worldwide.

Acknowledgments

I am very grateful to Christophe Letellier for suggesting this writing, and for many suggestions for its scope and presentation, and further to my colleagues Jean-Marc

Ginoux, Lev Lerman, Christian Mira, Miguel A. F. Sanjuan, Charles Tresser, and Yoshisuke Ueda for very abundant historical details and editorial advice. This is my most collaborative writing project ever, and I am both grateful and honored by their participation.

References

1. R. H. Abraham & Y. Ueda, *The Chaos Avant-garde: Memories of the Early Days of Chaos Theory*, Singapore: World Scientific, 2000.
2. H. Poincaré, *La Science et l'Hypothèse*, Flammarion, Paris (1902), English translation by W. J. Greenstreet: *Science and Hypothesis*, London : Walter Scott Pub. Co., 1905.
3. H. Poincaré, *La Valeur de la Science*, Flammarion, Paris (1905), Translated by G. B. Halsted (1907), Lancaster, Pa.: Science Press, 1946.
4. H. Poincaré, *Science et Méthode*, Flammarion, Paris (1908). Translated by F. Maitland (1914), London : T. Nelson
5. H. Poincaré, Sur les courbes définies par une equation différentielle, *Comptes-Rendus de l'Académie des Sciences* (Paris), **90**, 673-675 (1880).
6. H. Poincaré, Sur les courbes définies par les equations différentielles, *Comptes-Rendus de l'Académie des Sciences* (Paris), **93**, 951-952 (1881).
7. H. Poincaré, Mémoire sur les courbes définies par une equation différentielle (1ère partie), *Journal de Mathématiques Pures et Appliquées*, **7**, 375-422 (1881).
8. H. Poincaré, Mémoire sur les courbes définies par une equation différentielle (2nde partie), *Journal de Mathématiques Pures et Appliquées*, **8**, 251-296 (1882).
9. H. Poincaré, Sur les courbes définies par les equations différentielles (3ème partie), *Journal de Mathématiques Pures et Appliquées*, **4**, 167-244 (1885).
10. H. Poincaré, Sur les courbes définies par les equations différentielles, *Journal de Mathématiques Pures et Appliquées*, **2**, 151-217 (1886).
11. H. Poincaré, *Sur les propriétés des fonctions définies par des équations aux différences partielles*, Ph.D. Thesis, Paris, Gauthier-Villars (1879).
12. J. Barrow-Green, *Poincaré and the Three Body Problem*, Providence, RI: American Mathematical Society (1997).
13. H. Poincaré, Sur la télégraphie sans fil, *La lumière Électrique*, **4**, 259-266, 291-297, 323-327, 355-359, 387-393 (1908).
14. J.-M. Ginoux & L. Petitgirard, Poincaré's forgotten conferences on wireless telegraphy. *International Journal of Bifurcation and Chaos*, **20** (11), 3617-3626 (2010).
15. H. Poincaré, Sur le problème des trois corps et les équations de la dynamique, *Acta Mathematica*, **13**, 1-270 (1890).
16. H. Poincaré, *Les méthodes nouvelles de la mécanique céleste*, Vol III, Gauthier-Vilars, Paris (1899). English translation, *New methods of celestial mechanics*, Volume III, NASA (1967).
17. C. Mira, *Some historical aspects concerning the theory of dynamic systems*, In: Diner, pp. 250-261 (1986).
18. G. D. Birkhoff, Proof of Poincaré's geometric theorem, *Transactions of the American Mathematical Society*, **14**, 14-22 (1913).
19. G. D. Birkhoff, *Dynamical Systems*. New York: American Mathematical Society (1927).
20. G. D. Birkhoff & P. A. Smith, Structure analysis of surface transformations, *Journal de Mathématiques Pures et Appliquées*, IX, **7**, 345-379, (1928).

21. R. H. Abraham, & C. D. Shaw, *Dynamics: the Geometry of Behavior*, Second Edition. Redwood City, CA: Addison-Wesley (1992).
22. G. D. Birkhoff, *G. D. Birkhoff, Collected Mathematical Papers*, 3 volumes, New York: American Mathematical Society (1950).
23. N. Minorsky, *Introduction to Non-linear Mechanics: Topological Methods, Analytical Methods, Non-linear Resonance, Relaxation Oscillations*, Ann Arbor, MI: J. W. Edwards (1947).
24. A. A. Andronov, & S. E. Khaikin, *Theory of Oscillations* (1937). Translated by. S. Lefschetz Princeton, NJ: Princeton University Press (1949).
25. S. Lefschetz, *Lectures on Differential Equations*, Princeton, NJ: Princeton University Press (1946).
26. A. Andronov & L. Pontryagin, Systèmes grossiers, *Doklady Akademii Nauk SSSR*, **14**, 247-251 (1937).
27. S. Smale, Stable manifolds for differential equations and diffeomorphisms, *Annali della Scuola Normale Superiore di Pisa*, III, **17**, 97-116 (1963).
28. S. Smale, Finding a horseshoe on the beaches of Rio, In *The Chaos Avant-garde: Memories of the Early Days of Chaos Theory*, Singapore: World Scientific (Abraham & Ueda, Eds), pp. 7-22 (2000).
29. S. Batterson, *Stephen Smale: The mathematician who broke the dimension barrier*, Providence, RI: American Mathematical Society (2000).
30. S. Smale, *The Mathematics of Time*, Berlin: Springer-Verlag (1980).
31. L. M. Osen, *Women in Mathematics*, Cambridge, MA: MIT Press (1974).
32. D. H. Kennedy, *Little Sparrow: A Portrait of Sophia Kovalevsky*, Athens, OH: Ohio University Press, (1983).
33. S. Kovalevskaya, *A Russian Childhood* (1890), Translated by B. Stillman, Berlin: Springer-Verlag (1978).
34. S. Kovalevsky, Mémoire sur un cas particulier du problème de la rotation d'un corps pesant autour d'un point fixe, où l'intégration s'effectue l'aide des fonctions ultra-elliptiques du temps, *Mémoires présentés par divers savants à l'Académie des Sciences de l'Institut National de France*, **31**, 1-62 (1890).
35. A. Boutroux, *Vingt ans de ma vie, simple vérité* (1912-1913), (L. Rollet, Ed), Paris: Hermann (2012).
36. J. Moser, *Stable and random motions in dynamical systems with special emphasis on celestial mechanics*, Princeton, NJ: Princeton University Press, (1973).
37. N. Krylov & N. Bogoliubov, *Introduction to nonlinear mechanics*, Translated by S. Lefschetz, Princeton University Press (1943).
38. C. Bissell, The role of A. A. Andronov in the development of Russian control engineering, *Automation and Remote Control*, **62**, 863-874 (2001).
39. C. Mira, History, Part 1, In *Chaos in Discrete Dynamical Systems: A Visual Introduction in 2 Dimensions* (A. Abraham, C. Gardini & C. Mira, Eds) New York: Springer Verlag, (1997).
40. V. S. Afraimovich, L.M. Lerman & S. V. Gonchenko, L. P. Shilnikov-75, *Regular and Chaotic Dynamics*, **15** (2-3), 101-106 (2010).
41. C. Hayashi, *Nonlinear oscillations in physical systems*, Mc Graw-Hill, 1953.
42. L. P. Shilnikov, A contribution to the problem of the structure of an extended neighborhood of a rough equilibrium state of saddle-focus type, *Math. USSR Sbornik*, **10**, 91-102 (1970).
43. C. Tresser, Un théorème de Shilnikov en $C^{1,1}$. *Comptes-Rendus de l'Académie des Sciences* (Paris), I, **296**, 545-548 (1983).
44. C. Tresser, About some theorems of L. P. Shilnikov. *Annales de l'Institut Henri Poincaré*, **40**, 441-461 (1984).

45. D. Ruelle, *Elements of differentiable dynamics and bifurcation theory*, London: Academic Press (1989).
46. N. Metropolis, M. L. Stein & P. R. Stein, On finite limit sets for transformations on the unit interval, *Journal of Combinatorial Theory A*, **15**, 25-44 (1973).
47. M. J. Feigenbaum, Quantitative universality for a class of nonlinear transformation, *Journal of Statistical Physics*, **19** (1), 25-52 (1978).
48. P. Coullet & C. Tresser, Itérations d'endomorphismes et groupe de renormalisation, *Journal de Physique* (Colloque C5 — supplément), **8** (39), C5-25 (1978).
49. J. J. Stoker, *Nonlinear Vibrations*. New York: Interscience, (1950).
50. Y. Ueda, *The Road to Chaos*, Santa Cruz, CA: Aerial Press (1992).

Chapter 3

A Toulouse research group in the "prehistoric" times of chaotic dynamics

Christian Mira

19 rue d'Occitanie
31130 Quint
France

The first researches in the nonlinear dynamics field began in Toulouse from the year 1958. This paper relates the first period (i.e. the "prehistoric" one ending in 1976, the emergence of the word chaos due to May, and the beginning of the "historic" times) of the Toulouse group activity in complex nonlinear dynamics. This period was characterized by a close collaboration with Igor Gumowski, at that time, the western scientist who had the largest and the most profound information and understanding from the results of the schools on nonlinear dynamics in the former Soviet Union. Without using the non existent words like "chaos", and "fractal", complex dynamical behaviors were explained, in particular, from models in the form of recurrence (map).

Contents

1. Introduction. Birth of the Group. Approach of Dynamic Problems 39
2. Basin Boundaries of Two-dimensional Noninvertible Maps (1963-1975) 42
   2.1. Genesis of the results 42
   2.2. Complex organization of basins (1968-1973) 44
3. Chaotic Attractors of Two-Dimensional Noninvertible Maps (1968-1975) 45
4. Normal Forms for Resonant Bifurcations (1969-1974) 45
5. Two-Dimensional Conservative Maps (1970-1975) 47
6. Study of One-Dimensional Non-Invertible Maps (from 1972) 48
7. Applications 49
8. Two Exhibitions of Chaotic Images (1973 and 1975) 52
9. Conclusion 54
References 57

1. Introduction. Birth of the Group. Approach of Dynamic Problems

A two-year stay of Igor Gumowski from October 1958, in the laboratory of my first research efforts, is at the origin of the birth of the *Toulouse research group* (collective name gathering together four denominations, adopted here for greater convenience) and my orientation toward the "nonlinear world". Igor Gumowski arrived as Professor Assistant from "*Université Laval*" (Québec). From this encounter followed the creation of the group, which until 1999 worked on nonlinear dynamics and its applications in Toulouse.

As far as I know, Igor Gumowski was at that time one of the three western scientists who had the largest and the most profound information and understanding from the results of the schools on nonlinear dynamics in the former Soviet Union.[1-3] Thanks to him, I had access from 1958 to exceptionally wide information, rather unknown in western countries at that time. This unawareness occurred in spite of a text due to J.-P. La Salle and S. Lefschetz[4] who wrote in 1961:

> "In USSR the study of differential equations has profound roots, and in this subject the USSR occupies incontestably the first place. One may also say that Soviet specialists, far from working in vacuum, are in intimate contact with applied mathematicians and front rank engineers. This has brought great benefits to the USSR and it is safe to say that USSR has no desire to relinquish these advantages."

I. Gumowski also drew my attention to the high interest of the Japanese results of the "*Hayashi' School*" and their applications to electric and electronic circuits. This important contribution to nonlinear dynamics is recapitulated in the Hayashi's books.[5,6] It must be noted that papers on what would be latter called "*chaotic dynamics*" were published by Hayashi with his disciples Ueda and Kawakami, long before (1976) the appearance of this denomination.[7] This first meeting with I. Gumowski was followed by a long and friendly collaboration, source of fruitful results based on the *qualitative methods*, and the *analytical methods* of nonlinear dynamics.

About the *qualitative* methods it is reminded that their "strategy" is based on the characterization of the complex transcendental non tabulated solutions of continuous (resp. discrete) nonlinear models by their singularities. For continuous (resp. discrete) nonlinear models the singularities are equilibrium points (resp. fixed points), or periodical solutions (resp. cycles), phase trajectories (resp. invariant curves), stable and unstable manifolds, boundary (or separatrix) of the influence domain (domain of attraction, or basin) of a stable (attractive) stationary state, homoclinic, or heteroclinic singularities, more complex singularities of fractal, or non-fractal type. The qualitative methods concern the identification of two spaces associated with a dynamic system. The first one, called phase space, identifies these singularities (location, nature). The second one deals with the evolution of these singularities when the system parameters vary, or in the presence of a continuous structure modification of the system (study of bifurcation sets in a parameter space, or in a function space). The second approach corresponds to the *analytical methods*. Here, the above mentioned complex transcendental functions are defined by convergent, or at least asymptotically convergent series expansions, or in "the mean". The *method of the Poincaré's small parameter*, the *asymptotic methods of Krylov-Bogoliubov-Mitropolski* are analytical. So are the *averaging method*, and the *method of harmonic linearization* in the theory of nonlinear oscillations. These approaches constitute two relatively independent branches of the nonlinear oscillations theory. They have the same aims: construction of mathematical tools for the solution of concrete problems, improvement and development of the *dynamic systems theory*.

In the study of complex dynamics, especially chaotic dynamics, analytical methods quickly attain limits, which restrict their use to some limited local behavior. It is the reason why the basic tools, at the origin of the results described in this text, are essentially those of the qualitative methods, which achieved their highest level of evolution in the framework of the *Andronov's school*.[3] This approach has been the background of theses preparation for many researchers of the group I managed in Toulouse. The important contribution (from 1903 to 1918) of Samuel Lattès, a Professor of the Faculté des Sciences de Toulouse, is less known in the field of discrete nonlinear dynamics. Lattes published fundamental texts on nonlinear recurrence relationships (or maps), in particular in relation with the stable and unstable manifolds of fixed points and cycles.[8–11] These results provided another set of fundamental bases for the Toulouse group studies in *nonlinear discrete models*, and their complex behaviors, with the Julia and Fatou publications.

The first contact with complex dynamics (1958-1960) was the study of the parallel electronic integrator. When the circuit behaves as an integrator, its equation leads to a *critical case in the Ljapunov sense*. It results that the linear approximation is not sufficient to explain the circuit behavior, and the circuit may generate phenomena of complex, and very sensitive, oscillations for certain parameter choices. At that time, studying a transistor amplifier, I. Gumowski showed in 1959[12–17] that a reliable description of the dynamics implies a model containing not only a small parasitic capacitance, but also a time lag which has a nonlinear dependence with the input amplitude. Here the lag is due to the transit time of the charge carriers, an increase of the electric field at the collector junction leading to a decrease of the diffusion distance.

Such studies had the advantage of drawing attention to complex phenomena generated by nonlinear characteristics, and motivated the orientation of future researches in nonlinear dynamics. The Toulouse group was born from the numerous questions induced by these electronic circuits, and the related discussions with I. Gumowski.

The present text is an abbreviated version of chapter 8 of the book *"The Chaos Avant-Garde. Memories of the early days of the chaos theory"*.[18] It relates the first period (i.e. the "prehistoric" one ending in 1976, the emergence date of the word *chaos* and the beginning of the "historic" times) of the Toulouse group activity in complex nonlinear dynamics, and its collaboration with I. Gumowski. During this period, the most part of the available papers were essentially published in French, which explains that they have remained relatively unknown, and that some of their results were rediscovered. Each of the following sections is devoted to a topic which led to observation of complex dynamics, basins and attractors of two-dimensional noninvertible maps, resonant bifurcations, two-dimensional area-preserving map, one-dimensional noninvertible maps, and several applications in engineering. Figure 1 gives a simplified view of the research fields, when the group began its activity.

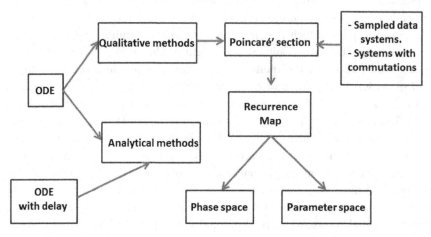

Fig. 1.   1958-1975 Research areas of the 'Toulouse group.'

2. Basin Boundaries of Two-dimensional Noninvertible Maps (1963-1975)

2.1. *Genesis of the results*

After two years of military service, my researches were oriented toward *nonlinear sample data control systems*, and some control and electronic systems using different types of modulations. The corresponding models were discrete in the form of recurrence relationships (equivalent designations: maps, point mapping, iteration, substitution). More particularly the studies were concentrated on the class of models described by *two-dimensional noninvertible maps* (nonlinear by nature). In this field, the determination of the basin boundary of a stable stationary state is of prime importance for practical problems. Indeed it gives the domain of perturbations (classed as changes of initial conditions) that the system can undergo without changing its qualitative behavior.

In 1963 the results on nonlinear maps were in an underdeveloped state with respect to those related to ordinary differential equations. Our first basic reference was the book *Leçons sur les récurrences et leurs applications*,[19] giving an account of the French school results on iteration. This school (Auguste Grévy, Dénes Koenigs, L. Leau, Ernest Lémeray, Jacques Hadamard, Samuel Lattès, Gaston Julia, Pierre Fatou) was the most active one in this field from the end of the 19th century to around 1930.

With respect to authors of the iteration French School, I. Gumowski and I were more particularly attracted by the "global" study given in the Julia and Fatou's papers related to one-dimensional maps with a complex variable.[20,21] Such maps concern a restricted class of two-dimensional noninvertible maps, that of two-dimensional maps with real variables, defined by two functions satisfying the Cauchy-Riemann conditions. In spite of this limitation, Julia and Fatou offered a starting point for studies of basin boundaries generated by general two-dimensional noninvertible maps with real variables. The results obtained by these authors can be summarized as follows. The critical points, and the set E of all the repulsive cycles, play a prime role for defining the map properties. The set $E\prime$, derived from E, contains E, and E is perfect. In page 49 of Ref.[20] it is said that "*la structure de E' est la même dans toutes ses parties*", which means that the $E\prime$ structure is self similar, called *fractal* from 1977. "Nonlinear analysts" were not interested by such questions when we began such researches. Now, it is well known that the set $E\prime$ (called *Julia set* since the years 1980) is a *basin boundary*, when the map generates an attractor.

Paul Montel (Director of the "*Institut Poincaré*" in Paris) and Julia were still alive at that time. It was not the Fatou's case. So at first Gumowski and I met Montel in 1963 for information about the existence of eventual results on basin boundaries, when they are generated by two-dimensional real maps not satisfying the particular Cauchy-Riemann conditions. He told us the lack of such an extension, as far as he knew. Then I contacted Julia, approximately 70 years old at that time. Julia told me that, due a health state partially related to his old war wounds, he has left his mathematical activities since many years. Nevertheless after having confirmed Montel's opinion, he added that more general cases might present high difficulties associated with a lot of possible different qualitative situations. Such difficulties emerged gradually and laboriously in the Toulouse group researches.

From this contact with Montel and Julia, with I. Gumowski we decided to begin studies on basin boundaries of two-dimensional real noninvertible maps by collecting many very simple generic examples, illustrating different qualitative situations. They essentially concerned quadratic and cubic maps, or piecewise linear ones, a phase plane point having at most three real rank-one preimages. A first basic tool was obtained with the introduction of a new singularity, the *critical curve* (a *critical set* for higher dimensions) as locus of points having two merging rank-one preimages (inverses).[22] As for the *critical point* related to a one-dimensional map, a critical curve separates regions the points of which have a different number of rank-one preimages. A second tool was given by the determination of the stable manifold of a saddle fixed point, or cycle, from the series expansions defined by Lattès. This permits to obtain a "germ", the inverse iterates of which lead to get the *stable manifold of a saddle point*.

Following this way, before approaching more complex situations as in Refs.,[23,24] a first step was the study of several "regular" smooth, or non smooth, examples given in[25,26] and references therein, for identifying different structures of *simply connected basin boundaries*, with a situation on the boundary (see p. 11, Sec. 4.6 in Ref.[23]), at that time not well identified, described as a *"cycle of very high period, perhaps infinite"*. After having understood many "regular" examples, this method led from 1968 to three more complex examples of basins, one having properties of self-similarity, generated by two-dimensional real maps not satisfying the Cauchy-Riemann conditions.

2.2. *Complex organization of basins (1968-1973)*

The starting point *"fractal* structure" (at that time the word did not exist) of *non connected basins* was the understanding of a very simple quadratic map generating a basin made up of only two non connected parts. This illustrated the basic global bifurcation *"simply connected basin* ↔ *non-connected basin"*[36] (see also pp. 243-244 in Ref.[22]). Imbedding this map into a cubic one leads to a *fractalization of the basin*. This imbedding is a cubic "perturbation" of the initial map. As far as I know, this imbedding into a cubic map[24] has given the first example of a fractal non-connected basin for maps which do not satisfy the Cauchy-Rieman conditions. At that time the word "fractal" did not exist, but the description given in Ref.[24] shows clearly the self similarity properties of the basin structure.

A case of boundary fractalization is also described in Alain Giraud's thesis,[27] after a *non classical global bifurcation* corresponding to the creation of a whole segment made up of unstable period two cycles. After the bifurcation the mechanism of period doubling was identified for the cycles initially located on this segment. Details are given pages 252-255 of Ref.,[22] and pages 399-416 of Ref.[28] Another example of such a non classical bifurcation is also given in Ref.[22] when a critical parameter value gives rise to a circle made up of period 2 cycles, one of the two eigen-values (S_1, S_2) (multipliers) being $S_1 = +1$. (details pages 255-260 in Ref.,[22] and 437-447 in[28]). Crossing through this parameter value, the boundary separating two basins undergoes a very strong modification. It is worth noting that such a bifurcation issued from a one-dimensional set made up of cycles with a multiplier + 1; or -1, even now seems quasi-unknown.

A *multiply connected basin* corresponds to a dual situation, with respect to the non-connected basin one. It seems that the example given in Ref.[29] (1973) is the first one published for noninvertible maps.

In these examples the role of the critical curve appears essential for existence of one of the three basic basin situations: *simply connected, nonconnected, multiply connected*, and the related bifurcations. A detailed explanation of the fundamental mechanisms leading to the different cases of Section 2.2 can be found, in English, in the relatively recent book.[28]

3. Chaotic Attractors of Two-Dimensional Noninvertible Maps (1968-1975)

From 1968, the study of some maps showed what has been after called *chaotic attractor*, or *strange attractor*, and also *strange repeller*.

A paper published in *Automatica* (1969)[26] described an attractor, generated by a piecewise linear map. This attractor was a period two area, without stable cycle, called *attractive limit set* in this paper, *stochastic area* in Ref.,[29] and *chaotic attractor*, or *chaotic area* when this word was introduced in 1975. The Toulouse group also used "*Pulkin's phenomenon*" associated with such behaviors. This designation was inspired from a Russian paper published in 1950,[30] showing that if a one-dimensional noninvertible map generates infinitely many unstable cycles and increasing classes of their limit points, this may lead to complex iterated sequences. The example of Ref.[26] was considered in[29] with more details, indicating the role of critical curves and related bifurcations. It permitted to define the boundary of complex attractors generated by two-dimensional non-invertible maps. This boundary is made of an arc of the critical set and its successive images until a certain rank (more details are given in the 1996 book[28]). It was numerically observed that the contact of the attractor boundary with the basin one leads to the destruction of this attractor. This bifurcation was studied more completely in 1978.[31]

Another interesting example was obtained by approximating the piecewise linear map by a smooth one. This was made by replacing a neighborhood of the non smoothness points with a polynomial of degree 2.[29] The chaotic attractor generated by this map presents a different "internal structure" with respect to the above nonsmooth map. Indeed arcs of the successive images of the critical curve correspond to higher density of iterated points for the smooth case, and separation of areas with different densities of iterated points in the nonsmooth case. Such "internal structures" were explained later in 1977[32,33] considering the rank-m images of critical curves.

Another type of attractors, in particular generated by maps resulting from a perturbation of conservatives cases, were presented in Refs.,[22,29] and also published in the Proceedings of IFIP (Stockholm, 1974).[33]

4. Normal Forms for Resonant Bifurcations (1969-1974)

These normal forms played a central role for the study of some complex dynamical behaviors, in the framework of the Toulouse group. In particular they were firstly applied to resonant Hamiltonian bifurcations generated by area-preserving maps. Here details about this topic, and the list of the old related references, are not given, but they can be obtained from chapter 3 of Ref.,[22] its annexes C, D (pp. 425-440) of Ref.,[22] and chapter 5 of Ref.[34]

A first set of studies was made for a fixed point of a two-dimensional map, in the following situations of critical cases in the Ljapunov's sense (i.e. the fixed point nature cannot be defined from the linear approximation, but by the map nonlinear terms):

- The fixed point multipliers (eigenvalues) are $S_1 = \pm 1$, $|S_2| \neq 1$. Related bifurcations by crossing through such critical cases were considered in the general case.
- The fixed point multipliers are $S_1 = S_2 = +1$, or $S_1 = -S_2 = 1$, or $S_1 = -S_2 = -1$, i) when the linear approximation matrix is reducible to the diagonal form, ii) when it is not reducible to the diagonal form. Related bifurcations by crossing through such critical cases were considered.

A second set of results were formulated for a fixed point, the multipliers (eigenvalues) of which are $S_{1,2} = \exp(\pm j\varphi)$. For this purpose a variant of the *Cigala's method* was used. This method was described in Ref.[35] (1905) for solving the particular problem of stability in two-dimensional conservative diffeomorphisms. Extended to dissipative two-dimensional maps,[36,37] having a sufficiently high degree of smoothness, the reduction to a normal form consists in identifying and isolating the "dominating terms" of the nonlinear parts. This objective is achieved by means of a sequence of nonlinear transformations of almost-identity type, which successively remove the non dominating terms.

This critical case leads to two essentially different situations. The most complex one, called *exceptional case*, occurs when the angle φ is commensurable with 2π, $\varphi = \frac{2k\pi}{q}$. It leads after a certain number of applications of Cigala's transformations to a problem of denominators which cancel, and small denominators near this angle value. Then at this level the normal form is obtained, because it is impossible to continue the process of successive transformations. The second situation, called *non-exceptional case*, is related to the angle φ, either non-commensurable with 2π, or if it is commensurable it does not lead to a problem of small denominators.

Bifurcation by crossing through an exceptional case, or a non-exceptional case, from a parameter variation, is very often wrongly called "Hopf bifurcation". Since 1969 the Toulouse group has always called this qualitative change *Neïmark's bifurcation*. Indeed the first contribution is due to Ju Neïmark (1963-1968) when φ is non-commensurable with 2π, in the particular case of the crossing through a *complex focus of multiplicity one*, giving rise to only one invariant closed curve.[38–41] Moreover the crossing through the exceptional cases $\varphi = \frac{2\pi}{3}$, $\varphi = \frac{2\pi}{4}$, was considered by this author in Ref.[41] The Neïmark's results furnished some of the basic tools for the Toulouse group studies.

In the *non-exceptional case* the Cigala's method permitted to extend Neïmark's results to the crossing through the situation of a *complex focus of multiplicity m*, $m \geq 1$, when $S_{1,2} = \exp(\pm j\varphi)$. This bifurcation may give rise to s, $0 \leq s \leq m$, invariant closed curves.[42–44]

The *exceptional cases* lead to new singular points of different types, those defined locally by $2q$ asymptotes (complex saddle) giving the resonant situations.[43,44]

Such studies of exceptional and non-exceptional cases were followed by their equivalent related to a four-dimensional map, a fixed point having two pairs of complex multipliers in this case. Due to the presence of two angles φ_1 and φ_2 the exceptional case gives rise to 8 basic situations (Ref.[45] and appendix C of Ref.[22]).

All papers of this section were written in French at that time, but their content was after reproduced in English in the book.[34]

5. Two-Dimensional Conservative Maps (1970-1975)

From 1966 to 1976 I. Gumowski occupied a position of Senior Physicist in the *"European Organization for Nuclear Research"* in Geneva. In this framework he had particularly to study the problem of *"stochastic"* instability in accelerators and storage rings. Such instabilities appear to increase with the amount of non-linearity. Physically this amount increases with the self-fields, and the latter increase with beam intensity. Here "stochastic", adjective also used in the Geneva research center, is related to what was called "chaotic" from 1975. Nevertheless here "stochastic" will be used in preference to chaotic in order to remain in the vocabulary context of that time. Gumowski's new position led to a collaboration with the Toulouse group on a new topic: two-dimensional conservative (or Hamiltonian) maps. Such maps are area-preserving. Considering the results obtained in the framework of this collaboration, it must be noted that Gumowski's part was more important than the Toulouse group one. Gumowski showed that a map describing the transverse motion of a particle in an accelerator, or a storage ring, can be found in a straightforward manner if the assumption is made that the ring is composed of localized elements only, and can be reduced to a two-dimensional conservative one.

This led to the following steps

- Study of unbounded quadratic map and their bifurcations
- Study of unbounded cubic map.
- Study of map with bounded non-linearity, introducing new features leading to classify cycles into at least four distinct families : internal ones surrounding the main center fixed point, internal ones surrounding a secondary center, external ones surrounding both the two centers, mixed ones, characterized by the property that all their points are located inside the areas bounded by the intersections of the two manifolds of a main saddle fixed point. Points of both internal and external cycles converge separately to the homoclinic points of the saddle manifolds, and the convergence law was identified. Moreover situations of Birkoff's *"unstable centers"* were found in the maps study.

Many new results were obtained for such families of maps. Details with their related references (not given here) are given in Chapter 6 of,[22] and Chapter 3 of Ref.[46]

6. Study of One-Dimensional Non-Invertible Maps (from 1972)

The bases of this study are:

- Pulkin's paper,[30] published in 1950. In the more general case of piecewise continuous maps, or continuous ones, Pulkin was the first to show that the presence of unstable cycles, and their limit sets, may generate complex oscillating iterated sequences (i.e. in this case chaos). His publication deals with one-dimensional maps generating infinitely many unstable cycles, giving rise to what he named *completely invariant sets*. Such sets are related to the existence of limit sets of different classes. So infinitely many limit points of the unstable cycles set, when their period tends toward infinity, lead of *class 1 limit set*. The limit sets of class 1 generate limit points of *class 2*, and so on until limit points of *class* ∞. The increasing rank classes of limit points leads to what was called 25 years after *fractal structure*.
- Myrberg's[47–50] and Sharkovskji's[51] results drew attention to the prime importance of their contribution directly related to the Toulouse group preoccupations. The results of these authors are complementary. They gave from 1972 the reference points which permitted, step by step, to identify the global fractal bifurcation organization generated by smooth maps defined by a function with only one extremum. This organization was called *box-within-the-box bifurcation structure* [a].

Before famous Li & Yorke's paper "*Period three implies chaos*",[58] *Sharkovskij's cycles ordering*[51] implied this assertion saying that period 3 leads to the existence of infinitely many cycles of any period k. Indeed in general these cycles are unstable (Pulkin situation), except some of them which may be stable. Likewise this situation is implicitly related to a case which leads either to *stable chaos*, or to *unstable chaos*.

From the Pulkin's results, the Sharkovskij's cycles ordering compared with the Myrberg's one was the germinal point of a thinking on cycles classification generated by an unimodal map. Indeed the Sharkovskij's ordering is general in the sense that it concerns general forms of continuous of one-dimensional maps with any extremums, but cannot discern between the cycles having the same period k, the number of which drastically increases with the period growth (see p. 96 in Ref.[34]). For example, in the case of unimodal maps (i.e. with only one extremum) the number of cycles of period $k = 30$ is $N_k = 35790267$, and the number of bifurcation values generating them is $N_\lambda(k) = 17895679$. Myrberg's ordering is limited to the particular case of unimodal maps, but it permits to differentiate cycles of same period by their *rotation sequence* (permutation order of their points), which leads to a sharp classification based on two indexes k (cycle period), j the cycle

---

[a] "*Structure boîtes-emboîtées*" in French [52–56] by the Toulouse group, and "*embedded boxes*" by Guckenheimer when he quoted these results,[57] more particularly saying "*the decomposition of non-wandering sets of maps, studied by Leo Jonker, David Rand, can be explained in different terms making use of Mira's concept of the embedded boxes*".

numbering in the Myrberg ordering of rotation sequences related to cycles having the same period k. Considering these two basic results as a starting point, the guide line, adopted to go further in the study of unimodal maps, was to introduce the set of *critical points* of rank $r = 1, 2, 3, ...$, (i.e. the increasing sequence of images of the map extremum). Such a set is made up of "non-classical" singularities introduced by the map non-invertibility. This leads to identify *non-classical bifurcations* corresponding to the merging of two singular points of different nature: a critical point and an unstable cycle. In Myrberg' sense a *spectrum* is the infinite sequence of bifurcations by period doubling related to a basic cycle $(k; j)$ created from a fold bifurcation (multiplier $S = +1$). The introduction of this non-classical bifurcation permitted a new classification of *Myrberg' spectra*, through the identification of the fractal box-within-the-box bifurcation structure (or embedded boxes).

All these fundamental results have been passed over in silence in most of contemporary papers dealing with this subject, which has a very large vogue since 1978. The most part of these results have been very often attributed to authors who rediscovered them after using another forms of quadratic map such as the logistic map, or maps of the unit interval. So the publication[56] (1976) describes the situation called *"cycle en valeur moyenne"* near one of the two boundaries of each box, which was rediscovered in 1980 as *intermittency phenomenon*. The other box boundary, called in [56] *"segment stochastique cyclique en valeur moyenne"*, was rediscovered in 1982 under the name *chaotic attractor in crisis*, or *boundary crisis*. All the results of this section were published in French before 1976, but they are presented in English in Ref.[34] with more details.

7. Applications

After a study of a nonlinear control sampled data system,[23,24] the group researches were oriented toward rectifiers *"alternating current → direct current"* using thyristors with voltage feedback, or current feedback, the load being made up of a resistance and an inductance (1967-1971). A discrete model in the form of a recurrence relationship was constructed without any approximation, the discrete states of this systems being determined at two consecutive switching times of the thyristors. The analytical form of this model is very complex and bulky, but can be programmed in a computer without any difficulty. It is of implicit and parametric type, including several relations.[22,27,59–61] If it might be analytically solved, the explicit form corresponding to the determination having a physical sense would be :

$$X_{n+1} = F(X_n, t_{n+1}), \qquad g(X_n, t) = 0, \qquad X(t_n) = X_n$$

where X is the state vector, t_{n+1} is the smallest root, larger than t_n (switching time), of the relation $g(X_n, t) = 0$. For a voltage feedback X is a one-dimensional vector. With a current feedback X is a two-dimensional vector. In such a problem a stable fixed point corresponds to the user specifications about what is called *residual ripple* of the rectifier output, the mean value of which is associated with the system input. Figures 2-4 illustrate some of the results obtained at that time. More details are given in Giraud's thesis,[27] also in pages 441-460 of Ref.,[22] a shorter version with figures being accessible in Chapter 8 of Ref.[18]

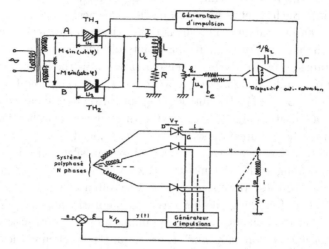

Fig. 2. Rectifier alternating current → direct current: 2 phases and N phases types, with "saturation" protection. The link "AC" corresponds to a voltage feedback. The link "BC" corresponds to a current feedback. A period k cycle is related to a rank-k subharmonic of the specified frequency $\frac{N\omega}{2\pi}$ (residual ripple). The subharmonic amplitude is all the higher as the rank-k is high.

It is important to note that Giraud's thesis (1969) (see page 190 of Ref.[27]) introduced the expression "*cycle stable d'ordre élevé*" [b] (Fig. 3e) which was used for convenience sake in order to qualify a steady state without any periodicity (obtained in the framework of this study), called "*pseudo-periodic*" by this researcher. Such a denomination gives an example of the fluctuations in the vocabulary choice, revealing the group perplexity in presence of phenomena identified as *chaotic* after 1975. At a time of very bad performances of computers, Giraud defined exactly the basin of the specified residual ripple, in presence of another attractor (a period 3 chaotic one), and bifurcation curves in the parameter plane (K gain of the control amplifier, τ time constant of the load). Before the Toulouse group studies, made in the years 1967-1971, the manufacturers of rectifiers thought that subharmonic and complex behaviors in such systems were essentially due to asymmetries of the phases of the secondary winding, and "antenna effects". The Toulouse group results have shown that, even with a perfect phases symmetry, the thyristors nonlinearities generate these non-wanted behaviors. Moreover a correct choice of the gain K for a given τ can lead to the global stability of the residual oscillation specified by the users.

---

[b]Large period stable cycle.

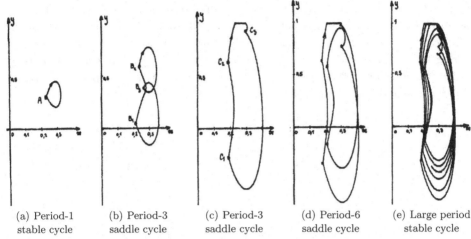

(a) Period-1 stable cycle
(b) Period-3 saddle cycle
(c) Period-3 saddle cycle
(d) Period-6 saddle cycle
(e) Large period stable cycle

Fig. 3. Continuous steady states of the rectifier, the switch curve is made up of periodic straight line segments. Phases number $N = 2$, reduced amplifier gain $K = 6$, reduced time constant of the load $\tau = \frac{L\omega}{R} = 10$, input reduced value $E = 0.3$. Two attractors coexist. One (case (a) "*cycle 1 stable*") is related to the specified residual ripple (fixed point P, i.e. period 1 cycle for the associated recurrence relationship). Case (e) denoted "*cycle stable d'ordre élevé*" was presented in Ref.[27] as a *state without any periodicity* (which after has appeared chaotic). Cases (b) and (c) correspond to two different period 3 saddles. Case d corresponds to a period 6 saddle (period doubling of the case (c)). According to Giraud's thesis[27] (1969).

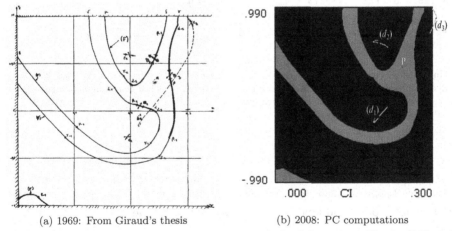

(a) 1969: From Giraud's thesis
(b) 2008: PC computations

Fig. 4. Boundary of the basins of the two attractors obtained from the stable manifold of the period 3 saddle (Fig. 3b). Here $x = \frac{RI}{M}$, y is a reduced value of the amplifier output (cf. Fig. 2). Here the current I is positive (physical sense). The basin of the fixed point P is non connected. Figure (a) was obtained in 1969 after very long computer work and Figure (b) was obtained in less than one minute with an ordinary PC. In the latter, the read area (light grey) is the basin of the fixed point P. The blue (dark) area is the basin of a period 3 chaotic attractor d_i ($i = 1, 2, 3$), containing the period 3 saddle (Fig. 3c) and the period 6 saddle (Fig. 3d), made up of arcs folded back into themselves infinitely many times.

It is worth noting that the above study was preceded in Ref.[27] by a theory of what was called at that time the "*systèmes à commutations*" [c] whose one example is shown in Figure 5, which has appeared to belong to what has been called after "*Hybrid Systems*". The "systèmes à commutations" theory was elaborated for preparing the study of rectifiers "*alternating current* → *direct current*". A more extended presentation of these results (1966-1969) was given later in pages 441-445 in Ref.[22] and pages 366-370 in Ref.[61]

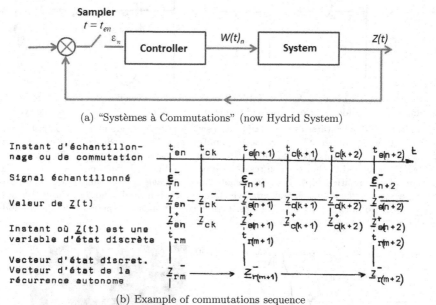

(b) Example of commutations sequence

Fig. 5. In hybrid system (a), $t = t_{ck}$ is a commutation time related to changes in the controller, or to $Z(t)$ discontinuities, or changes for the system mathematical model. The commutation time t_{ck} is solution to $g_k[Z(t), t] = 0$, and the sampling time t_{en} is solution to $g_n[Z(t), t] = 0$. (According to Giraud's thesis.[27])

A second application related to the longitudinal motion of particles in an accelerator is due to I. Gumowski.[46,62] Other applications giving rise to complex dynamics have been described for systems of satellite attitude controlled by a frequency modulator of second kind,[63] and in a satellite with inertia being a periodic function of time.[64]

8. Two Exhibitions of Chaotic Images (1973 and 1975)

The International Colloquium "*Transformations Ponctuelles et leurs Applications* [d] held in the framework of "Laboratoire d'Automatique et d'Analyse des Systèmes"

---

[c]Switching systems.
[d]"*Point mapping and Applications*".

(LAAS), in Toulouse, from Sept. 10 to 14, 1973. Prof. Lagasse, director of the LAAS, was its chairman. I. Gumowski led the scientific committee. I was vice-chairman. Eleven papers were devoted to "stochastic" (chaotic) behaviors, among the thirty ones published in the Proceedings. Let us note that Joe Ford, John H. Bartlett, Marek Kuczma, Igor Gumowski, M. Urabe, Michel Hénon, Boris V. Chirikov, and V. K. Melnikov were authors of some of them.

The colloquium was accompanied by an exhibition of chaotic images (Fig. 6) produced by equations

$$\begin{cases} x_{n+1} = y_n + \mu x_n + \dfrac{2(1-\mu)x_n^2}{1+x_n^2} + \alpha \left(1 - \beta y_n^2\right) y_n \\ y_{n+1} = -x_n + \mu x_{n+1} + \dfrac{2(1-\mu)x_{n+1}^2}{1+x_{n+1}^2} \end{cases} \quad (1)$$

studied by the Toulouse group. It is in this framework that I announced this exhibition of "stochastic" images (the word *chaotic* did not yet exist), quoting David Birkhoff's papers dealing with the laws of aesthetic (see pp. 320-364 in Ref.[66]), and Poincaré dealing with the aesthetic emotion which can be communicated by mathematics in the following terms:[65]

"Le savant digne de ce nom, le géomètre surtout, éprouve en face de son œuvre la même impression que l'artiste; sa jouissance est aussi grande et de même nature. Si je n'écrivais pour un public amoureux de la Science, je n'oserais pas m'exprimer ainsi; je redouterais l'incrédulité des profanes. Mais ici, je puis dire toute ma pensée. Si nous travaillons, c'est moins pour obtenir ces résultats positifs auxquels le vulgaire nous croit uniquement attachés, que pour ressentir cette émotion esthétique et la communiquer à ceux qui sont capables de l'éprouver." [e]

I took the liberty of saying that these images had begun to manifest such an emotion in a form opened not only to specialists as Poincaré said, but also to a broad audience (page 27 in Ref.[67]), due to the new possibilities offered by numerical simulations. The same exhibition, entitled *"Morphogénèse et Mathématiques"*, was organized by Mr. Marcel Barthes, Director of the Alliance Française (Rio de Janeiro) in the *Centre Culturel de la Maison de France*, from May 8 to 30, 1975.

---

[e] "The scientist worthy of such a name, especially the surveyor, experienced in front of his work the same impression as an artist, his enjoyment is as great, and of the same nature. If I was not writing for an audience fond of science, I would not say so, I dread the incredulity of laymen. But here I can express my thoughts. If we work, it is less for these positive results which the vulgar believes us only attached, as to feel that aesthetic emotion and communicate it to those who are able to experience it."

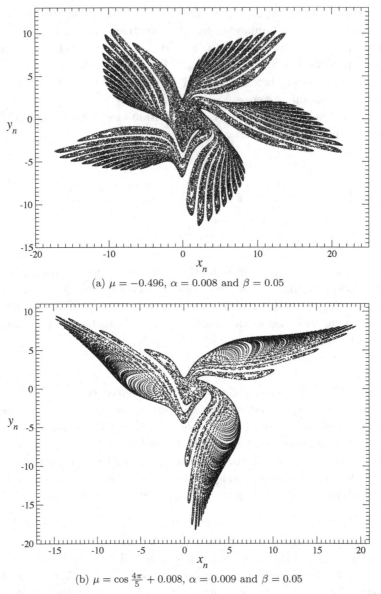

Fig. 6. Two "aesthetic" attractors produced by map (1) and presented in 1973 at the colloquium "Point Mapping and Applications".

9. Conclusion

The "history" of chaotic dynamics begins from around 1976, when the word "*Chaos*" was initially introduced in dynamics, with the well-known May's paper[68] [f] which

---

[f]The word "*Chaos*" appeared one year before in Li and Yorke's paper entitled Period three implies chaos.[58] Editor's note.

quotes the Toulouse group results.[52,53] This group continued a fruitful collaboration with I. Gumowski, who occupied a new Professor position in the Faculty of Sciences of Paul Sabatier University, till his retirement in 1987. The results obtained during the "prehistoric" times were the starting point of many developments of the problems considered in this text.

Among such developments the *fractal box-within-the-box bifurcation structure* (or *embedded boxes*) was extended to one-dimensional maps defined by functions with several extrema, to two-dimensional diffeomorphisms in a foliated parameter plane (i.e. considered as made up of sheets, each one being associated with a well defined cycle). Basic complex communications (*spring area, saddle area* and *crossroad area*) between these sheets were identified in a parameter plane, and their qualitative changes when a third parameter varies.[69–72] In the years 1986-1987, a six months stay of H. Kawakami (University of Tokushima) in the Toulouse group, led him to improve the properties of the *box-within-the-box structure*, this by introduction of the notion of "*adjoined cycles*" (pages 130-131 and 319-326 in Ref.[34]). It resulted a *symbolic phase plane* representation (pages 341-349 in Ref.[34]), a "*contracted representation of the parameter plane*" via the notions of "*connection chains*" and "*communication cells*" (pages 373-395 in Ref.[34]).

Using a Poincaré section, various types of models in the form of ordinary differential equations were studied from the above point of view of the foliated parameter plane and its communication areas.[72–74,76–80]

Another type of development of the *fractal box-within-the-box bifurcation structure* notion is its relations with the *Julia sets J* generated by a quadratic Dim1 complex map, that is a two-dimensional map defined by functions satisfying the Cauchy-Riemann conditions. In Ref.,[81] an identification of the J singularities and their localizations from the symbolism associated with this fractal organization was performed. In a companion paper,[82] the embedding of the quadratic map into a two-dimensional noninvertible map was investigated, with a study of the bifurcations when the embedding parameter tends toward zero. This approach permits to explain the bifurcations route toward a Julia set.

Now consider the situation out of the above local aspect. It is interesting to mention that the post-1976 developments of the Toulouse group have occurred in a new international scientific environment for researches in dynamics. Indeed nonlinear dynamics was not a favorite choice for pure mathematicians in western countries before 1976. This in spite of the highly significant Smale's contribution of prime interest, leading in particular to the "*horseshoe*" map properties (1963), and other rare interesting publications. In the field of nonlinear dynamics, the most part of publications out of SSSR, since the beginning of the last century and before 1977, concerned authors coming from engineering, physics, and applied mathematics. Therefore the Toulouse group publications were not written according to the standards of what can be called an "abstract theory of nonlinear dynamics", implicitly defined by certain pure mathematicians after 1976. The purpose of the

Toulouse group was to facilitate the communication with researchers in applied sciences. Indeed the simplest possible terminology has been adopted, in order not to add artificial vocabulary-hurdles to the intrinsic difficulty of the subject. Ars-pro-artis generalizations have been deliberately sacrificed in order not to obscure the dominating internal mechanisms of dynamic processes, and thus preserve a phenomenological transparency (cf. p.1 in Ref.[46] and preface of Ref.[34]). From this point of view the Toulouse group approach has been in agreement with Birkhoff (quoted by Marston Morse[83]), when he considered that

> "The systematic organization, or exposition, of a mathematical theory is always secondary in importance to its discovery, ... some of the current mathematical theories being no more than relatively obvious elaborations of concrete examples."

Moreover a relatively large part of the results obtained at those "prehistoric" times of "chaotic dynamics" were essentially attained via a numerical way (with classical checks of precision), from programs guided by the fundamental considerations of the qualitative and analytical theories of nonlinear dynamics. The resulting formulation of problems has therefore appeared primitive, even simplistic, the terminology old-fashioned, and the results non relevant for many abstractly inclined researchers. Some of them found occasions to feel irritated by such a *modus operandi*, for example through rejection of manuscripts. More precisely it is possible to say that nonlinear dynamics has been the action field of two sets of scientists since some 35 years (the "historic" times of chaotic dynamics). The first set is made up of pure mathematicians. The second one is formed with some applied mathematicians, and more essentially with scientists coming from Engineering, or Applied Sciences. If we except the Poincaré and Birkhoff cases and very rare others, the second set was historically the first which considered such questions since the beginning of the last century, in order to solve practical problems of oscillations in their own technical fields. The researchers of this set implicitly adopted as their own Joseph Fourier's statement:[84]

> " L'étude approfondie de la nature est la source la plus féconde des découvertes mathématiques. Non seulement, cette étude, en offrant aux recherches un but déterminé, a l'avantage d'exclure les questions vagues et les calculs sans issue; elle est encore un moyen assuré de former l'Analyse elle-même et d'en découvrir les éléments qu'il importe le plus de connaître et de conserver. Ces éléments fondamentaux sont ceux qui se reproduisent dans tous les effets naturels."

Indeed this text says that the study of the Nature (that is the real world) is the most productive source of mathematical discoveries, providing the advantage of excluding vague problems and unwieldy calculations. It is also a mean to isolate the most important aspects to be known in the Mathematical Analysis, these fundamental aspects being those which appear in all natural effects.

The approaches of each of the two sets work with different standards, have different purposes, but each approach has its own merits and interest in its proper field. This depends on the study context. Nevertheless this situation is the source of high difficulties of communication between the two sets. They lie in the fact that what is important for one set is not for the other, and inversely. For the first set only results in a sufficiently general abstract form are significant. They consider that analyses performed in a particular study, or without using their patterns, are without any relevance. Some members of the first set do not accept the "non-use" of their standards for papers writing, which increases the above difficulties of communication. For the second set, only papers written in the least abstract form, using the least sophisticated mathematical tools, can transmit information permitting to solve well defined practical problems. This last set considers that the following "proposition": *the more a result is formulated in an abstract form, the less it can be used for solving a practical problem*, is a result of experience and dictated by the common sense. My contacts with many researchers of this second set have shown that some of them, as "impure" mathematicians daring to set foot in the reserved field of a noble caste, have felt a kind of contempt from certain (fortunately not all) pure mathematicians, considering that only their approach is significant and meaningful.

Acknowledgement

Christophe Letellier revised and improved the presentation of this chapter.

References

1. C. Mira, Some historical aspects of nonlinear dynamics. Possible trends for the future, *International Journal of Bifurcation and Chaos*, **7** (9 & 10), 2145-2174 (1997) — *The Journal of the Franklin Institute B*, **334** (5/6), 1075-1113, (1997).
2. C. Mira, Chua's circuit and the qualitative theory of dynamical systems, *International Journal of Bifurcation and Chaos*, **7** (9&10), 1911-1916 (1997) — *The Journal of the Franklin Institute B*, **334** (5/6), 737-744 (1997).
3. E. S. Boïko, *The Academician A. A. Andronov' school* (in Russian), Ed. Nauka, Moscow (1983).
4. J.-P. La Salle & S. Lefschetz, Recent Soviet contributions to ordinary differential equations and nonlinear mechanics, *Journal of Mathematical Analysis and Applications*, **2**, 467-499 (1961).
5. Ch. Hayashi, *Nonlinear oscillations in physical systems*, Mc Graw-Hill, New York (1964).
6. Ch. Hayashi, *Selected papers on nonlinear oscillations*, Nippon Printing and Publishing Company (1991).
7. Ch. Hayashi, Y. Ueda & H. Kawakami, Periodic solutions of Duffing's equation with reference to doubly asymptotic solutions, *Proceedings of the 5th International Conference on Non-Linear Oscillations*, Kiev, pp. 235-255 (1969).
8. S. Lattes, Sur les équations fonctionnelles qui définissent une courbe ou une surface invariante par une transformation, *Annali di Matematica*, III, **13**, 1-137 (1906).

9. S. Lattes, Sur la convergence des relations de récurrence. Surface invariante par une transformation, *Comptes-Rendus de l'Académie des Sciences*, **150**, 1106-1109 (1910).
10. S. Lattes, Sur les formes réduites des transformations ponctuelles dans le domaine d'un point double, *Bulletin de la Société Mathématique*, **39**, 309-345 (1911).
11. S. Lattes, Sur les suites récurrentes non linéaires et sur les fonctions génératrices de ces suites", *Annales de la Faculté des Sciences de Toulouse*, III, **3**, 73-124 (1912).
12. I. Gumowski, Sur un effet non linéaire dans les amplificateurs à transistor avec réaction, *Comptes-Rendus de l'Académie des Sciences*, **249**, 1959, 2514-2517 (1960) — **250**, 822-825 (1960).
13. I. Gumowski, J. Lagasse & Y. Sevely, Mise en équation d'un amplificateur à transistor non linéaire, *Comptes-Rendus de l'Académie des Sciences*, **250**, 1995-1998 (1960).
14. I. Gumowski, Sur le comportement d'un amplificateur à transistor non linéaire au voisinage de sa limite de stabilité, *Comptes-Rendus de l'Académie des Sciences*, **250**, 3142-3145 (1960).
15. I. Gumowski, Calcul de la réponse en fréquence d'un amplificateur à transistor non linéaire au voisinage de sa limite de stabilité, *Comptes-Rendus de l'Académie des Sciences*, **250**, 4322-4325 (1960).
16. I. Gumowski, Sur la réponse transitoire des amplificateurs à transistor avec réaction, *Comptes-Rendus de l'Académie des Sciences*, **253**, 1671-1674 (1961).
17. I. Gumowski, Calcul de la réponse transitoire à une onde carrée d'un amplificateur à transistor non linéaire, *Comptes-Rendus de l'Académie des Sciences*, **253**, 2207-2210 (1961).
18. C. Mira, I. Gumowski and a Toulouse research group in the "prehistoric" times of chaotic dynamics, In The Chaos Avant-Garde. Memories of the early days of the chaos theory, R. Abraham & Y. Ueda (eds), *World Scientific Series on Nonlinear Science*, Series A, **39** pp. 95-197 (2000).
19. P. Montel, *Leçons sur les récurrences et leurs applications*, (recueillies et rdiges par Jacques Dufresnoy & Éloi Lefebvre) Gauthier-Villars, Paris (1957).
20. G. Julia, Mémoire sur l'itération des fonctions rationnelles, *Journal de Mathématiques Pures & Appliquées*, VII, **4** (1), 47-245 (1918).
21. P. Fatou, Mémoire sur les équations fonctionnelles, *Bulletin de la Société Mathématique*, **47**, 161-271 (1919) — **48**, 33-94 and 208-314 (1920).
22. I. Gumowski & C. Mira, *Dynamique chaotique. Transition ordre-désordre*, Ed. Cépadués, Toulouse (1980).
23. C. Mira & F. Roubellat, Cas où le domaine de stabilité d'un ensemble limite attractif d'une récurrence du deuxième ordre n'est pas simplement connexe, *Comptes-Rendus de l'Académie des Sciences A*, **268**, 1657-1660 (1969).
24. F. Roubellat, *Contribution à l'étude des solutions des récurrences non linéaires et applications aux systèmes à données échantillonnées*, Thèse de Doctorat ès-sciences Physiques, **364**, Faculté des Sciences de l'Université de Toulouse (1969).
25. C. Mira, Etude de la frontière de stabilité d'un point double stable d'une récurrence non linéaire autonome du deuxième ordre, *Proceedings of International Pulse Symposium*, Budapest, D43-7/II, 1-28 (1968).
26. I. Gumowski & C. Mira, Sensitivity problems related to certain bifurcations in non-linear recurrences relations, *Automatica*, **5**, 303-317 (1969).
27. A. Giraud, *Application des récurrences à l'étude des systèmes de commande*, Thèse de Docteur-Ingènieur, **205**, Faculté des Sciences de Toulouse (1969).
28. C. Mira, L. Gardini, A. Barugola & J.-C. Cathala, *Chaotic Dynamics in two-dimensional noninvertible maps*, World Scientific, Series A on Nonlinear Sciences (Ed. L. Chua), **20** (1996).

29. J. Bernussou, Liu Hsu & C. Mira, Quelques exemples de solutions stochastiques bornées dans les récurrences autonomes du second ordre, Collected preprints of Colloque International du CNRS **229**, "Transformations Ponctuelles et Applications", Toulouse (September 1973), *Proceedings Editions du CNRS*, Paris, 195-226 (1976).
30. C. P. Pulkin, Oscillating iterated sequences (in Russian), *Dokl. Akad. Nauk SSSR*, **76** (6), 1129-1132 (1950).
31. I. Gumowski & C. Mira, Bifurcation déstabilisant une solution chaotique d'un endomorphisme du second ordre, *Comptes-Rendus d'Académie des Sciences A*, **286**, 427-431 (1978).
32. I. Gumowski & C. Mira, Solutions chaotiques bornées d'une récurrence ou transformation ponctuelle du second ordre à inverse non unique, *Comptes-Rendus de l'Académie des Sciences A*, **285**, 477-480 (1977).
33. I. Gumowski & C. Mira, Point sequences generated by two-dimensional recurrences, *Proceedings of Information Processing 74* (IFIP), Stockholm (1974). North-Holland Publishing Company, pp. 851-855 (1974).
34. C. Mira, *Chaotic dynamics. From the one-dimensional endomorphism to the two-dimensional diffeomorphism*, World Scientific, Singapor (1987).
35. A. R. Cigala, Sopra un criterio di instabilita, *Annali di Matematica*, III, **11**, 67-75 (1905).
36. C. Mira, Etude d'un premier cas d'exception pour une récurrence, ou transformation ponctuelle, du deuxième ordre, *Comptes-Rendus de l'Académie des Sciences A*, **269**, 1006-1009 (1969).
37. C. Mira, Etude d'un second cas d'exception pour une récurrence, ou transformation ponctuelle, du deuxième ordre, *Comptes-Rendus de l'Académie des Sciences A*, **270**, 332-335 (1970).
38. Yu. I. Neïmark, Method of maps in the nonlinear oscillations theory (in Russian), *Int. Conf. on Nonlinear Oscillations* (ICNO), Kiev (1961), *Proceedings Ukr. Acad. Nauk.* **2**, 285-298 (1963).
39. Yu. I. Neïmark, Stability of a fixed point of a map for a critical case (in Russian), *Radiofisika*, **3** (2), 342-343 (1960).
40. Yu. I. Neïmark, *Method of maps in the nonlinear oscillations theory* (in Russian), Ed. Nauka, Moscow (1972).
41. L. O. Barsuk, N. M. Belosludstiev, Yu. I. Neïmark & N. M. Salganskaja, Stability of a fixed point in a critical case. Bifurcations (in Russian), *Radiofisika*, **11** (11), 1632-1641 (1968).
42. J.-P. Babary & C. Mira, Sur un cas critique pour une récurrence autonome du deuxième ordre, *Comptes-Rendus de l'Académie des Sciences A*, **268**, 129-132 (1969).
43. C. Mira, Traversée d'un cas critique pour une récurrence autonome du deuxième ordre, sous l'effet d'une variation de paramètre, *Comptes-Rendus de l'Académie des Sciences A*, **268**, 621-624 (1969).
44. I. Gumowski & C. Mira, Bifurcation pour une récurrence du deuxième ordre par traversée d'un cas critique avec deux multiplicateurs complexes conjugués, *Comptes-Rendus de l'Académie des Sciences A*, **278**, 1591-1594 (1974).
45. C. Mira, Cas critique d'une récurrence, ou transformation ponctuelle, du quatrième ordre avec multiplicateurs complexes, *Comptes-Rendus de l'Académie des Sciences A*, **272**, 1727-1730, (1971).
46. I. Gumowski & C. Mira, Recurrences and discrete dynamic systems — An introduction, *Lecture Notes in Mathematics*, **809**, Springer-Verlag (1980).
47. P. J. Myrberg, P. J., 1958. "Iteration von Quadratwurzeloperationen. I". *Annales AcademiæScientiarum FennicæA*, **256**, 1-10 (1958).

48. P. J. Myrberg, Iteration von Quadratwurzeloperationen. II, *Annales AcademiæScientiarum FennicæA*, **268**, 1-10 (1959).
49. P. J. Myrberg, Sur l'itération des polynômes réels quadratiques, *Journal de Mathématiques Pures et Appliquées* IX, **41**, 339-351 (1962).
50. P. J. Myrberg Iteration von Quadratwurzeloperationen. III, *Annales AcademiæScientiarum FennicæA*, **336**, 1-10 (1963).
51. A. N. Sharkovskij, Coexistence of cycles of a continous map of a line into itself, *Ukrain. Mat. Journal*, **16** (1), 61-71 (1964).
52. I. Gumowski & C. Mira, Sur les récurrences, ou transformations ponctuelles du premier ordre avec inverse non unique, *Comptes-Rendus de l'Académie des Sciences A*, **280**, 905-908, (1975).
53. I. Gumowski & C. Mira, Accumulation de bifurcations dans une récurrence, *Comptes-Rendus de l'Académie des Sciences A*, **281**, 45-48 (1975).
54. C. Mira, Accumulation de bifurcations et structures boîtes emboîtées dans les récurrences et transformations ponctuelles, *7th International Conference on Nonlinear Oscillations* (ICNO), Berlin (September 1975), Proceedings, Akademic Verlag, Berlin (1977).
55. C. Mira, Sur la notion de frontière floue de stabilité, *Proceedings of the third Brazilian Congress of Mechanical Engineering*, Rio de Janeiro, **D-4**, 905-918 (1975).
56. C. Mira, Sur la double interprétation, déterministe et statistique, de certaines bifurcations complexes, *Comptes-Rendus de l'Académie des Sciences A*, **283**, 911-914 (1976).
57. J. Guckenheimer, The bifurcation of quadratic functions, *Annals of the New York Academy of Science*, **75** (1), 343-347 (1980).
58. T.-Y. Li & J. Yorke, Period three implies chaos, *American Mathematics Monthly*, **82**, 985-992 (1975).
59. R. Valette, *Etude du comportement à fort signal de systèmes de commande comprenant un redresseur polyphasé*, Thèse de Docteur-Ingénieur, **295**, Faculté des Sciences de Toulouse (1971).
60. R. Prajoux, Etude des générateurs utilisant un redresseur polyphasé et un filtre dynamique en tant que systèmes de commande, Thèse de Docteur ès Sciences Physiques, **462**, Université Paul Sabatier, Toulouse (1971).
61. C. Mira, Systèmes asservis non linéaires, Hermes, Paris (1990).
62. I. Gumowski, *Some properties of large amplitude solutions of conservative dynamic systems*, Part 1: Quadratic and cubic non-linearities, and Part 2: Bounded nonlinearities, Rapport CERN/SI/Int. BR/72-1, Genève (1972).
63. J. Bernussou, Contribution à l'étude des solutions des récurrences non linéaires. Application à l'étude de certains systèmes à modulation, Thèse de Doctorat ès-Sciences Physiques, **596**, Université Paul Sabatier de Toulouse (1974).
64. Liu Hsu, Contribution à l'étude des solutions des récurrences non linéaires. Application aux systèmes dynamiques conservatifs, Thèse de Doctorat ès-Sciences Physiques, **597**, Université Paul Sabatier de Toulouse (1974).
65. H. Poincaré, Note sur Halphen, *Journal de l'Ecole Polytechnique*, Cahier LX, 137-161 (1890).
66. G. D. Birkhoff, *Collected Mathematical Papers*, Dover Publications, New York, Vol. **3**, (1968).
67. C. Mira, Exposé d'Introduction, Colloque International du CNRS **229** *Transformations Ponctuelles et Applications*, (Toulouse September 1973), Proceedings Editions du CNRS, Paris, pp. 19-27 (1976).
68. R. M. May, Simple mathematical models with complicated dynamics, *Nature*, **261**, 459-470 (1976).

69. C. Mira & J. P. Carcassès, On the crossroad area-saddle area and spring area – crossroad area transitions, *International Journal of Bifurcation and Chaos*, **1** (3), 641-655, (1991).
70. J. C. Cathala, C. Mira & H. Kawakami, Singular points with two multipliers $S_1 = -1$, $S_2 = 1$ in the bifurcation curves of maps, *International Journal of Bifurcation & Chaos*, **2** (4), 1001-1004 (1992).
71. R. Allam & C. Mira, Crossroad area — dissymmetrical spring area — symmetrical spring area, and double crossroad area — double spring area transitions, *International Journal of Bifurcation & Chaos*, **3** (2), 429-435 (1993).
72. C. Mira, H. Kawakami & R. Allam, The dovetail bifurcation structure and its qualitative changes, *International Journal of Bifurcation & Chaos*, **3** (4), 903-919 (1993).
73. J. P. Carcasses & C. Mira, An autonomous ordinary differential equation generating alternating isoordinal cascades of cusps in a bifurcation plane, *Proceeding of International Conference on Iteration Theory* (ECIT'89), World Scientific, pp. 29-49 (1991).
74. C. Mira & L. Djellit, Bifurcation structure in a model of a frequency modulated CO^2 laser, *International Journal of Bifurcation & Chaos*, **3** (1), 97-111 (1993).
75. C. Mira & M. Qriouet, On a crossroad area — spring area transition occurring in a Duffing-Rayleigh equation with a periodic excitation, *International Journal of Bifurcation & Chaos*, **3** (4), 1029-1037 (1993).
76. M. Qriouet & C. Mira, Fractional harmonic synchronization in the Duffing-Rayleigh differential equation, *International Journal of Bifurcation & Chaos*, **4** (2), 411-426 (1994).
77. C. Mira, M. Touzani-Qriouet & H. Kawakami, Bifurcation structures generated by the non autonomous Duffing equation, *International Journal of Bifurcation & Chaos*, **9** (7) 1363-1379 (1999).
78. M. Touzani-Qriouet, C. Mira & H. Kawakami, Reducible fractional harmonics generated by the non autonomous Duffing-Rayleigh equation. Pockets of reducible harmonics and Arnold's tongues, *International Journal of Bifurcation & Chaos*, **10** (6), 1345-1366 (2000).
79. H. Khammari, C. Mira & J. P. Carcasses, Part I Behavior of harmonics generated by a Duffing equation with a nonlinear damping, *International Journal of Bifurcation & Chaos*, **15** (10), 3181-3221 (2005).
80. H. Khammari & C. Mira, Part II Behavior of harmonics generated by a Duffing equation with a nonlinear damping, *International Journal of Bifurcation & Chaos*, **19** (4) 1227-1254 (2009).
81. C. Mira & L. Gardini, From the box-within-a-box bifurcation organization to the Julia set. Part I: Revisited properties of the sets generated by a quadratic map with a real parameter, *International Journal of Bifurcation & Chaos*, **19** (1), 281-327 (2009).
82. C. Mira, A. Agliari & L. Gardini, From the box-within-a-box bifurcation organization to the Julia set. Part II: Bifurcation routes to different Julia sets from an inderect embedding of a quadratic complex map, *Internationl Journal of Bifurcation & Chaos*, **19** (10), 3235-3282 (2009).
83. M. Morse, George David Birkhoff and his mathematical work, *Bulletin of the American Mathematical Society*, **52** (5), 357-391 (1946).
84. J. Fourrier, *Théorie Analytique de la Chaleur*, Firmin Didot Ed., Paris, (1822). Reprinted by J. Gabay (1988).

Chapter 4

Can we trust in numerical computations of chaotic solutions of dynamical systems ?

René Lozi

Laboratoire J.A. Dieudonné — UMR CNRS 7351
Université de Nice Sophia-Antipolis
Parc Valrose
06108 Nice Cedex 02
France

Since the famous paper of Edward Lorenz in 1963 numerical computations using computers play a central role in order to display and analyze solutions of non-linear dynamical systems. By these means new structures have been emphasized like hyperbolic and/or strange attractors. However theoretical proofs of their existences are very difficult and limited to very special linear cases. Computer aided proofs are also complex and require special interval arithmetic analysis. Nevertheless, numerous researchers in several fields linked to chaotic dynamical systems are confident in the numerical solutions they found using popular software and publish without checking carefully the reliability of their results. In the simple case of discrete dynamical systems (e.g. Hénon map) there are concerns about the nature of what a computer find out : long unstable pseudo-orbits or strange attractors? The shadowing property and its generalizations which ensure that pseudo-orbits of a homeomorphism can be traceable by actual orbits even if rounding errors are not inevitable are not of great help in order to validate the numerical results. Continuous dynamical systems (e.g. Chua, Lorenz, Rössler) are even more difficult to handle in this scope and researchers have to be very cautious to back up theory with numerical computations. We present a survey of the topic based on these, only few, but well studied models.

Contents

1. Introduction 64
2. Continuous and Discrete Chaotic Dissipative Dynamical Systems: a Paradigm for Possibly Flawed Computations 65
   2.1. Some classes of dynamical system 65
   2.2. Poincaré map: a bridge between continuous and discrete dynamical system 66
3. Collapsing Effects 67
   3.1. Undesirable chaotic transient 67
   3.2. Enigmatic computations for the logistic map (1838) 69
   3.3. Collapsing orbit of the symmetric tent map 73
   3.4. Statistical properties 74
4. Shadowing and Parameter-shifted Shadowing Property of Mappings of the Plane 76
   4.1. Hénon map (1976) found by mistake 76

    4.2. Lozi map (1977) a tractable model 80
    4.3. Shadowing, parameter-shifted shadowing and orbit-shifted shadowing properties . 83
5. Continuous Models 85
    5.1. Lorenz attractor (1963) 85
    5.2. Geometric Lorenz attractor 87
    5.3. Lorenz map 89
    5.4. Rössler attractor (1976) 91
    5.5. Chua attractor (1983) 92
6. Conclusion 95
References 96

1. Introduction

Since the famous paper of Edward Lorenz in 1963,[1] numerical computations using computers play a central role in order to display and analyze solutions of nonlinear dynamical systems. By these means new structures have been emphasized like hyperbolic and/or strange attractors. However theoretical proofs of their existence are very difficult and limited to very special linear cases.[2] Computer aided proofs are also complex and require special interval arithmetic analysis.[3,4] Nevertheless, numerous researchers in several fields linked to chaotic dynamical systems are confident in the numerical solutions they found using popular software and publish without checking carefully the reliability of their results. In the simple case of discrete dynamical systems (*e.g.* Hénon map[5]) there are concerns about the nature of what a computer find out : long unstable pseudo-orbits or strange attractors?[6] The shadowing property which ensures that pseudo-orbits of a homeomorphism can be traceable by actual orbits, even if rounding errors are not inevitable, is not a great help in order to validate the numerical results.[7] Continuous dynamical systems (*e.g.* Chua, Lorenz, Rössler) are even more difficult to handle in this scope and researchers have to be very cautious to back up theory with numerical computations.[8,9] We present a survey of the topic based on these, only few, but well studied models.[10]

In Sec. 2 we define the paradigm for possibly flawed computations: continuous and discrete chaotic dissipative dynamical systems. In Sec. 3, some undesirable collapsing effects are highlighted for an example of strange attractor and for 1-dimensional discrete dynamical systems (logistic and tent maps). The shadowing properties (parameter-shifted, orbit-shifted shadowing) are presented in Sec. 4, together with the classical mappings of the plane into itself: the Hénon and Lozi maps. Finally in Sec. 5 the case of the seminal Lorenz model and its following metaphors: Rössler and Chua equations are examined.

2. Continuous and Discrete Chaotic Dissipative Dynamical Systems: a Paradigm for Possibly Flawed Computations

2.1. *Some classes of dynamical system*

Dynamical systems are involved in the modeling of phenomena which evolve in time. Their theory attempts to understand, or at least describe in form of mathematical equations, the changes over time that occur in biological, chemical, economic, financial, electronic, physical or artificial systems. Examples of such systems include the long-term behavior of solar system (sun and planets) or galaxies, the weather, the growth of crystals, the struggle for life between competing species, the stock market, the formation of traffic jams, etc.

Dynamical systems can be continuous *vs* discrete, autonomous *vs* non-autonomous, conservative *vs* dissipative, linear *vs* nonlinear, etc. When differential equations are used, the theory of dynamical systems is called continuous dynamical systems. Instead when difference equations are employed the theory is called discrete dynamical systems. Some situations may also be modeled by mixed operators such as differential-difference equations. It is the theory of hybrid systems. In mathematics, an autonomous system or autonomous differential equation is a system of ordinary differential equations which does not explicitly depend on the independent variable, if it is not the case the system is called non-autonomous. Classical mechanics deals with dynamical systems without damping or friction as the ideal pendula, the solar system. In this case the dynamical systems involved are conservative dynamical systems. When damping or friction occur the dynamical systems are dissipative. Linear dynamical systems can be solved in terms of simple functions and the behavior of all orbits classified, on the contrary there are in general no explicit solutions of non linear dynamical systems which model more complex phenomena. Before the advent of fast computing machines, solving a non linear dynamical system required sophisticated mathematical techniques and could be accomplished only for a small class of dynamical systems. Numerical methods implemented on computers have simplified the task of determining the orbits of a dynamical system. However when chaotic dynamical systems are studied the crucial question is: can we rely on these solutions ?

Chaos theory studies the behavior of dynamical systems that are highly sensitive to initial conditions, an effect which is popularly referred to as the butterfly effect. Small differences in initial conditions (such as those due to rounding errors in numerical computation) yield widely diverging outcomes for chaotic systems, rendering long-term prediction impossible in general. This happens even though these systems are deterministic, meaning that their future behavior is fully determined by their initial conditions, with no random elements involved. In other words, the deterministic nature of these systems does not make them predictable. The first example of such chaotic continuous system in the dissipative case was pointed out by the meteorologist Lorenz in 1963.[1]

In this article we limit our study to chaotic (hence non linear) dissipative dynamical systems, either continuous or discrete, autonomous or non-autonomous. The case of linear system is not relevant because most of the solutions are given by closed formulas. The case of conservative system is much more difficult to handle as the lack of friction (dissipation of energy) leads to exponential increasing of rounding errors. Dedicated techniques are necessary to obtain reliable solutions.[11]

Although there exist peculiar mathematical tools in order to study non-autonomous dynamical systems, they can be easily transformed in autonomous systems increasing by one the dimension of the space variable. Then we only consider autonomous sytems. We focus our study to the most popular models: for the discrete case: logistic and tent map in 1-dimension, Hénon and Lozi map in 2-dimension; for the continuous case: Lorenz, Rössler and Chua model.

2.2. Poincaré map: a bridge between continuous and discrete dynamical system

Generally (*i.e.* when there exist periodic solutions) the Poincaré map allows us to build a correspondence between continuous and discrete dynamical systems. If we consider a 3-dimensional continuous dynamical system (*i.e.* a system of three differential autonomous equations):

$$\begin{cases} \dot{x}_1 = f_1(x_1, x_2, x_3) \\ \dot{x}_2 = f_2(x_1, x_2, x_3) \\ \dot{x}_3 = f_3(x_1, x_2, x_3) \end{cases} \quad (1)$$

the solution of such a system can be seen as a parametric curve $(x_1(t), x_2(t), x_3(t))$ in the space $\mathbb{R}^3$. A periodic solution (also called a cycle) is no more than a loop in this space as shown on Fig. 1 for the solution starting from, and arriving to, the same initial condition X_0.

Poincaré map defined in a neighborhood of this cycle is the map φ of the plane $\Sigma = \mathbb{R}^2$ into itself which associates to the initial point belonging to this plane, the first return point of the solution starting from this very initial point $\varphi : X \in \Sigma \to \varphi(X) \in \Sigma$ (*e.g.* on Fig. 1. the first intersection point X_2 of the plane Σ with the solution starting from X_1, $\varphi : X_1 \in \Sigma \to \varphi(X_1) = X_2 \in \Sigma$). Then the study of n-dimensional continuous system is equivalent to the study of $(n-1)$-dimensional discrete system.

Figs. 2 and 3 display the discrete periodic orbit $\{X_0, X_1, X_2, X_3, X_4 = X_0\}$ associated to the continuous periodic orbit of period 4:

$$\varphi^{(4)}(X_0) = \varphi \circ \varphi \circ \varphi \circ \varphi(X_0) = X_0$$

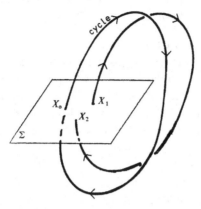

Fig. 1. Solutions of a differential system in $\mathbb{R}^3$.

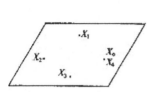

Fig. 2. Discrete periodic orbit associated to the continuous periodic orbit of period 4.

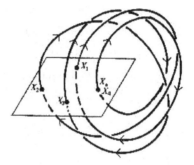

Fig. 3. Continuous period 4 orbit.

3. Collapsing Effects

3.1. *Undesirable chaotic transient*

In 2008, Zeraoulia Elhadj and Clinton Sprott[12] introduced a two-dimensional discrete mapping with C^∞ multifold chaotic attractors. They studied a modifed Hénon map given by:

$$f(x_n, y_n) = \begin{pmatrix} x_{n+1} \\ y_{n+1} \end{pmatrix} = \begin{pmatrix} 1 - a\sin(x_n) + by_n \\ x_n \end{pmatrix} \quad (2)$$

where the quadratic term x^2 in the Hénon map (see Sec. 4) is replaced by the nonlinear term $\sin x$. They studied this model for all values of a and b. The essential motivation for this work was to develop a C^∞ mapping that is capable of generating chaotic attractors with "multifolds" via a period-doubling bifurcation route to chaos which has not been studied before in the literature.

They prove the following theorem on the existence of bounded orbits.

Theorem 1. *(Elhadj & Sprott)* *The orbits of the map (2) are bounded for all $a \in \mathbb{R}$, and $|b| < 1$, and all initial conditions $(x_0, x_1) \in \mathbb{R}^2$.*

The existence of chaotic attractors is only inferred numerically. They display four examples, convincing at first glance, of what they call "chaotic multifold attractors". Unfortunately one obtained for $a = 4.0$ and $b = 0.9$ (Fig. 4) is no more than a long transient regime which collapses to a trivial and degenerate behavior: a periodic orbit of period 6 when the computation is done for sufficently large value of n. When programming in Language C (Borland® compiler), using a computer with Intel DuoCore2 processor and computing with double precision numbers, from any initial points after a transient regime for approximatively 110,000 iterates (actually the length of the transient regime depends upon the initial value) the orbit is trapped to the period-6 attractor given by:

$$x_{120003} = 8.95855079898761453 = x_{120009} = x_{120015} = \cdots,$$
$$x_{120004} = 10.96940289429559630 = x_{120010} = x_{120016} = \cdots,$$
$$x_{120005} = 13.06132591362670500 = x_{120011} = \cdots,$$
$$x_{120006} = 8.97249334266406962 = \cdots,$$
$$x_{120007} = 11.90071070225514802 = \cdots,$$
$$x_{120008} = 13.07497800339342220 = \cdots.$$

This attractor is shown on Fig. 5.

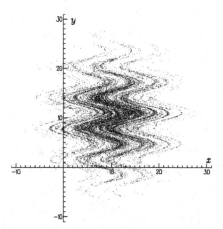

Fig. 4. Chaotic multifold attractor of the map (2) for $a = 4.0$ and $b = 0.9$ (Elhadj & Sprott[12]).

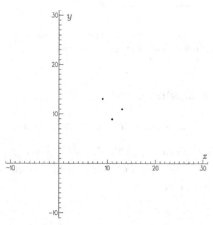

Fig. 5. Period 6 attracting orbit of the map (2) for $a = 4.0$ and $b = 0.9$ (Elhadj & Sprott[12]). The six points on the graph are magnified.

Remark 4.1. It is obvious that the phase space (x_n, x_{n+1}) on which the points $(x_n, x_{n+1}) \in \mathbb{R}^2$ are computed is finite when finite arithmetic replaces continum state spaces and one can object that every orbit of a mapping must be periodic in this finite space. However with double precision numbers for each component, it is generally possible (in presence of chaotic attractor) to obtain periodic orbit with period as long as 10^{11} (see Lozi map in Sec. 4.2) which could be a good approximation for the attractor. Instead an attracting period-6 orbit has a very different behavior than a strange attractor because such orbit does not possess the sensitivity to initial condition property, which is a minimal necessary condition (but not a sufficient one) of existence of chaos.

This example of long chaotic transient regime hiding a periodic attractor with a very short period is not unique in scientific literature, numerous researchers in several fields linked to chaotic dynamical systems are confident in the numerical solutions they found using popular software and publish without checking carefully the reliability of their results. Most of the time they compute only few iterates (i.e. few means less than 10^9) of mapping and falsely conclude the existence of chaotic regimes upon these numerical clues.

3.2. *Enigmatic computations for the logistic map (1838)*

In 1838 the belgium mathematician Pierre François Verhulst[13] introduced the differential equation

$$\begin{cases} \dfrac{dp}{dt} = mp - np^2 \\ p(0) = p_0 \end{cases} \quad (3)$$

modelling the growth of population in a simple demographic model, as an improvement of the Malthusian growth model, in which some resistence to the natural increase of population is added. The function $p(t)$ being the size of the population of the mankind. He later, in 1845,[14] called logistic function the solution of this equation. Putting

$$x(t) = \frac{n}{m} p(t) \quad (4)$$

Eq. (3) is equivalent to

$$\frac{dx}{dt} = mx(1-x). \quad (5)$$

In 1973, the biologist Robert May introduced the nonlinear, discrete time dynamical system

$$x_{n+1} = rx_n(1 - x_n) \quad (6)$$

as a model for the fluctuations in the population of fruit flies in a closed container with constant food.[15] Due to the similarity of both equations although one is a continuous dynamical system, and the other a discrete one, he called Eq. (6) logistic equation. The logistic map $f_r : [0,1] \to [0,1]$

$$f_r(x) = rx(1-x) \quad (7)$$

associated to Eq. (6) and generally considered for $r \in [0, 4]$ is often cited as an archetypal example of how complex, chaotic behaviour can arise from very simple non-linear dynamical equations.

This dynamical system which has excellent ergodic properties on the real interval $[0, 1]$ has been extensively studied especially by May,[16] and Mitchell Feigenbaum[17] who introduced what is now called the Feigenbaum's constant

$$\delta = 4.6692016091029906718532038204662016172581855777...$$

explaining by a new theory (period doubling bifurcation) the onset of chaos.

For every value of r there exist two fixed points: $x = 0$ which is always unstable and $x = \frac{r-1}{r}$ which is stable for $r \in]1, 3[$ and unstable for $r \in]0, 1[\cup]1, 4[$ (see Fig. 6).

When $r = 4$, the system is chaotic. The set

$$\left\{ \frac{5 - \sqrt{5}}{8}, \frac{5 + \sqrt{5}}{8} \right\} = \{0.3454915028, 0.9045084972\}$$

is the period-2 orbit. In fact there exist infinity of periodic orbits and infinity of periods (furthermore several distinct periodic orbits having the same period can coexist). This dynamical system possesses an invariant measure

$$\mu(x) = \frac{1}{\pi \sqrt{x(1-x)}}$$

(see Fig. 7).

It is quite surprising that a simple quadratic equation can exhibit such complex behaviour. If the logistic equation with $r = 4$ modelled the growth of fruit flies, then their population would exhibit erratic fluctuations from one generation to another.

In the coordinate sytem

$$\begin{cases} X = 2x - 1 \\ Y = 2y - 1 \end{cases} \quad (8)$$

the map of Eq. (7) written in the equivalent form (for $r = 4$)

$$g(X) = 1 - 2X^2 \quad (9)$$

was studied in the interval $[-1, 1]$ by Stan Ulam and John von Neumann well before the modern chaos era. They proposed it as a computer random number generator.[18]

In order to compute periodic orbits whose period is longer than 2 the use of computer is required, as it is equivalent to find roots of polynomial equation of degree greater than 4, for which Galois theory teaches that no closed formula is available. However, numerical computation on computer uses ordinarily double precision numbers (IEEE-754) so that the working interval contains roughly 10^{16} representable points. Doing such a computation[19] in Eq. (6) with $r = 4$ with 1,000 randomly chosen initial guesses, 596, i.e., the majority, converge to the unstable fixed point $x = 0$, and 404 converge to a cycle of period 15,784,521 (Tab. 1)

Table 1. Coexisting periodic orbits with 1,000 initial guesses for the logistic map.

Period	Orbit	Relative Basin Size
1	{0} (unstable fixed point)	596 over 1,000
15,784,521	Scattered over the interval	404 over 1,000

In an experimental work,[20] Oscar Lanford III, does the same search of numerical periodic solution of the logistic equation under the form of Eq. (9). The precise discretization studied is obtained exploiting evenness of this equation to fold the interval $[-1, 0]$ to $[0, 1]$, i.e. replacing Eq. (9) by

$$G(X) = |1 - 2X^2| \qquad (10)$$

on $[0, 1]$. It is not difficult to see that the folded mapping has the same set of periods as the original one. In order to avoid the particular discretization of this interval when the standard IEEE-754 is used for double precision numbers, the working interval is then shifted from $[0, 1]$ to $[1, 2]$ by translation, and the iteration of the translated folded map is programmed in straightforward way. Out of 1,000 randomly chosen initial points, 890, i.e., the overwhelming majority, converged to the fixed point corresponding to the unstable fixed point $x = -1$ in the original representation of Eq. (9), 108 converged to a cycle of period 3,490,273, the remaining 2 converged to a cycle of period 1,107,319 (Table 2).

Table 2. Coexisting periodic orbits with 1,000 initial guesses for the logistic map in the form of Eq. (9).

Period	Orbit	Relative Basin Size
1	{0} (unstable fixed point)	890 over 1,000
1,107,319	Scattered over the interval	2 over 1,000
3,490,273	Scattered over the interval	108 over 1,000

Thus, in both cases at least, the very long-term behaviour of numerical orbits is, for a substantial fraction of initial points, in flagrant disagreement with the true behaviour of typical orbits of the original smooth logistic map.

In other numerical experiments we have performed,[21,22] the computer working with fixed finite precision is able to represent finitely many points in the interval in question. It is probably good, for purposes of orientation, to think of the case where the representable points are uniformly spaced in the interval. The true logistic map is then approximated by a discretized map, sending the finite set of representable points in the interval to itself.

Describing the discretized mapping exactly is usually complicated, but it is roughly the mapping obtained by applying the exact smooth mapping to each of the discrete representable points and "rounding" the result to the nearest representable point. In our experiments, uniformly spaced points in the interval with several orders of discretization (ranging from 9 to 2,001 points) are involved. In each experiment the questions addressed are:

- how many periodic cycles are there, and what are their periods?
- how large are their respective basins of attraction, i.e., for each periodic cycle, how many initial points give orbits with eventually land on the cycle in question?

Table 3 shows coexisting periodic orbits for the discretization with regular meshes of $N = 9$, 10 and 11 points. There are respectively 3, 2 and 2 cycles. Table 4 displays cases $N = 99$, 100 and 101 points, there are exactly 2, 4 and 3 cycles. Table 5 stands for regular meshes of $N = 1999$, $N = 2000$ and $N = 2001$ points. It seems that the computation of numerical approximations of the periodic orbits leads to unpredictable and somewhat enigmatic results. As says Lanford III,[20] "The reason is that because of the expansivity of the mapping the growth of roundoff error normally means that the computed orbit will remain near the true orbit with the chosen initial condition only for a relatively small number of steps typically of the order of the number of bits of precision with which the calculation is done. It is true that the above mapping like many 'chaotic' mappings satisfies a *shadowing theorem (see Sec. 4.3 in this article)* which ensures that the computed orbit stays near to some true orbit over arbitrarily large numbers of steps. The flaw in this idea as an explanation of the behavior of computed orbits is that the shadowing theorem says that the computed orbit approximates *some* true orbit but not necessarily that it approximates a *typical* one." He adds, "This suggests the discouraging possibility that this problem may be as hard of that of non equilibrium statistical mechanics."

Table 3. Computation on regular meshes of N points.

N	Period	Orbit	Basin Size
9	1	$\{0\}$	3 over 9
9	1	$\{6\}$	2 over 9
9	2	$\{3,7\}$	4 over 9
10	1	$\{0\}$	2 over 10
10	2	$\{3,8\}$	8 over 11
11	1	$\{0\}$	3 over 11
11	4	$\{3,8,6,9\}$	8 over 11

The existence of very short periodic orbit (Tables 1 and 2), the existence of a non constant invariant measure (Fig. 7) and the easily recognized shape of the function in the phase space avoid the use of the logistic map as a Pseudo Random Number Generator. However, its very simple implementation in computer programs led some authors to use it as a base of cryptosytem.[23,24]

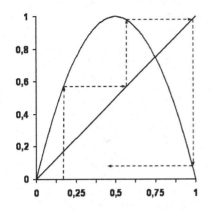

Fig. 6. First-return map for the logistic map $f_4(x)$.

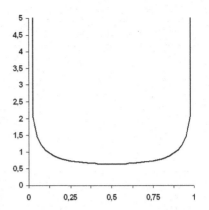

Fig. 7. Invariant measure $\mu(x) = \frac{1}{\pi\sqrt{x(1-x)}}$ of the logistic map $f_4(x)$.

Table 4. Computation on regular meshes of 99, 100 and 101 points.

N	Period	Orbit	Basin Size
99	1	$\{0\}$	3 over 99
99	10	$\{3, 11, 39, 93, 18, 58, 94, 15, 50, 97\}$	96 over 99
100	1	$\{0\}$	2 over 100
100	1	$\{74\}$	2 over 100
100	6	$\{11, 39, 94, 18, 58, 96\}$	72 over 100
100	7	$\{7, 26, 76, 70, 82, 56, 97\}$	24 over 100
101	1	$\{0\}$	3 over 101
101	1	$\{75\}$	2 over 101
101	3	$\{19, 61, 95\}$	96 over 101

3.3. Collapsing orbit of the symmetric tent map

Another often studied discrete dynamical system is defined by the symmetric tent map

$$x_{n+1} = T_a(x_n) \tag{11}$$

$$T_a(x) = 1 - a|x| \tag{12}$$

on the interval $J = [-1, 1]$. Despite its simple shape (see Fig. 8), it has several interesting properties. First, when the parameter value $a = 2$, the system possesses chaotic orbits. Because of its piecewise-linear structure, it is easy to find those orbits explicitly. More, owing to its simple definition, the symmetric tent map's shape under iteration is very well understood. The invariant measure is the Lebesgue measure. Finally, and perhaps the most important, the tent map is conjugate to

Table 5. Computation on regular meshes of 1,999, 2,000 and 2,001 points.

N	Period	Orbit	Basin Size
1999	1	{0}	3 over 1999
1999	4	{554, 1601, 1272, 1848}	990 over 1999
1999	8	{3, 11, 43, 168, 615, 1702, 1008, 1997}	1006 over 1999
2000	1	{0}	2 over 2000
2000	1	{1499}	14 over 2000
2000	2	{691, 1808}	138 over 2000
2000	3	{376, 1221, 1900}	6 over 2000
2000	8	{3, 11, 43, 168, 615, 1703, 1008, 1998}	1840 over 2000
2001	1	{0}	5 over 2001
2001	1	{1500}	34 over 2001
2001	2	{691, 1809}	92 over 2001
2001	8	{3, 11, 43, 168, 615, 1703, 1011, 1999}	608 over 2001
2001	18	{35, 137, 510, 1519, 1461, 1574, ···}	263 over 2001
2001	25	{27, 106, 401, 1282, 1840, 588, ···}	1262 over 2001

the logistic map, which in turn is conjugate to the Hénon map (see Sec. 4.1) for small values of b.[25]

However the symmetric tent map is dramatically numerically instable: Sharkovskiĭ's theorem applies for it.[26] When $a = 2$ there exists a period three orbit, which implies that there is infinity of periodic orbits. Nevertheless the orbit of almost every point of the interval J of the discretized tent map eventually wind up to the (unstable) fixed point $x = -1$ (this is due to the binary structure of floating points) and there is no numerical attracting periodic orbit.[27]

The numerical behaviour of iterates with respect to chaos is worse than the numerical behaviour of iterates of the approximated logistic map. This is why the tent map is never used to generate numerically chaotic numbers.

3.4. *Statistical properties*

Many others examples could be given, but those given may serve to illustrate the intriguing character of the results, the outcomes proves to be extremely sensitive to the details of the experiment, but the findings all have a similar flavour: a relatively small number of cycles attract near all orbits, and the lengths of these significant cycles are much larger than one but much smaller than the number of representable points. Patrick Diamond *et al*,[28,29] suggest that statistical properties of the phenomenon of computational collapse of discretized chaotic mapping can be modelled by random mappings with an absorbing centre. The model gives results which are very much in line with computational experiments and there appears to be a type of universality summarised by an *Arcsin* law. The effects are discussed with special reference to the family of dynamical systems

$$x_{n+1} = 1 - |1 - 2x_n|^l, \quad 0 \leqslant x \leqslant 1, \quad 1 \leqslant l \leqslant 2 \tag{13}$$

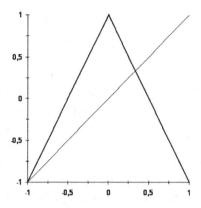

Fig. 8. Tent map $T_2(x)$.

Computer experiments show close agreement with prediction of the model. However these results are of statistical nature, they do not give accurate information on the exact nature of the periodic orbits (e.g. length of the shortest or the greatest one, size of their basin of attraction ...). Following this work, Guo-cheng Yuan and James Yorke[30] study precisely the collapse to the repelling fixed point $x = -1$ of the iterates of the one dimensional dynamical system

$$x_{n+1} = 1 - 2|x_n|^l, \quad -1 \leqslant x \leqslant 1, \quad l > 2 \tag{14}$$

for a large fraction of initial conditions when calculated on a mesh of N points equally spaced from -1 to $+1$ (as we have done in Sec. 3.2). The map associated to this system

$$T_{a,l}(x) = 1 - a|x|^l$$

being an usual nonlinear generalization of the map $T_a(x)$ of Eq. (12). They not only prove rigorously that the collapsing effect does not vanish when arbitrarily high numerical precision is employed, but also they give a lower bound of the probability for which it happens. They define $P_{collapse}$ which is the probability that there exists n such that $x_n = -1$ and they give the proof that $P_{collapse}$ depends only on the numbers N and l. The lower bound of $P_{collapse}$ is given by

$$\liminf_{N \to \infty} P_{collapse} \geqslant \sqrt{\pi} K' \left[1 - \mathrm{erf}(K')\right] \exp(K')^2 > 0 \tag{15}$$

$$K' = lK^{-1} 2^{-1/.2 - 1/l} \left[1 + \left((2l)^{1/l} - 1\right)^{-2}\right]^{-1/2}$$

$$K = \left(\sum_{i=1}^{\infty} k_i^2\right)^{1/2}$$

$$k_i = l\left[(i+1/2)^{1/l} - (i-1/2)^{1/l}\right]$$

with

$$\text{erf}(x) = \frac{2}{\sqrt{\pi}} \int_0^x e^{-t^2} dt. \qquad (16)$$

They plot the curve given by Eq. (15) along with the numerical results (Figure 9). Each numerical datum is obtained as follows. For each l they sample 10,000 pairs of $(\bar{l}, x)$ from $(l - 0.01l, l + 0.01l) \times (-1, 1)$ with uniform distribution. For each sample they keep iterating the dynamical system defined by Eq. (14) with the initial condition x until the numerical trajectory repeats. Then they calculate the portion of trajectories that eventually map to -1. The deviation is clear since Eq. (15) only gives a lower bound. Nonetheless the theoretical curve reveals the fact that $P_{collapse}$ is already substantial for $l = 3$ and it predicts that $P_{collapse}$ increases as l increases and that $\lim_{l \to \infty} P_{collapse} = 1$.

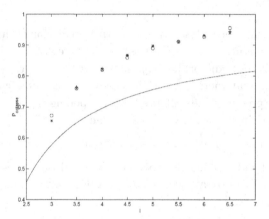

Fig. 9. $P_{collapse}$ as a function of l. The curve in this figure is the lower bound computed from Eq. (15). Numerical results are obtained by using different numerical precisions are also shown in this figure "+" -double precision "o" -single precision "×"-fixed precision 10^{-12} (from[30]).

4. Shadowing and Parameter-shifted Shadowing Property of Mappings of the Plane

4.1. *Hénon map (1976) found by mistake*

The chaotic behavior of mapping of the real line is relatively simple compared to the behavior of mapping on the plane. These mappings could be seen as a simple expansion in a phase space with increased dimension of both models (logistic and tent) we have introduced in the previous section. However they have been discovered not in this "ascending" way, instead they come in "descending way" as metaphor of Poincaré map of 3-D continuous dynamical systems.

Fig. 10. Observatory of Nice, the office of Michel Hénon was located inside the building.

In order to study numerically the properties of the Lorenz attractor,[1] Michel Hénon an astronomer of the observatory of Nice, France (see Fig. 10) introduced in 1976[5] a simplified model of the Poincaré map of this attractor. The Lorenz attractor being in dimension 3, the corresponding Poincaré map is a map from the plane $\mathbb{R}^2$ to $\mathbb{R}^2$. The Hénon map is then also defined in dimension 2 as

$$H_{a,b} : \begin{pmatrix} x \\ y \end{pmatrix} = \begin{pmatrix} y + 1 - ax^2 \\ bx \end{pmatrix} \qquad (17)$$

It is associated to the dynamical system

$$\begin{cases} x_{n+1} = y_n + 1 - ax_n^2 \\ y_{n+1} = bx_n \end{cases} \qquad (18)$$

For the parameter value $a = 1.4$, $b = 0.3$, M. Hénon pointed out numerically that there exists an attractor with fractal structure. This was the first example of strange attractor (previously introduced by David Ruelle and Floris Takens[31]) for a mapping defined by an analytic formula.

As highlighted in the sequence of Figs. 11-14, the like-Cantor set structure in one direction orthogonal to the invariant manifold in this simple mapping was a dramatic surprise in the community of physicists and mathematicians. Nowadays hundreds of research papers have been published on this prototypical map in order to fully understand its innermost structure. However as for 1-dimensional dynamical systems, there is a discrepancy between the mathematical properties of this map in the plane and the numerical computations done using (IEEE-754) double precision numbers.

If we call Megaperiodic orbits,[6] those whose length of the period belongs to the interval of natural numbers $[10^6, 10^9[$ and Gigaperiodic orbits, those whose length of the period belongs to the interval $[10^9, 10^{12}[$, Hénon map possesses both Mega and Gigaperiodic orbits. On a Dell computer with a Pentium IV microprocessor running at 1.5 Gigahertz frequency, using a Borland C compiler and computing with ordinary (IEEE-754) double precision numbers, one can find for $a = 1.4$ and $b = 0.3$ one attracting period of length $3,800,716,788$, *i.e.* two hundred forty times longer than the longest period of the one-dimensional logistic map (Table 1). This periodic orbit (we call it here Orbit 1) is numerically slowly attracting. Starting with the initial value

$$(x_0, y_0)_1 = (-0.35766, 0.14722356)$$

one winds up:

$$(x_{11,574,730,767}, y_{11,574,730,767})_1 = (x_{15,375,447,555}, y_{15,375,447,555})_1$$
$$= (1.27297361350955662, -0.0115735710153616837)$$

The length of this period is obtained subtracting

$$15,375,447,555 - 11,574,730,767 = 3,800,716,788$$

However this Gigaperiodic orbit is not unique: starting with the other initial value

$$(x_0, y_0)_2 = (0.4725166, 0.25112222222356)$$

the following Megaperiodic orbit (Orbit 2) of period $310,946,608$ is computed

$$(x_{12,935,492,515}, y_{12,935,492,515})_2 = (x_{13,246,439,123}, y_{13,246,439,123})_2$$
$$= (1.27297361350865113, -0.0115734779561870744).$$

Remark 4.2. This second orbit can be reached more rapidly starting form the other initial value

$$(x_0, y_0) = (0.881877775591, 0.0000322222356)$$

then

$$(x_{4,459,790,707}, y_{4,459,790,707}) = (1.27297361350865113, -0.0115734779561870744)$$

Remark 4.3. It is possible that some other periodic orbits coexist with both Orbit 1 and Orbit 2. However there is no peculiar method but the brute force to find it if any.

Remark 4.4. The comparison between Orbit 1 and Orbit 2 gives a perfect idea of the sensitive dependence on initial conditions of chaotic attractors:

Orbit 1 passes through the point

$$(\mathbf{1.27297361350}955662, \mathbf{-0.0115735}710153616837)$$

and Orbit 2 passes through the point

$$(\mathbf{1.27297361350}865113, \mathbf{-0.0115734}779561870744)$$

The same digits of the coordinates of these points are bold printed, they are very close.

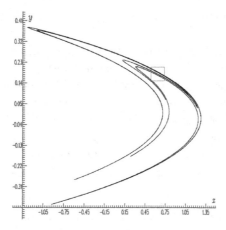

Fig. 11. Hénon strange attractor, 10000 successive points obtained by iteration of the mapping.

The discovery of a strange attractor for maps of the plane boosted drastically the research on chaos in the years '70. This is miraculous considering that if M. Hénon tried to test rigorously his model nowadays with powerful computers he should only find long periodic orbits. In 1976, M. Hénon used the electronic pocket calculator built by Hewlett-Packard company HP-65 (Fig. 15) and one of the only two computers available at the university of Nice an IBM 7040 (Fig. 16), the other one was an IBM 1130, slower. The HP-65 first introduced in 1974 was worth 750 € (equivalent to $3,500$ € nowadays), the IBM 7040, was worth 1,100,000 € (equivalent to $7,800,000$ € in the year 2012).

He concluded,[5]

> "The simple mapping (17) appears to have the same basic properties as the Lorenz system. Its numerical exploration is much simpler: in fact most of the exploratory work for the present paper was carried out with a programmable pocket computer (HP-65). For the more extensive computations of Figures 2 to 6, we used a IBM 7040 computer, with 16-digit accuracy. The solutions can be followed over a much longer time than in the case of a system of differential equations. The accuracy is also increased since there are no integration errors. Lorenz (1963) inferred the Cantor-set structure of the attractor from reasoning, but could not observe it directly because the contracting ratio after one 'circuit' was too small: 7×10^{-5}. A similar experience was reported by Yves Pomeau (1976). In the present mapping, the contracting ratio after one iteration is 0.3, and one can easily observe a number of successive levels in the hierarchy. This is also facilitated by the larger number of points. Finally, for mathematical studies the mapping (4) might also be easier to handle than a system of differential equations."

The large number of points was in fact very few:[5]

"The transversal structure (across the curves) appears to be entirely different, and much more complex. Already on Figures 11 a number of curves can be seen, and the visible thickness of some of them suggests that they have in fact an underlying structure. Figure 12a is a magnified view of the small square of Figure 11: some of the previous 'curves' are indeed resolved now into two or more components. The number n of iterations has been increased to 10^5, in order to have a sufficient number of points in the small region examined. The small square in Figure 12a is again magnified to produce Figure 12b, with n increased to 10^6: again the number of visible 'curves' increases. One more enlargement results in Fig. 12c, with $n = 5 \times 10^6$: the points become sparse but new curves can still easily be traced."

In 1976, the one million €, IBM 7040 took several hours to compute five millions points, today a basic three hundred euros laptop, runs the same computation in one hundredth of a second.

With so few iterations nowadays the following claim:[5] "These figures strongly suggest that the process of multiplication of 'curves' will continue indefinitely, and that each apparent 'curves' is in fact made of an infinity of quasi-parallel curves. Moreover, Figures 4 to 6 indicate the existence of a hierarchical sequence of 'levels', the structure being practically identical at each level save for a scale factor. This is exactly the structure of a Cantor set." should be obviously untrue. One can assert that Hénon map was found not by chance (the reasoning of M. Hénon was straightforward) but by mistake. It is a great luck for the expansion of chaos studies!

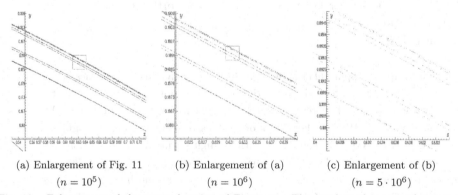

(a) Enlargement of Fig. 11 ($n = 10^5$)   (b) Enlargement of (a) ($n = 10^6$)   (c) Enlargement of (b) ($n = 5 \cdot 10^6$)

Fig. 12. Enlargement of the squared region of Figure 11. The number of computed points is increased as specified in the subcaption.

4.2. *Lozi map (1977) a tractable model*

Using one of the first desktop electronic calculator HP-9820, he usually employed to initiate mathematics teacher trainees to computer sciences, René Lozi found out, on June 15, 1977, (on the scientific campus of the university of Nice, few kilometers

Fig. 13. Electronic pocket calculator HP-65.   Fig. 14. Computer IBM 7040.

apart of the observatory of this town of the French Riviera where Michel Hénon worked), that the linearized version of the Hénon map displayed numerically the same structure of strange attractors, but the curves were replaced by straight lines. He published this result in the proceedings of a conference on dynamical systems which held in Nice in July of the same year,[32] in a very short paper of one and an half pages.

The aim of Lozi, a numerical analyst, was to allow algebraic computations on an analog of the Hénon map, such direct computation being untractable on the original model due to the square function. He changed this function with an absolute value, defining the map

$$L_{a,b} : \begin{pmatrix} x \\ y \end{pmatrix} = \begin{pmatrix} y + 1 - a\,|x| \\ bx \end{pmatrix} \qquad (19)$$

and the associate dynamical system

$$\begin{cases} x_{n+1} = y_n + 1 - a\,|x_n| \\ y_{n+1} = bx_n \end{cases} \qquad (20)$$

He then found out on the plotter device attached to his small primitive computer that for the value $a = 1.7$, $b = 0.5$, the same Cantor-set like structure was apparent. Due to the linearity of this new model, it is possible to compute explicitly any periodic orbit solving a linear system. Moreover in december 1979, Michal Misiurewicz[2] gave a talk in the famous conference organized in december 1979 at New-York by the New-York academy of science in which he etablished rigorously the proof that what he call "Lozi map" has a strange attractor for some parameter values (including the genuine one $a = 1.7$, $b = 0.5$). Since that time, hundreds of papers have been published on the topic.

Remark 4.5. The open set $\mathcal{M}$ of the parameter space (a, b) where the existence of strange attractor is proved is now called the *Misiurewicz domain*. It is defined as

$$0 < b < 1, \ a > b + 1, \ 2a + b < 4, \ a\sqrt{2} > b + 2, \ b < \frac{(a^2 - 1)}{(2a + 1)} \quad (21)$$

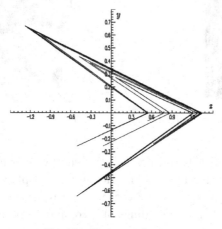

Fig. 15. Lozi map $L_{1.7, 0.5}$.

Today, in the same conditions of computation as for Hénon map (see Sec. 4.1), running the computation during nineteen hours, one can find a Gigaperiodic attracting orbit of period $436, 170, 188, 959$ more than one hundred times longer than the period of Orbit 1 found for the Hénon map.

Starting with

$$(x_0, y_0) = (0.88187777591, 0.0000322222356)$$

one obtains

$$(x_{686,295,403,186}, y_{686,295,403,186}) = (x_{250,125,214,227}, y_{250,125,214,227})$$
$$= (1.34348444450739479, -2.07670041497986548.10^{-7})$$

There is a transient regime before the orbit is reached. It seems that there is no periodic orbit with a smaller length. This could be due to the quasihyperbolic nature of the attractor. However, the parameter-shifted shadowing property, the orbit-shifted shadowing property of Lozi map (and generalized Lozi map), which are new variations of the shadowing the property which ensures that pseudo-orbits of a homeomorphism can be traceable by actual orbits even if rounding errors in computing are not inevitable, has been recently proved.[7,33]

4.3. Shadowing, parameter-shifted shadowing and orbit-shifted shadowing properties

The shadowing property is the property which ensures that pseudo-orbits (i.e. orbits computed numerically using finite precision number) of a homeomorphism can be traceable by actual orbit. The concept and primary results of shadowing property for uniformly hyperbolic diffeomorphisms were introduced by Dmitri Anosov[34] and Rufus Bowen,[35] who proved that such diffeomorphisms have the shadowing property. Various papers associated with shadowing property and uniform hyperbolicity were presented by many authors. Comprehensive expositions on these works were included in books by Ken Palmer[36] or Serguei Pilyugin.[37] The precise definition of this property is[38]

Definition 4.1. *(Shadowing)* Let (X,d) a metric space, $f : X \to X$ a function and let $\delta > 0$. A sequence $\{x_k\}_{k=p}^{q}$, $(p,q) \in \mathbb{N}^2$ of points is called a δ-pseudo-orbit of f if $d(f(x_k), x_{k+1}) < \delta$ for $p \leq k \leq q-1$ (e.g. a numerically computed orbit is a pseudo-orbit). Given $\epsilon > 0$, the pseudo-orbit $\{x_k\}_{k=p}^{q}$, is said ϵ – *shadowed* by $x \in X$, if $d(f^k(x), x_k) < \epsilon$ for $p \leq k \leq q$. One says that f has the *shadowing property* if for $\epsilon > 0$ there is $\delta > 0$ such that every δ-pseudo-orbit of f can be ϵ – *shadowed* by some point.

However, if one relaxes the uniform hyperbolicity condition on diffeomorphisms slightly (Lozi map, for example, has not such a property), then many problems concerning the shadowing property are not solved yet. Nowadays, it is widely supposed that many diffeomorphisms producing chaos **would not have the shadowing property**. In fact, Yuan and Yorke[39] found an open set of absolutely nonshadowable C^1 maps for which nontrivial attractors supported by partial hyperbolicity include at least two hyperbolic periodic orbits whose unstable manifolds have different dimensions. Moreover, Flavio Abdenur and Lorenzo Díaz[40] showed recently that the shadowing property does not hold for diffeomorphisms in an open and dense subset of the set of C^1-robustly nonhyperbolic transitive diffeomorphisms.

In order to fix this issue the notion of the parameter-shifted shadowing property (for short *PSSP*) was introduced by Ethan Coven, Ittai Kan and James Yorke.[41] In fact, they proved that the tent map (Eq. (12)) has the parameter-fixed shadowing property for almost every slope a in the open interval $I = (\sqrt{2}, 2)$, but do not have it for any a in a certain dense subset of I. Moreover, they proved that, for any $a \in I$, the tent map T_a has *PSSP*. The definition of this variant property being[33]

Definition 4.2. *(PSSP)* Let $\{f_a\}_{a \in J}$ be a set of maps on $\mathbb{R}^2$ where J is an open interval in $\mathbb{R}$.

- For $\delta > 0$, the sequence $\{x_n\}_{n \geq 0} \subset \mathbb{R}^2$ is called δ-pseudo-orbit of f_a if $\|f_a(\mathbf{x}_n) - \mathbf{x}_{n+1}\| \leq \delta$ for any integer $n \geq 0$.
- For $a \in J$, we say that f_a has the *parameter – shifted shadowing property* if, for any $\epsilon > 0$, there exist $\delta = \delta(a, \epsilon) > 0$, $\tilde{a} = \tilde{a}(a, \epsilon) \in J$ such that any

δ-pseudo-orbit $\{x_n\}_{n\geq 0}$ of f_a can be ϵ-shadowed by an actual orbit of $f_{\tilde{a}}$, that is, there exists a $y \in \mathbb{R}^2$ such that $\|f_{\tilde{a}}^n(y) - x_{n+1}\| \leq \epsilon$ for any $n \geq 0$.

For Lozi map, Shin Kiriki and Teruhiko Soma proof that[33]

Theorem 2. *There exists a nonempty open set $\mathcal{O}$ of the Misiurewicz domain $\mathcal{M}$ such that, for any $(a,b) \in \mathcal{O}$, the Lozi map $L_{a,b}$ has the parameter-shifted shadowing property in a one-parameter family $\{L_{\tilde{a},b}\}_{\tilde{a} \in J}$ fixing b, where J is a small open interval containing a.*

However they note that "The problem of parameter-fixed shadowability is still open even for the Lozi family as well as the Hénon family."

Another variation of shadowability is the orbit-shifted shadowing property (for short *OSSP*) recently introduced by Akio Sakurai[7] in order to study generalized Lozi maps introduced by Lai-Sang Young.[42] The limited extend of this article does not allow us to recall the complex definition of these generalized maps.

Definition 4.3. For $\delta_0 > 0$, $\delta_1 > 0$, and $\delta > 0$, with $\delta_0 \leq \eta_0$, a sequence $\{x_n\}_{n \geq 0}$ in $R = [0,1] \times [0,1]$ is an $(\delta_0, \delta_1) - shifted\ \delta - pseudo - orbit$ of f if, for any $n \geq 1$, x_n and x_{n+1} satisfy the following conditions.
   (i) $\|f(x_n) - x_{n+1}\| \leq \delta$ if $B_{\delta_0}(x_{n-1}) \cap (S_1 \cup S_2) = \emptyset$.
   (ii) $\|f(x_n) - (\delta_1, 0) - x_{n+1}\| \leq \delta$ if $B_{\delta_0}(x_{n-1}) \cap S_1 \neq \emptyset$.
   (iii) $\|f(x_n) + (\delta_1, 0) - x_{n+1}\| \leq \delta$ if $B_{\delta_0}(x_{n-1}) \cap S_2 \neq \emptyset$.
where S_1 and S_2 are the two components of the essential singularity set of f.

Definition 4.4. *(OSSP)* We say that f has the orbit-shifted shadowing property if, for any $\epsilon > 0$ with $\epsilon \leq \eta_0$, there exist $\delta_0, \delta_1, \delta > 0$ such that any $(\delta_0, \delta_1) - shifted\ \delta - pseudo - orbit$ $\{x_n\}_{n \geq 0}$ in R of f can be ϵ-shadowed by an actual orbit of f, that is, there exists $z \in R$ such that $\|f^n(z) - x_n\| \leq \epsilon$ for any $n \geq 0$.

Theorem 3. *Any generalized Lozi map f satisfying the conditions (2.1)-(2.4) given in[7] has the orbit-shifted shadowing property. More precisely, for any $0 < \epsilon \leq \eta_0$, there exists $\delta > 0$ such that any $(\epsilon, \epsilon/2)$-shifted δ-pseudo-orbit of f in R is ϵ-shadowed by an actual orbit.*

For sake of simplicity, we do not give the technical conditions (2.1)-(2.4), however one can prove that any original Lozi map $L_{a,b}$ indicated in Theorem 4.1 satisfies these conditions.

In conclusion to both Secs. 3 and 4, it appears that if, owing to the introduction of computers since forty years, it is easy to compute orbits of mapping in 1 or 2-dimension (and more generally in finite dimension), it is not unproblematic to obtain reliable results. The mathematical study of what is actually computed (for the orbits) is a rather difficult and still challenging problem. Trusty results are obtained under rather technical assumptions. Considering a new mathematical tool, the Global Orbit Pattern (*GOP*) of a mapping on finite set, R. Lozi and Clarisse Fiol[19,21] obtain some combinatorial results in 1-dimension.

There is not place in this paper to consider other computations than the computations of orbits. However there is a need for several other reliable numerical results as the Lyapunov numbers,[43] the fractal dimensions (correlation, capacity, Haussdorff,...) and Lyapunov spectrum ,[44] topological entropy,[45] extreme value laws.[46,47] ...The researches are very active in these fields. One can mention as an example the relationship between the expected value of the period scales and the correlation dimension for the case of fractal chaotic attractors of D-dimensional maps.[48]

Theorem 4. *The expected value of the period scales with round off ϵ is*

$$\overline{m} = \epsilon^{-D/2} \qquad (22)$$

where D is the correlation dimension of the chaotic attractor. The periods m have substantial statistical fluctuation. That is, P(m), the probability that the period is m, is not strongly peaked around $\overline{m}$. This probability is given by

$$P(m) = \left(\frac{1}{\overline{m}}\right) F\left(\frac{\sqrt{\frac{\pi}{8m}}}{\overline{m}}\right) \qquad (23)$$

where

$$F(x) = \sqrt{\frac{\pi}{8}}\left[1 - \sqrt{\frac{\pi}{2}} erf\left(\frac{x}{\sqrt{2}}\right)\right] \qquad (24)$$

and erf(x) is defined by Eq. (16)

5. Continuous Models

We have highlighted the close relationship between continuous and discrete models via Poincaré map. We have also pointed out that Michel Hénon constructed his model in order to compute more easily Poincaré map of the Lorenz model which is crunching too fast for direct numerical computations. It is time to study directly these initial equations together with the metaphoric ones that follows naturally: the Rössler and the Chua equations.

5.1. *Lorenz attractor (1963)*

The following non linear system of differential equations was introduced by Edward Lorenz[a] in 1963.[1] As a crude model of atmospheric dynamics, these equations led

---

[a]Edward Lorenz, born on May, 23, 1917 at west Hartford (Connecticut), dead on April, 16, 2008 at Cambridge (Massachussets), studied the Rayleigh-Bénard problem (*i.e.* the motion of a flow heated from below, as an approach of the atmospheric turbulence) using a primitive Royal McBee LPG-30 computer. He first considered a series of 12 differential equations when he discovered the "butterfly effect". Then he simplied his model from 12 to only three equations. As for Hénon model, his discovery was made by mistake due to rounding errors.

Lorenz to the discovery of sensitive dependence of initial conditions — an essential factor of unpredictability in many systems (*e.g.* meteorology).

$$\begin{cases} \dot{x} = -\sigma(x+y) \\ \dot{y} = \rho x - y - xz \\ \dot{z} = xy - \beta z \end{cases} \tag{25}$$

Numerical simulations for an open neighbourhood of the classical parameter values

$$\sigma = 10, \ \rho = 28, \ \beta = \frac{8}{3} \tag{26}$$

suggest that almost all points in phase space tend to a strange attractor the *Lorenz attractor*. One can note that the system is invariant under the transformation

$$S(x, y, z) = (-x, -y, +z). \tag{27}$$

This means that any trajectory that is not itself invariant under S must have a symmetric "twin trajectory." For $\rho > 1$ there are three fixed points: the origin and the two "twin points"

$$C^{\pm} = \left(\pm\sqrt{\beta(\rho-1)}, \pm\sqrt{\beta(\rho-1)}, \rho-1\right)$$

For the parameter values generally considered, $C^{\pm}$ have a pair of complex eigenvalues with positive real part, and one real, negative eigenvalue. The origin is a saddle point with two negative and one positive eigenvalues satisfying

$$0 < -\lambda_3 < -\lambda_3 < \lambda_1 < -\lambda_2.$$

Thus the stable manifold of the origin $W^s(0)$ is two dimensional and the unstable manifold of the origin $W^u(0)$ is one dimensional. As indicated by M. Hénon[5] in his initial publication the flow contracts volumes at a significant rate (see Sec. 4.1). As the divergence of the vector field is $-(\sigma+\beta+1)$ the volume of a solid at time t can be expressed as

$$V(t) = V(0)e^{-(\sigma+\beta+1)t} \approx V(0)e^{-13.7t}$$

for the classical parameter values. This means that the flow contracts volumes almost by a factor *one million* per time unit which is quite extreme. There appears to exist a forward invariant open set U containing the origin but bounded away from $C^{\pm}$. The set U is a double torus (one with two holes), with its holes centered around the two excluded fixed points. If one lets φ denote the flow of Eq. (25), there exists the maximal invariant set

$$\mathcal{A} = \bigcap_{t \geq 0} \varphi(U, t)$$

Due to the strong dissipativity of the flow, the attracting set $\mathcal{A}$ must have zero volume. As indicated by Warwick Tucker,[49] it must also contain the unstable

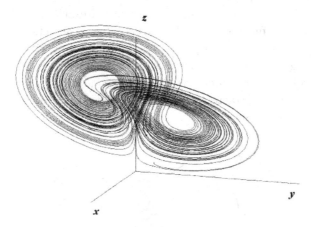

Fig. 16. Lorenz attractor.

manifold $W^u(0)$ of the origin, which seems to spiral around $C^\pm$ in a very complicated non periodic fashion (see Fig. 16).

In particular $\mathcal{A}$ contains the origin itself, and therefore the flow on $\mathcal{A}$ can not have a hyperbolic structure. The reason is that fixed points of the vector field generate discontinuities for the return maps (Poincaré maps), and as a consequence, the hyperbolic splitting is not continuous. Apart from this, the attracting set appears to have a strong hyperbolic structure.

5.2. *Geometric Lorenz attractor*

As it was very difficult to extract rigorous information about the attracting set $\mathcal{A}$ from the differential equations themselves, Hénon constructed his famous model (Sec. 4.1) in order to understand numerically its inner structure. In another way of research, a *geometric model* of the Lorenz flow was introduced by John Guckenheimer and Robert Williams[50] at the end of the seventies (see Fig. 17). This model has been extensively studied and it is well understood today. The question whether or not the original Lorenz system for such parameter values has the same structure as the geometric Lorenz model has been unsolved for more than 30 years. By combination of normal form theory and rigorous computations, Tucker[51] answered this question affirmatively, that is, for classical parameters, the original Lorenz system has a robust strange attractor which is given by the same rules as for the geometric Lorenz model. From these facts, it is known that the geometric Lorenz model is crucial in the study of Lorenz dynamical systems. Oddly enough the original equations introduced by Lorenz have remained a puzzle until the same author proved in 1999 in his Ph.D Thesis[49] the following theorem

Theorem 5. *For the classical parameter values, the Lorenz equations support a robust strange attractor $\mathcal{A}$. Furthermore the flow admits a unique SRB measure μ_X with $supp(\mu_X) = \mathcal{A}$.*

Remark 4.6. SRB measures, are an invariant measures introduced by Yakov Sinai, David Ruelle and Rufus Bowen in the 1970's. These objects play an important role in the ergodic theory of dissipative dynamical systems with chaotic behavior. Roughly speaking, SRB measures are the invariant measures most compatible with volume when volume is not preserved. They provide a mechanism for explaining how local instability on attractors can produce coherent statistics for orbits starting from large sets in the basin.

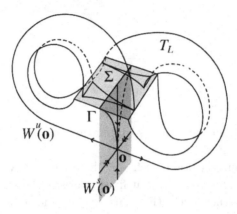

Fig. 17. Geometric Lorenz model from.[52]

In addition, Tucker indicates, "In fact, we prove that the attracting set is a singular hyperbolic attractor. Almost all nearby points separate exponentially fast until they end up on opposite sides of the attractor. This means that a tiny blob of initial values soon smears out over the entire attractor. It is perhaps worth pointing out that the Lorenz attractor does not act quite as the geometric model predicts. The latter can be reduced to an interval map which is everywhere expanding. This is not the case for the Lorenz attractor: there are large regions in the attracting set that are contracted in all directions under the return map. However, such regions only require a few more iterations before accumulating expansion. This corresponds to the interval map being eventually expanding, and does not lead to any different qualitative long time behaviour. Apart from this, the Lorenz attractor is just as the geometric model predicts: it contains the origin, and thus has a very complicated Cantor book structure."

One of the many ingredients required for the proof of this theorem is a rigourous arithmetics on a trapping region consisting in 350 adjacent rectangles belonging to the return plate $z = 27 = \rho - 1$. Warwick Tucker says

> "One major advantage of our numerical method is that we totally eliminate the problem of having to control the global effects of rounding errors due to the computer's internal floating point representation. This is achieved by using a high

dimensional analogue of *interval arithmetic*. Each object Ξ (*e. g.* a rectangle or a tangent vector) subjected to computation is equipped with a maximal absolute error Δ_Ξ, and can thus be represented as a box $\Xi \pm \Delta_\Xi = [\Xi_1 - \Delta_{\Xi_1}, \Xi_1 + \Delta_{\Xi_1}] \times \cdots \times [\Xi_n - \Delta_{\Xi_n}, \Xi_n + \Delta_{\Xi_n}]$. When following an object from one intermediate plane to another, we compute upper bounds on the images of $\Xi_i + \Delta_{\Xi_i}$, and lower bounds on the images of $\Xi_i - \Delta_{\Xi_i}$, $i = 1, \cdots, n$. This results in a new box $\tilde{\Xi} \pm \Delta_{\tilde{\Xi}}$, which strictly contains the exact image of $\Xi \pm \Delta_\Xi$. To ensure that we have strict inclusion, we use quite rough estimates on the upper and lower bounds. This gives us a margin which is much larger than any error caused by rounding possibly could be. Hence, the rounding errors are taken into account in the computed box $\tilde{\Xi} \pm \Delta_{\tilde{\Xi}}$, and we may continue to the following intermediate plane by restarting the whole process.

As long as we do not flow close to a fixed point, the local return maps are well defined diffeomorphisms, and the computer can handle all calculations. Some rectangles, however, will approach the origin (which is fixed point), and then the computations must be interrupted as discussed in the previous section *(i.e. the normal form theory)*."

The work of Tucker is a big step toward the reliability of numerical solutions of chaotic continuous dynamical systems. However it has been obtained after a great deal of effort (he published two revised versions of the proof anytime he detected a mistake in the code) which can not be afforded to any new continuous chaotic model.

5.3. *Lorenz map*

In contrast the analysis of the Poincaré map associated to the geometric model of the Lorenz flow allows the use of the parameter-shifted shadowing property.[52] This first return map on a Poincaré cross section of a geometric Lorenz flow is called a Lorenz map $L : \Sigma \backslash \Gamma \to \Sigma$, where

$$\Sigma = \{(x, y) \in \mathbb{R}^2; |x|, |y| \leqslant 1\}$$

and $\Gamma = \{(0, y) \in \mathbb{R}^2; |y| \leqslant 1\}$. More precisely it is defined as follows

Definition 4.5. *(Lorenz map)* Let $\Sigma_\pm$ denote the components of $\Sigma \backslash \Gamma$ with $\Sigma_\pm \ni (\pm 1, 0)$ (Fig. 17). A map $L : \Sigma \backslash \Gamma \to \Sigma$, is said to be a Lorenz map if it is a piecewise C^1 diffeomorphism which has the form

$$L(x, y) = (\alpha(x), \beta(x, y))$$

where $\alpha : [-1, 1] \backslash \{0\} \to [-1, 1]$ is a piecewise C^1-map with symmetric property $\alpha(-x) = -\alpha(x)$ and satisfying

$$\begin{cases} \lim_{x \to 0+} \alpha(x) = -1, & \alpha(1) < 1 \\ \lim_{x \to 0+} \alpha'(x) = \infty, & \alpha'(x) > \sqrt{2} \text{ for any } x \in (0, 1] \end{cases}$$

(Fig. 18a), and $\beta : \Sigma \backslash \Gamma \to [-1,1]$ is a contraction in the y-direction. Moreover, it is required that the images $L(\Sigma_+)$, $L(\Sigma_-)$ are mutually disjoint cusps in Σ, where the vertices $\mathbf{v}_+$, $\mathbf{v}_-$ of $L(\Sigma_+)$ are contained in $\{\mp 1\} \times [-1,1]$ respectively (Fig. 18b).

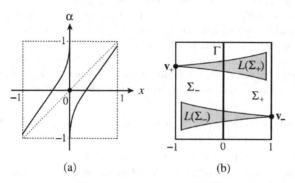

Fig. 18. Lorenz map from.[52]

Then, Kiriki and Soma proved the parameter-shifted shadowing property (*PSSP*) for Lorenz maps.[52]

Theorem 6. *There exists a definite set $\mathcal{L}$ of Lorenz maps satisfying the following condition:*

For any $L \in \mathcal{L}$ and any $\epsilon > 0$, there exist $\mu > 0$ and $\delta > 0$ such that any δ – pseudo orbit of the Lorenz map L_μ with $L_\mu(x,y) = L(x,y) - (\mu x, 0)$ is ϵ–shadowed by an actual orbit of L.

Theorem 7. *Any geometric Lorenz flow controlled by a Lorenz map $L \in \mathcal{L}$ has the parameter-shifted shadowing property.*

Due to the lack of space, we refer to[52] for the strict description of $\mathcal{L}$.

Besides this work on the Lorenz map, there are many some others rigorous results on Lorenz equations. Zbigniew Galias and Piotr Zygliczyński,[53,54] for example, are basing their result on the notion of the topological shifts maps (TS-maps). Let $\Sigma = \{(x, y, z) \in \mathbb{R}^3; z = \rho - 1\}$ which is the standard choice for the Poincaré section. Let $\mathbf{P}$ be a Poincaré map generated on the plane by Eq. (25), they prove that:

Theorem 8. *For all parameter values in a sufficiently small neighborhood of parameter value given by Eq. (26), there exists a transversal section $I \subset \Sigma$ such that the Poincaré map $\mathbf{P}$ induced by Eq. (25) is well defined and continuous on I. There exists a continuous surjective map $\pi : \text{Inv}(I, \mathbf{P}^2) \to \Sigma_2$, such that*

$$\pi \circ \mathbf{P}^2 = \sigma \circ \sigma$$

The preimage of any periodic sequence from Σ_2 contains periodic points of $\mathbf{P}^2$.

Remark 4.7. The maximal invariant part of I (with respect to $\mathbf{P}^2$) is defined by
$$Inv\left(I, \mathbf{P}^2\right) = \bigcap_{i \in \mathbb{Z}} \mathbf{P}_{|I}^{-2i}(I)$$
and $\Sigma_2 = \{0, 1, \cdots, K-1\}^{\mathbb{Z}}$ is a topological space with the Tichonov topology; $0, 1, \cdots, K-1$ being symbols characterizing periodic infinite sequences of TS-maps (see[4] for more details).

In this case the Poincaré map is issued directly from Lorenz equations not from the geometric model. Due to the limited extend of this article, we do not cite all the results on computer aided proof. We refer the reader to.[4]

5.4. Rössler attractor (1976)

We have seen that from the seminal discovery of the Lorenz attractor, several strategies have been developped in order to study this appealing and complex mathematical object: geometric Lorenz model, Lorenz map, Hénon map (and Lozi map as linearized version). In 1976 Otto E. Rössler followed a different direction of research: instead of simplify the mathematic equation (25) and considering that, due to extreme simplification, there is no actual link between this equation and the Rayleigh-Benard problem from which it originated, he turned his mind to the study of a chemical multi-vibrator. He started to design some three-variable oscillator based on a two-variable bistable system coupled to a slowly moving third-variable. The resulting three-dimensional system was only producing limit cycles at the time. At an international congress on rhythmic functions held on September 8-12, 1975 in Vienna, he met Art Winfree, a theoretical biologist who challenged him to find a biochemical reaction reproducing the Lorenz attractor. Rössler failed to find a chemical or biochemical reaction producing the Lorenz attractor but, he instead found a simpler type of chaos in a paper he wrote during the 1975 Christmas holidays.[55] The obtained chaotic attractor (see Fig. 19)

$$\begin{cases} \dot{x} = -y - z \\ \dot{y} = x + ay \\ \dot{z} = b + z(x - c) \end{cases} \tag{28}$$

with

$$a = 0.2, \ b = 0.2, \ c = 5.7 \tag{29}$$

does not have the rotation symmetry of the Lorenz attractor (defined by Eq. (27)), but it is characterized by a map equivalent to the Lorenz map. The reaction scheme (see Fig. 20) leading to Eq. (28) is meticulously analyzed by Christophe Letellier and Valérie Messager.[56]

The structure of the Rössler attractor is simpler than the Lorenz's one. However even if hundreds of papers has been written on it, the rigorous proof of its existence is not yet established as done for Lorenz equations. In 1997, Zygliczyński using

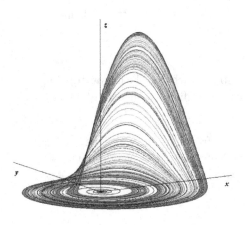

Fig. 19. Rössler attractor.

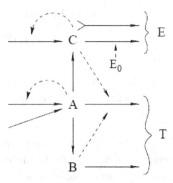

Fig. 20. Combination of an Edelstein switch with a Turing oscillator in a reaction system producing chaos. E = switching subsystem, T = oscillating subsystem; constant pools (sources and sinks) have been omitted from the scheme as usual (From,[56] adapted from[55]).

reduced Rössler equations to two parameters instead of three (i.e. stating that $a = b$ in Eq. (28)) gave, using computer assisted proof similar results as he did for Lorenz equations.[4]

Let $\Theta = \left\{(x,y,z) \in \mathbb{R}^3; x = 0,\ y < 0,\ \dot{x} > 0\right\}$

Theorem 9. *For all parameter values in a sufficently small neighborhood of* $(a,c) = (0.2, 5.7)$ *there exists a transversal section* $N \subset \Theta$ *such that the Poincaré map* $\mathbf{P}$ *induced by Eq. (28) is well defined and continuous. There exists a continuous surjective map* $\pi : Inv\,(N, \mathbf{P}) \to \Sigma_3$, *such that*

$$\pi \circ \mathbf{P} = \sigma \circ \pi$$

$\Sigma_A \subset \pi\,(Inv\,(N, \mathbf{P}))$, *where*

$$A = \begin{bmatrix} 0 & 1 & 1 \\ 0 & 1 & 1 \\ 1 & 0 & 0 \end{bmatrix}$$

The preimage of any periodic sequence from Σ_A *contains periodic points of* $\mathbf{P}$.

As this result is related to the Poincaré map of Rössler equations, nowadays there is still a need to apply the method developed by Tucker to prove the true existence of a strange attractor.

5.5. *Chua attractor (1983)*

Before he discovered his equations, Rössler in collaboration with Friedrich Seelig, inspired by a little book entitled *Measuring signal generators, Frequency Measuring Devices and Multivibrators from the Radio-Amateur Library*,[57] began to "translate" electronic systems into nonlinear chemical reaction systems. The know-how he developed led eventually in to his chaotic model. Few years later, in October 1983, visiting Takeshi Matsumoto at Waseda University, L. O. Chua found an electronic circuit (see Fig. 21) mimicking directly on an oscilloscope screen a chaotic signal (see Fig. 22). As we have seen, only two autonomous systems of ordinary differential equations were generally accepted then as being chaotic, the Lorenz equations and the Rössler equations. The nonlinearity in both systems is a function of two variables namely, the product function which is very difficult to build in an electronic circuit. Leon Chua[58] says,

> "I was to have witnessed a live demonstration of presumably the world's first successful electronic circuit realization of the Lorenz Equations, on which Professor Matsumoto's research group had toiled for over a year. It was indeed a remarkable piece of electronic circuitry. It was painstakingly breadboarded to near perfection, exposing neatly more than a dozen IC components, and embellished by almost as many potentiometers and trimmers for fine tuning and tweaking their incredibly sensitive circuit board. There would have been no need for inventing a more robust chaotic circuit had Matsumoto's Lorenz Circuit worked. **It did not.** The fault lies on the dearth of a critical nonlinear IC component with a near-ideal characteristic and a sufficiently large dynamic range; namely, the analog multiplier. Unfortunately, this component was the key to building an autonomous chaotic circuit in 1983."

He adds,

> "Prior to 1983, the conspicuous absence of a reproducible functioning chaotic circuit or system seems to suggest that chaos is a pathological phenomenon that can exist only in mathematical abstractions, and in computer simulations of contrived equations. Consequently, electrical engineers in general, and nonlinear circuit theorists in particular, have heretofore paid little attention to a phenomenon which many had regarded as an esoteric curiosity. Such was the state of mind among the nonlinear circuit theory community, circa 1983. Matsumoto's Lorenz Circuit was to have turned the tide of indifference among nonlinear circuit theorists. Viewed from this historical perspective and motivation, the utter disappointment that descended upon all of us on that uneventful October afternoon was quite understandable. So profound was this failure that the wretched feeling persisted in my subconscious mind till about bedtime that evening. Suddenly it dawned upon me that since the main mechanism which gives rise to chaos, in both the Lorenz and the Rössler Equations, is the presence of at least two unstable equilibrium points — 3 for the Lorenz Equations and 2 for the Rössler Equations — it seems only prudent to design a simpler and more robust circuit having these attributes.
>
> Having identified this alternative approach and strategy, it becomes a simple exercise in elementary nonlinear circuit theory to enumerate systematically all such circuit candidates, of which there were only 8 of them, and then to systematically eliminate those that, for one reason or another, can not be chaotic."

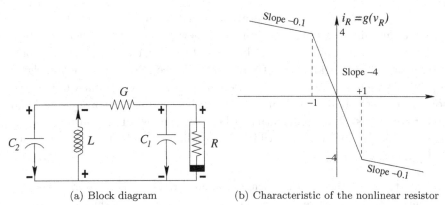

(a) Block diagram (b) Characteristic of the nonlinear resistor

Fig. 21. Chua's circuit.

The Chua's equations

$$\begin{cases} \dot{x} = \alpha\left(y - \Phi\left(x\right)\right) \\ \dot{y} = x - y + z \\ \dot{z} = -\beta y \end{cases} \quad (30)$$

where

$$\Phi\left(x\right) = x + g\left(x\right) = m_1 x + \frac{1}{2}\left(m_0 - m_1\right)\left[|x+1| - |x-1|\right] \quad (31)$$

and with

$$\alpha = 15.60, \ \beta = 28.58, \ m_0 = -\frac{1}{7}, \ m_1 = \frac{2}{7} \quad (32)$$

were soon carefully analyzed by Matsumoto.[59] The main mathematical idea behind the invention of this circuit is the same one as simplifying the Hénon map by linearizing the parabola with an absolute value as done in Lozi map six years before (even if L. O. Chua did not know these maps at that time).

The "nonlinear" characteristic which is in fact only piecewise linear allows some exact computations. Henceforth, L. O. Chua et al. proved very soon that the mechanism of chaos exists in this attractor.[60] However in spite it is possible to obtain closed formula for the solution of Eqs. (30) in every subspace $\{x \leq -1\}, \{-1 < x < 1\}, \{1 \leq x\}$, due to the transcendental nature of the equation allowing the computation at the matching boundaries $\{x = -1\}, \{x = 1\}$, of the global solution,[9] it is not possible to compute it explicitely. As only numerically computed solutions are avaliable, it remains the gnawing problem of what is really computed. To that problem may be added the fact that the electronic realization of the Chua's circuit is not exactly governed by Eqs. (30), due to the instability of the electronics components, the parameter value (Eq. (32)) is randomly fluctuating around its mean value. It is very difficult to analyze the experimental observed chaos.[61]

One possible way to perform this analysis may be the use of a new mathematical tool: *confinor* instead of *attractor*.[62] The confinor theory, when applied to Chua's circuit allowed the discovery of coexisting chaotic regimes.[8,63]

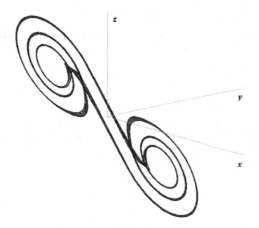

Fig. 22. Chua attractor.

6. Conclusion

We have shown, in the limited extent of this article, on few but well known examples,[10] that it is very difficult to trust in numerical solution of chaotic dynamical dissipative systems. In some cases one can even prove that it is never possible to obtain reliable results. We have focused our survey on models related to the seminal Lorenz model which is the most studied model of strange attractor. These models are not the only ones studied today. Countless models presenting chaos, arising in every fields concerned with dynamics such as ecology, biology, chemistry, economy, finance, electronics ..., are developped since forty years. However their study is less carefully done as those presented here, conducting sometimes to hasty results and flawed theories in these sciences. In addition the disturbing phenomenom of "ghost solution" appears when discretization of nonlinear differential equation by central difference scheme is used (several examples of such ghost solutions when central difference scheme is used sole or in combination with Euler's scheme, are given by Masaya Yamaguti *et al.*[64]). Recently Aleksandr Sharkovskji and S. A. Berezovsky[65] pointed out the notion of "numerical turbulence" which appears, due to incorrectness of calculation method stipulated by discreteness. In conclusion one can say that there is room for more study of the relationship between numerical computation and theoretical behavior of chaotic solutions of dynamical systems.

References

1. E. N. Lorenz, Deterministic nonperiodic flow, *Journal of Atmospheric Science*, **20**, 130-141 (1963).
2. M. Misiurewicz, Strange attractors for the Lozi mappings, *Annals of the New-York Academy of Science*, **357** (1), 348-358 (1980).
3. W. Tucker, The Lorenz attractor exists, *Comptes-Rendus de l'Académie des Sciences* (Paris), I, **328** (12), 1197-1202 (1999).
4. P. Zgliczyński, Computer assisted proof of chaos in the Rössler equations and the Hénon map, *Nonlinearity*, **10** (1), 243-252 (1997).
5. M. Hénon, A Two-dimensional mapping with a strange attractor, *Communications in Mathematical Physics*, **50**, 69-77 (1976).
6. R. Lozi, Giga-periodic orbits for weakly coupled tent and logistic discretized maps, In *Modern Mathematical Models, Methods and Algorithms for Real World Systems*, A. H. Siddiqi, I. S. Duff & O. Christensen (eds), Anamaya Publishers, New Delhi, India, pp. 80-124 (2006).
7. A. Sakurai, Orbit shifted shadowing property of generalized Lozi maps, *Taiwanese Journal of Mathematics*, **14** (4), 1609-1621 (2010).
8. R. Lozi & S. Ushiki, Coexisting chaotic attractors in Chua's circuit, *International Journal of Bifurcation & Chaos*, **1** (4), 923-926 (1991).
9. R. Lozi & S. Ushiki, The theory of confinors in Chua's circuit: accurate analysis of bifurcations and attractors, *International Journal of Bifurcation & Chaos*, **3** (2), 333-361 (1993).
10. D. Blackmore, New models for chaotic dynamics, *Regular & Chaotic dynamics*, **10** (3), 307-321 (2005).
11. J. Laskar & P. Robutel, The chaotic obliquity of the planets, *Nature*, **361**, 608-612 (1993).
12. Z. Elhadj & J. C. Sprott, A two-dimensional discrete mapping with C^∞ multifold chaotic attractors, *Electronic Journal of Theoretical Physics*, **5** (17), 1-14 (2008).
13. P.-F. Verhulst, Notice sur la loi que la population poursuit dans son accroissement, *Correspondance mathématique et physique de l'observatoire de Bruxelles*, **4** (10), 113-121 (1838).
14. P.-F. Verhulst, Recherches mathématiques sur la loi d'accroissement de la population, *Nouveaux Mémoires de l'Académie Royale des Sciences et Belles-Lettres de Bruxelles*, **18**, 1-42 (1845).
15. R. M. May, *Stability and Complexity of Models Ecosystems*, Princeton University Press, Princeton, NJ (1973).
16. R. M. May, Biological populations with nonoverlapping generations: stable points, stable cycles, and chaos, *Science*, **186** (4164), 645-647 (1974).
17. M. J. Feigenbaum, The universal metric properties of nonlinear transformations, *Journal of Statistical Physics*, **21**, 669-706 (1979).
18. S. M. Ulam & J. von Neumann, On combination of stochastic and deterministic processes, *Bulletin of the American Mathematical Society*, **53** (11), 1120 (1947).
19. R. Lozi & C. Fiol, Global orbit patterns for dynamical systems on finite sets, *Conference Proceedings of the American Institute of Physics*, **1146**, 303-331 (2009).
20. O. E. Lanford III, Some informal remarks on the orbit structure of discrete approximations to chaotic maps, *Experimental Mathematics*, **7** (4), 317-324 (1998).
21. R. Lozi & C. Fiol, Global orbit patterns for one dimensional dynamical systems, *Grazer Mathematische Berichte*, **354**, 112-144 (2009).
22. R. Lozi, Complexity leads to randomness in chaotic systems, In *Mathematical Methods,*

    Models, and Algorithms in Science and Technology, A. H. Siddiqi, R. C. Singh and P. Manchanda (eds), World Scientific Publishing, (Singapore), pp. 93-125 (2011).
23. M. S. Baptista, Cryptography with chaos, *Physics Letters A*, **240**, 50-54 (1998).
24. M. R. K. Ariffin & M. S. M. Noorani, Modified Baptista type chaotic cryptosystem via matrix secret key, *Physics Letters A*, **372**, 5427-5430 (2008).
25. K. T. Alligood, T. D. Sauer & J. A. Yorke, *Chaos. An introduction to dynamical systems*, Textbooks in Mathematical Sciences, Springer, New-York (1996).
26. A. N. Sharkovskiĭ, Coexistence of cycles of a continuous map of the line into itself, *International Journal Bifurcation & Chaos*, **5** (5), 1263-1273 (1995).
27. J. C. Sprott, *Chaos and Time-Series Analysis*, Oxford University Press, Oxford, UK (2003).
28. P. Diamond, P. Kloeden, A. Pokrovskii, & A. Vladimorov, Collapsing effects in numerical simulation of a class of chaotic dynamical systems and random mappings with a single attracting centre, *Physica D*, **86**, 559-571 (1995).
29. P. Diamond & A. Pokrovskii, Statistical laws for computational collapse of discretized chaotic mappings, *International Journal of Bifurcation & Chaos*, **6** (12A), 2389-2399 (1996).
30. G. Yuan & J. A. Yorke, Collapsing of chaos in one dimensional maps, *Physica D*, **136**, 18-30 (2000).
31. D. Ruelle & F. Takens, On the nature of turbulence, *Communications in Mathematical Physics*, **20**, 167-192 (1971).
32. R. Lozi, Un attracteur étrange du type attracteur de Hénon, *Journal de Physique*, **39** (C5), 9-10 (1978).
33. S. Kiriki & T. Soma, Parameter-shifted shadowing property of Lozi maps, *Dynamical Systems*, **22** (3), 351-363 (2007).
34. D. V. Anosov, Geodesic flows on closed Riemann manifolds with negative curvature, *Proceedings of the Steklov Institute of Mathematics*, **90** (1967). English translation, *American Mathematical Society Transactions,* Providence, R.I. (1969).
35. R. Bowen, ω-limit sets for Axiom A diffeomorphisms, *Journal of Differential Equations*, **18**, 333-339 (1975).
36. K. Palmer, *Shadowing in Dynamical Sytems: Theory and Appplications*, Kluwer Academic Publications (2000).
37. S. Y. Pilyugin, Shadowing in dynamical systems, *Lecture Notes in Mathematics*, **1706**, (1999).
38. K. Sakai, Diffeomorphisms with the shadowing property, *Journal of the Australian Mathematical Society A)*, **61**, 396-399 (1996).
39. G.-C. Yuan & J. A. Yorke, An open set of maps for which every point is absolutely non-shadowable, *Proceedings of the American Mathematical Society*, **128**, 909-918 (2000).
40. F. Abdenur & L. J. Díaz, Pseudo-orbit shadowing in the C^1 topology, *Discrete and Continuous Dynamical Systems*, **17** (2), 223-245 (2007).
41. E. M. Coven, I. Kan & J. A. Yorke, Pseudo-orbit shadowing in the family of tent maps, *Transactions of the American Mathematical Society*, **308**, 227-241 (1988).
42. L.-S. Young, A Bowen-Ruelle measure for certain piecewise hyperbolic maps, *Transactions of the American Mathematical Society*, **287**, 41-48 (1985).
43. D. Delenau, Dynamic Lyapunov indicator: a practical tool for distinguishing between ordered and chaotic orbits in discrete dynamical systems, *Proceedings of the 13th WSEAS International Conference on Mathematical Methods*, N. Gavriluta, R. Raducanu, M. Iliescu, H. Costin, N. Mastorakis (eds), Iasi, Romania, pp. 117-122 (2011).
44. P. Grassberger, On Lyapunov and dimension spectra of 2D attractors, with an application to the Lozi map, *Journal of Physics A*, **22**, 585-590 (1989).

45. Y. Ishii & D. Sands, Monotonicity of the Lozi family near the tent-maps, *Communications in Mathematical Physics*, **198**, 397-406 (1998).
46. M. P. Holland, R. Vitolo, P. Rabassa, A. E. Sterk, & H. W. Broer, Extreme value laws in dynamical systems under physical observables, *Physica D*, **241** (5), 497-513 (2012).
47. D. Faranda, V. Lucarini, G. Turchetti, & S. Vaienti, Extreme Value distribution for singular measures, *arXiv:1106.2299*, submitted (2012).
48. C. Grebogi, E. Ott, & J. A. Yorke, Roundoff-induced periodicity and the correlation dimension of chaotic attractors, *Physical Review A*, **38** (7), 3688-3692 (1988).
49. W. Tucker, *The Lorenz attractor exists*, PhD thesis, Uppsala University, Sweden (1999).
50. J. Guckenheimer & R. F. Williams, Structural stability of Lorenz attractors, *Publications mathmatiques de l'IHÉS*, **50**, 59-72 (1979).
51. W. Tucker, A rigorous ODE solver and Smale's 14th problem, *Foundations of Computational Mathematics*, **2**, 53-117 (2002).
52. S. Kiriki & T. Soma, Parameter-shifted shadowing property for geometric Lorenz attractors, *Transactions of the American Mathematical Society*, **357** (4), 1325-1339 (2004).
53. Z. Galias & P. Zygliczyński, Chaos in the Lorenz equations for classical parameter values a computer assisted proof, *Universitatis Iagellonicae Acta Mathematica*, **36**, 209-210 (1998).
54. Z. Galias & P. Zygliczyński, Computer assisted proof of chaos in the Lorenz system, *Physica D*, **115**, 165-188 (1995).
55. O. E. Rössler, Chaotic behavior in simple reaction system, *Zeitschrift für Naturforschung A*, **31**, 259-264 (1976).
56. Ch. Letellier & V. Messager, Influences on Otto Rössler's earliest paper on chaos, *International Journal of Bifurcation & Chaos*, **20** (11), 1-32 (2010).
57. H. Sutaner, *Meßsender Frequenzmesser un Multivibratoren, Radio-Praktiker Bucherei*, **128/130**, Franzis Verlag, Munich, (1966).
58. L. O. Chua, The genesis of Chua's Circuit, *Archiv für Elektronik und Übertragungstechnik*, **46** (4), 250-257 (1992).
59. T. Matsumoto, A chaotic attractor from Chua's circuit, *IEEE Transactions on Circuit and Systems*, **31** (12), 1055-1058 (1984).
60. L. O. Chua, M., Kumoro, & T. Matsumoto, The Double Scroll Family, *IEEE Transactions on Circuit and Systems*, **32** (11), 1055-1058 (1984).
61. L. O. Chua, L. Kocarev, K. Eckert, & M. Itoh, Experimental chaos synchronization in Chua's circuit, *International Journal of Bifurcation & Chaos*, **2** (3), 705-708 (1992).
62. R. Lozi & S. Ushiki, Confinors and bounded-time patterns in Chua's circuit and the double-scroll family, *International Journal of Bifurcation & Chaos*, **1** (1), 119-138 (1991).
63. S. Boughaba & R. Lozi, Fitting trapping regions for Chua's attractor. A novel method based on Isochronic lines, *International Journal of Bifurcation & Chaos*, **10** (1), 205-225 (2000).
64. M. Yamaguti & S. Ushiki, Chaos in numerical analysis of ordinary differential equations, *Physica D*, **3** (3), 618-626 (1981).
65. A. N. Sharkovsky & S. A. Berezovsky Phase transitions in correct-incorrect calculations for some evolution problems, *International Journal of Bifurcation & Chaos*, **13** (7), 1811-1821 (2003).

Chapter 5

Chaos hierarchy — A review, thirty years later

Otto E. Rössler*

Institut für Physikalische und Theoretische Chemie
Auf der Morgenstelle 8, D-72076 Tübingen, Germany

Christophe Letellier

CORIA — Université de Rouen
Avenue de l'Université, BP 12
F-76801 Saint-Etienne du Rouvray cedex, France

An enriched analysis of the first steps into "chemical chaos" as tried under the influence of Winfree in 1975 is presented. This retrospective provides a bridge to the more mathematical context of the paper that follows in this collection of modern topological approaches to chaos.

Contents

1. Introduction . . . 99
2. A Short Autobiography . . . 100
3. Three Main Influences . . . 104
   3.1. The multivibrator by Andronov, Khaikin and Vitt . . . 104
   3.2. The Lorenz paper . . . 106
   3.3. The Li-Yorke theorem . . . 109
4. My Earliest Paper on Chaos . . . 111
   4.1. Phase space and chaotic attractor . . . 111
   4.2. First-return map to a Poincaré section . . . 111
   4.3. Qualitative properties of the expected dynamics . . . 112
   4.4. The equations and their chaotic solution . . . 113
   4.5. Topological analysis . . . 116
5. A Short Walk through Various Topologically Inequivalent Classes of Chaos . . . 118
6. Conclusion . . . 120
References . . . 122

1. Introduction

The first paper I wrote was about a theoretical system for the "Biogenesis".[1] Most of my following papers were about some kinds of chemical reactions interpreted in terms of electronic circuits.[2-4] The multivibrator[5,6] and the Bonhoeffer-van der Pol oscillator[3,7] were very often invoked. I started to publish about chaos in 1976. I

---
*Text written in the first-person perspective to add atmosphere.

then flooded the "chaos market" with various types of chaos with suggestive names like "spiral chaos,[8] "screw chaos" or "funnel chaos",[9] "sandwich chaos",[10] "walking-stick map",[11] "folded-towel map",[12] "superfat attractor"[13] among others. In this chapter, I choose to focus my attention, not on the most often quoted paper[14] but on my earliest paper on chaos,[8] since it contains conceptual insight related to the topology of chaos. With this latter paper, I published the second chaotic attractor displayed in the phase space.[15] I always tried to use informal (non-technical) terms and very suggestive pictures to force understanding on the reader (as Fichte said).

2. A Short Autobiography

I was born in 1940. My father, Otto Rössler (1907-1991), was a linguist recognized for having introduced a new system of Egypto-Semitic consonant correspondances and the term "Afro-semitic" languages.[16] As an adolescent, I built my own radio-transmitter and thus got acquainted with electronics while still in highschool at Tübingen. In 1957, I got an individual licence (DL9 KF) from the Deutscher Amateur Radio Club. I then studied medicine up to 1966 at the University of Tübingen. In 1966, I defended my inaugural dissertation — supervised by Erich Letterer (1895-1982) — for getting my grade of doctor in medicine.[17] Deeply interested in how Life could come from a "chemical soup", I exchanged letters with Carl-Friedrich von Weizäcker (1912-2007) and met him. Under his recommendation, I spent almost one year (1966-1967) at the Max-Planck Institute for the Physiology of Behavior (Seewiesen) with Horst Mittelstädt and Konrad Lorenz (1903-1989). I then spent two years at the University of Marburg where I was a medical assistant under the supervision of Reimara Waible who became my wife, one year later. During that period, I wrote — in German — a first paper entitled "Contributions to the theory of spontaneously evolving systems I: a simple model class" sent to the *Journal of Theoretical Biology*. The editor, Robert Rosen (1934-1998), who was reading German, accepted the paper for publication but required a translation into English. Not yet fluent in English, I never quite made it and the paper remained unpublished. Having caught interest because at this first paper, Rosen supported my application for a one-year position at the Center of Theoretical Biology (State University of New York at Buffalo).

Friedrich-Franz Seelig who had just obtained a chair ("Lehrstuhl") for Theoretical Chemistry at the University of Tübingen offered to join his new group with a stipend from the Deutsche Forschungsgemeinschaft (DFG). In the early 60s, Seelig had done his diploma work with Hans Kuhn and Fritz-Peter Schäfer to build an analog computer consisting of a network of electrical oscillators, connected to capacitors to solve the two-dimensional Schrödinger equation.[18] This system was excited by means of a radio frequency generator. In 1965, Seelig solved a two-dimensional Schrödinger equation with a digital computer (IBM 7090).[19] I had met Seelig via Hans Kuhn who was working on the origin of Life, my first research topic. Kuhn handed me down to Seelig as it were. Sharing an interest for the origin of life, in differential equations and in electronics (computers), we agreed that nonlinear systems like my evolutionary soup and electronic systems (without self-induction and without coupling condensers) were virtually isomorphic. This triggered a cooperation project between Seelig — a quantum chemist — and myself — a medical physiologist — to look for reaction-kinetic analogs to electronic circuits.

I joined Seelig at Tübingen on returning from Buffalo in 1970. After having been sent by the Division of Theoretical Chemistry to attend an EAI (Enterprise Application Integration) course on analog computing, I had to teach that topic, for which my radio-amateur past was useful. With the founding money obtained with his new position at Tübingen, Seelig bought (with 80 000 DM) an analog computer — a Dornier DO 240 (Fig. 1) — equipped with digital potentiometers, a digital clock and two function generators... As a Stipend-holder" of the DFG I was free in my research and started to study few-variable systems in parallel with Seelig. I started by investigating a chemical multivibrator.[3] To learn about the dynamics of such an electronic circuit, I read the textbook by Aleksandr Andronov, Sëmen Khaikin and Aleksandr Vitt in its 1966 English edition.[20]

Fig. 1. Analog computer Dornier DO 240 as bought by Seelig in 1970.

Inspired by a little book entitled "*Measuring-signal Generators, Frequency Measuring Devices and Multivibrators* from the Radio-Amateur Library,[21] Seelig and myself began to "translate" electronic systems into nonlinear chemical reaction systems (among them the RC-oscillator of Fig. 44 of that book as shown in Fig. 2). Morphogenetic reaction systems, devised by Nicholas Rashevsky[22] and Alan Turing,[23] fitted in, enabling the design of a chemical oscillator based on a chemical flip-flop, that is, a bistable multivibrator that has two stable states in a subsystem and hence and can be used as one bit of memory. I remained fascinated by the multivibrator[6] and the electronic Eccles-Jordan trigger as it is called.[4] This had led to the "flip-flop" study with Seelig.[2,24] I necessarily associated the multivibrator with the universal circuit introduced by Khaikin[25] and its description in phase space as described in[20] (see Sect. 3.1). Most of my early papers — say between 1971 and 1975 — were devoted to chemical reactions that reproduce the dynamics underlying some electronic circuits, and many of them explicitly discussed the multivibrator.[2,3,5,6,24] In 1972, with Dietrich Hoffmann, I provided "a first evidence that the Belousov-Zhabotinsky reaction is a Bonhoeffer oscillator, i.e. a special type of chemical hysteresis oscillators".[7] A link was explicitely made with relaxation oscillations as done by Bonhoeffer when he investigated a model for the excitation of nerve.[26] The model proposed by Bonhoeffer has all characteristics the so-called van der Pol equations have and the limit cycle drawn in the phase space by Bonhoeffer is very similar to the one published in.[27]

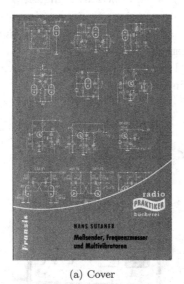

Fig. 2. Cover and Fig. 44 with Rössler's handwriting of Hans Sutaner's book published in 1966.

In this vein, I proposed with Seelig a two-cellular homogeneous chemical multi-vibrator.[2,24] An example of a homogeneous system involving a two-variable bistable system (switch) found on the analog computer was

$$\begin{cases} \dot{A} = -k_2 A - k_3 B \dfrac{A}{K+A} + k_1 + k_6 C \\ \dot{B} = -k_2 B - k_3 A \dfrac{B}{K+B} + k_1 + \beta_B \\ \dot{C} = k_4 B - k_5 C\,. \end{cases} \quad (1)$$

This system is still quoted as one of the very first chemical reactions system designed to implement logic circuits.[28] Computer output of this abtract chemical reaction (Fig. 2 in[6]) was compared to an electronic multivibrating device in.[29] Then starting to investigate the subsystem A-B by replacing term $k_6 C$ in the first equation with a constant term β_A, I commented in the paper submitted in 1971:[6]

> The equations of this partial system are well-known in electronics where they apply to the usual symmetrical RS flip-flop: the so-called Eccles-Jordan trigger;[30] only the nonlinear terms [...] are normally replaced by a more generally formulated class of functions (see Andronov et al,[20] p. 309, Eq. 5.61). However, the very system [A-B] is obtained, even in the electronic case, if n-channel field-effect transistors are employed as the active elements.[31]
>
> If the standard analytical techniques used in electronics (Andronov et al,[20] p. 310) are applied to the present special case, it is again found (a) that either equation, when set equal to zero, yields a curved nullcline; (b) that both nullclines intersect each other in either 1 or 3 steady states; (c) that the intermediate steady state is a saddle-point and the other ones (or the remaining one, respectively) are stable nodes; and (d) that the presence of additional limit sets (limit cycles) is excluded.

The contribution of Andronov's group not only framed my early studies on chemical reactions but also my first studies on chaos as I will explain in this chapter.

I thus started to design some three-variable oscillators based on a two-variable bistable system coupled to a slowly moving third-variable. The resulting three dimensional system was only producing limit cycles at the time. In this period, I also introduced *dynamical automata* as components for the building up of complex chemical reaction systems: in other words, I had in mind to build chemical reaction systems as complex as electronic circuits are.[3] At an international congress on *Rhythmic functions in biological systems* held on September 8-12, 1975 in Vienna, I met Art Winfree (1942-2002) again — a theoretical biologist who started his career as an engineering physicist and studied chemical waves,[32] circadian rhythm[33] and cardiac arrhythmia.[34] In 1975, he had invited me for a talk in Purdue University on my paper of ref.[31] Winfree — also an expert in computers — was regularly exchanging letters with me about oscillating chemical reactions and dynamical systems. At this conference in Vienna, Winfree challenged me in 1975 to find a biochemical reaction reproducing the Lorenz attractor. To stimulate me to the task, Art sent me a collection of reprints and preprints with a letter dated on October 7, 1975. The papers sent were:

(1) E. N. LORENZ, Deterministic Nonperiodic Flow, *Journal of the Atmospheric Sciences,*, **20** (1963) 130-141.
(2) R. MAY & G. F. OSTER. Bifurcations and dynamic complexity in simple ecological models,
(preprint later published[35]).
(3) F. C. HOPPENSTEADT & J. M. HYMAN. Periodic solutions of a logistic difference equation, (preprint later published[36]).
(4) T. Y. LI & J. A. YORKE. Period-three implies chaos, (preprint later published[37]).
(5) J. GUCKENHEIMER, G. F. OSTER & A. IPAKTCHI. Dynamics of density-dependent population models, (preprint later published[40]).

I was strongly impressed by Lorenz's paper. As it will be shown, Li and Yorke's paper[37] also had a strong influence and was crucial for providing a numerical proof of chaos. By these times, I failed to find a chemical or biochemical reaction producing the Lorenz attractor but I instead found a simpler type of chaos in a paper I wrote during the 1975 Christmas holidays.[8] This paper will be at the core of the present chapter. It is only much later that I discovered jointly with Peter Ortoleva a biochemical reaction scheme producing a Lorenz-like dynamics.[41] The obtained attractor does not have the rotation symmetry the Lorenz attractor has, but it is characterized by a map equivalent to the one published by Lorenz.[42] This type of chaos was later designated as "unimodal cut chaos" in reference.[43]

3. Three Main Influences

I briefly review three works which infuenced me while discovering my first chaotic system according to an abstract submitted on December 1st, 1975 for the 1976 *Biological Society Meeting* (Fig. 3). These influences are i) Lorenz's 1963 paper, ii) the Li-Yorke theorem and iii) the universal circuit for which a paper by Khaikin — whose content is discussed in[20] — is quoted (cf.[25]). But let us follow the chronology.

3.1. *The multivibrator by Andronov, Khaikin and Vitt*

As mentioned, I was quite acquainted with electronics. Since I was also attracted by dynamical systems, the textbook by Andronov, Khaikin and Vitt[20] became one of my favorite books in the early 70s. Indeed as soon as I realized that equations for describing life would be too complicated as an exclusive object of research, I concentrated my interest on the basic chemical elements that could be used to build complex chemical reactions. The very first elementary circuit investigated was a two-variable multivibrator.[5] According to Andronov, the Russian scientist and his co-workers quickly focused their interest on an intermediary circuit between a double *RC* circuit and a multivibrator.[20] This so-called *universal circuit* — a circuit as simple as possible with a wide variety of behaviors — was three-dimensional, thus

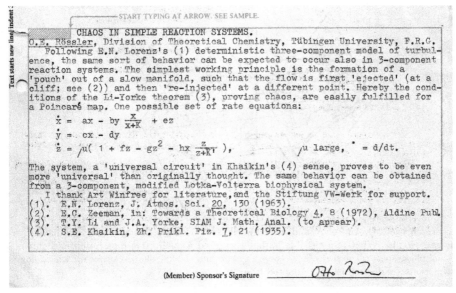

Fig. 3. Abstract I submitted on December 1st, 1975 for the Biological Society Meeting planned for 1976.

requiring three variables to describe the motion in phase space (Fig. 4). This is one of the most important methodological breakthroughs introduced by Andronov and co-workers. They thus described the trajectory drawn in the three-dimensional space spanned by two potentials measured on the circuit, potential u measured at one of the two triodes and potential v_1 measured at a condenser located between the two triodes, and the derivative $\dot{v}_1 = z$ of the second potential.

Oscillations were described as follows:[20]

> [...] the phase paths of "rapid" motion (jump) in the u, z, v_1 phase space recede away from the region $|u| \leq u_*$ of the surface F [...]. For $|u| \leq u_*$ only jumps of the voltage u are possible [...]. On the remaining part of the surface F (for $|u| > u_*$) [...] the paths of "rapid" motion approach the surface F [...] On the portion F^+ of F where $|u| \leq u_*$ there are "slow" motions along paths [...]. Outside F^+ $u \to \infty$, for $\mu \to +0$ but $\dot{z}$ and $\dot{v}_1$ remain finite, therefore there are "rapid" motions along the paths z =const, v_1 =const which lead to the surface F^+ where they pass into paths of "slow" motions. In due course all paths of "slow" motion pass into discontinuous jumps at $u = +u^*$ or at $u = -u^*$. It can easily be shown that all phase paths tend to a unique and stable limit cycle for $t \to +\infty$. Thus [...], whatever the initial conditions, discontinuous oscillations build up in the system.

In this description, Andronov and co-workers used a three-dimensional space to clearly distinguish "slow" and "fast" motions. They also explained why "jumps of the voltage" cannot be avoided. This therefore represents a dynamical analysis of the universal circuit that was provided.

The description used by Andronov combines analytical computations and qualitative properties of the trajectory in the phase space. The figure drawn was

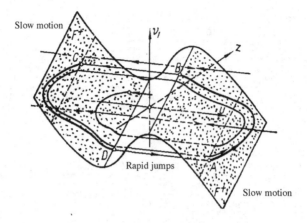

Fig. 4. Sketch used by Andronov and his co-workers to describe the trajectory produced by the universal circuit. What is noticeable is that they used a qualitative description in a three-dimensional phase space. The trajectory is organized around the "S"-shaped surface F where slow and fast motions are distinguished. (Adapted from[20]).

thus deeply conclusive to reach the conclusion that a stable limit cycle must exist. The unusual character of the description lies in combining physical properties of the system on the one hand and a representation of its evolution in abstract phase space on the other. Strictly speaking, they should have been led to observe oscillations more complicated than periodic ones. But only periodic behaviors were discussed in the literature by them.

This contribution kept my attention focused on the multivibrator and lead me to introduce an S-shaped two-dimensional surface to explain how to produce non trivial oscillating solutions. By "trivial" is here meant a limit cycle with a nearly constant speed. From a dynamical point of view, the background provided by Andronov and his co-workers influenced me a lot before 1975.

3.2. *The Lorenz paper*

Lorenz's paper also strongly influenced the way in which I wrote my first paper on a chaotic system was written. Edward Lorenz started his paper[42] by giving general definitions about trajectories in phase space. One clear breakthrough in the study of dynamical systems re-introduced by Lorenz was the use of a projection of phase space. He did not do that in an empirical way, but stated rather clearly that a system governed by the set of equations

$$\dot{X}_i = F_i(X_1, X_2, ..., X_M), (i = 1, ..., M) \qquad (2)$$

"may be studied by means of *phase space* — an M-dimensional Euklidean (sic) space Γ whose coordinates are $X_1, ..., X_M$". No doubt Lorenz was acquainted with such background through David Birkhoff's work. Birkhoff was Dean of the Faculty of Arts and Science at Harvard University where he taught since 1912. Lorenz got his AM in mathematics from Harvard University in 1940 and, he attended Birkhoff's lectures. Birkhoff is well known to be one of the continuators of Poincaré's work. The use of phase space was one of the key points in Lorenz's paper.

Lorenz also investigated some definitions about the stability of "nonperiodic flow", a mathematical background he inherited from Birkhoff. Some definitions given by him about stable and unstable points, periodic, quasi-periodic and nonperiodic solutions show this. He also stated that "two states differing by imperceptible amounts may eventually evolve into two considerably different states". As a consequence, "an acceptable prediction of an instantaneous state in the distant future may well be impossible". Such sensitivity to initial conditions was one of the relevant points highlighted by David Ruelle by the mid of the 70s to distinguish chaos from other qualitative types of dynamical behavior.[44]

To provide an idea of how the trajectory was organized in three-dimensional phase space $\mathbb{R}^3(x,y,z)$, Lorenz introduced "isopleths" that return the value of x as a smooth single-valued function of y and z. Isopleths allow to represent the "surface" on which the trajectory evolves (Fig. 5a). Lorenz was thus able to show that the trajectory "passes back and forth from one spiral to the other without intersecting itself". This surface was topologically equivalent to what is now called a branched manifold — or a template — on which all trajectories can be drawn (Fig. 5b). Such a manifold was used since the middle of 70s by Robert Williams for describing the Lorenz attractor[45,46] as "a picture already present in Lorenz' paper (compare Figs. 5a and 5b). As Williams wrote,[45] "a computer gives the same picture up to a smooth deformation when programmed to find the attractor of the system". The branched manifold proved important as a knot holder, that is, to synthesize the relative organization of unstable periodic orbits embedded within the attractor, as later shown by Joan Birman and Williams.[47]

Then Lorenz proposed a first-return map to maxima of the variable z in order to identify the possible periodic sequences that can be produced. It helped him to conclude that

> the periodic trajectories, whose sequences of maxima form a denumerable set, are unstable, and only exceptional trajectories, having the same sequences of maxima, can approach them asymptotically. The remaining trajectories, whose sequences of maxima form a nondenumerable set, therefore represent deterministic nonperiodic flow.

This argument was used to show that trajectories were actually nonperiodic since unstable periodic orbits were "exceptional" since the probability to have a trajectory remaining in the neighborhood of a periodic orbit is therefore nearly zero.

Lorenz then used a first-return map to describe the dynamics governing the transitions from one spiral to the other:

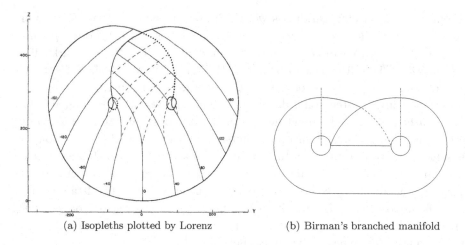

(a) Isopleths plotted by Lorenz (b) Birman's branched manifold

Fig. 5. (a) Isopleths of x as a function of y and z (thin solid curves). Where two values of x exist, the dashed lines are isopleths of the lower value. Heavy solid curve, and extensions as dotted curves, indicate natural boundaries of surfaces. (b) Representation of the associated branched manifold drawn by Williams. The two-component Poincaré section associated with the maxima of variable z is also drawn.

> ... the trajectory apparently leaves one spiral only after exceeding some critical distance from the center. Moreover, the extent to which this distance is exceeded appears to determine the point at which the next spiral is entered; this in turn seems to determine the number of circuits to be executed before changing spirals again.

In order to investigate that feature carefully, Lorenz used the successive maximum values of z. He thus plotted the value of the $(n+1)$th maximum value of z versus the nth maximum (Fig. 6). This is what is now called a *first-return map* to a Poincaré section. Lorenz introduced that tool for having "an empirical prediction scheme" allowing to predict the number of "circuits" (oscillations around one of the focus-type fixed points) described by the trajectory between two successive transitions from one spiral to the other. With such a map, it is possible to follow through how the trajectory visits the attractor using a simple geometric construction (Fig. 6). The increasing branch (left part of the map) corresponds to the successive oscillations around the one focus and the decreasing branch (right part) is associated with transition from one spiral to the other. For instance, as shown in Fig. 6, starting from point 1, there are thus two oscillations in the initial spiral (points 2 and 3), then one transition in the other spiral (point 4) and, finally, a return to the initial spiral (point 5) before new oscillations in the initial spiral occur, and so on.

To conclude, the most important points used by Lorenz were i) plotting the trajectory in plane projections of the phase space, ii) showing that the trajectory can be described as evolving on a surface and iii) using a first-return map (or a

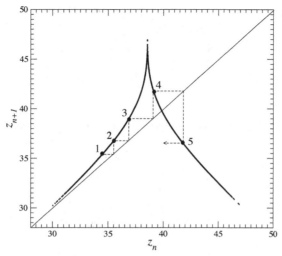

Fig. 6. Successive values of relative maximum plotted as z_{n+1} versus z_n as shown in.[42] The bisecting line has been added to make explicit the geometric construction (dashed line) that allows to track the evolution of the trajectory within the attractor.

Poincaré map) to show that the trajectory is nonperiodic with the help of periodic sequences.

3.3. *The Li-Yorke theorem*

The paper published in 1975 by Tien-Yien Li and James Yorke[37] remains highly reputed for i) having introduced the term chaos and ii) providing a theorem that can be understood as follows: as soon as a system has a period-3 orbit for a solution, then there is chaos. The term chaos was introduced for designating "complicated phenomena [that] may sometimes be understood in terms of a simple model".[37] In that sense, the word chaos was used in a quite adequate way since it traditionally designates the *"indescribable* state of Earth before creation". A simple difference equation (a second-order polynomial) may have "surprisingly complicated dynamic behavior", complicated meaning here: not actually understood. Similar conclusions could be obtained from the adverb "chaotically" used once in Ruelle and Takens' paper published in 1971.[38] James Yorke himself recently conceded that defining clearly what chaos is remains an open problem, particularly because it depends on the context in which it is used.[39] It has to be noted that the word "chaos" also appeared in the title of a paper published by May in 1974 in which he quotes Li and Yorke's preprint, and in the subsequent paper by Guckenheimer, Oster and Ipaktchi,[40] all of them in the collection sent by Winfree.

The most important theorem proved by Li and Yorke was

Let J be an interval and let $F : J \mapsto J$ be continuous. Assume there is a point

$a \in J$ for which the points $b = F(a)$, $c = F^2(a)$ and $d = F^3(a)$, satisfy

$$d \leq a < b < c \quad (\text{or } d \geq a > b > c).$$

Then

(1) for every $k = 1, 2, \ldots$ there is a periodic point in J having period k.
(2) Furthermore, there is an uncountable set $S \subset J$ (containing no periodic points), which satisfies the following conditions:

   (a) for every $p, q \in S$, with $p \neq q$,

   $$\lim_{n \to \infty} \sup |F^n(p) - F^n(q)| > 0$$

   and

   $$\lim_{n \to \infty} \inf |F^n(p) - F^n(q)| = 0.$$

   (b) for every $p \in S$ and periodic point $q \in J$,

   $$\lim_{n \to \infty} \sup |F^n(p) - F^n(q)| > 0$$

Consequently, when $q = a$, that is, when there is a periodic point with period 3, then there is period-k point for any $k = 1, 2, \ldots$ Furthermore, there is an uncountable subset of points x in J which are not even asymptotically periodic. This means that there is an uncountable subset of unstable periodic orbits.

Chaotic behaviors will be only encountered indefinitely when there is no stable periodic point. This last property is quite hard to prove and, for instance, a computer assisted proof for the chaoticity of the Lorenz attractor was only obtained in 1999.[48] René Lozi discusses this problem in chapter 4. It is nearly impossible to actually obtain an aperiodic orbit using numerical simulation. Consequently, when a map of the interval has a period-3 orbit, one can conclude that there is at least one orbit of each period and these orbits are nondenumerable. But it remains to prove that there is no stable periodic point embedded, something that remains non trivial to be shown in most cases.

Strictly speaking, showing that there is a period-3 orbit in a uni-dimensional map is not enough to prove that the behavior is chaotic. For scientists working with an experimental data set or with "computer experiments", there is no rigorous way to distinguish an arbitrarily long periodic orbit from a chaotic solution. Still, numerical experimentalists often use the existence of a period-3 orbit as a proof for an uncountable subset of non periodic points and implicitely assumed that the studied solution was chaotic.

4. My Earliest Paper on Chaos

4.1. *Phase space and chaotic attractor*

To me chaos became soon familiar since we can refer to Poincaré's work on two coupled oscillators, Arnold's map, Smale's horseshoe map and Ruelle and Takens' strange attractor introduced in the context of the route to turbulence. If we should discuss in which sense Chaos was known before 1976, for sure it was not very broadly known although the papers by May[49] and Li and Yorke[37] were already published with that word in the title. This term was also used by Guckenheimer, Oster and Ipaktchi.[40] To me, the relevant heritage left by Poincaré was the *Analysis situs*, that is, the topological analysis.

4.2. *First-return map to a Poincaré section*

I understood that the Poincaré map was a very useful tool to investigate dynamical systems. For rigorously speaking, a return-map to a Poincaré section is not built with the maxima of a given state variable only, but rather is obtained using the trajectorial intersections with a surface of section transverse to the flow. My simple spiral attractor was easier to carefully investigate than the Lorenz attractor. In particular, a Poincaré section was easy to define as I wrote:[14]

> Unexpectedly, the qualitative behavior of [the Lorenz equations] is still insufficiently understood, mainly because the usual technique for analyzing oscillations — to find a (Poincaré) cross-section through the flow which is a (auto-) diffeomorphism[50] is not applicable. A trick which exploits the inherent (although imperfect) symmetry between the two leaves of the [Lorenz] flow, so that in effect only a single leaf needs to be considered, has yet to be found.

To me it appeared some relation should exist between the Lorenz map and a "single leaf" attractor as I will discuss below. A more extended study will be discussed in chapter 10. This is why I asked the question

$$\text{"Lorenz map (return map to maxima)} \stackrel{?}{=} . \text{Poincaré map"}$$

I understood that an important role has to be played by the so-called "cap-shaped" map. "Cap" meaning "a soft, flat hat without a rim and usually with a peak". In a more mathematical way, we should speak about a smooth unimodal map. I am not a mathematician — since I never acquired a formal background in mathematics — but I was always fascinated by mathematics, from my first self-posed task (to write down the differential equation for Life), to the works which influenced me the most, namely that of Lorenz (who had graduated in mathematics), of Smale, of Li and Yorke, and of Andronov and co-workers... Being afraid of misrepresenting mathematical concepts, I avoided mistakes by using words like "cap-shaped" and many others, as can be found in all of my papers on chaos.

In his main paper, Lorenz first plotted the trajectory of his system in the corresponding phase space $\mathbb{R}^3(x, y, z)$, and then computed a first-return map built on the maximum values of variable z. This is the natural way for investigating a system. But I understood that reversing the Lorenz procedure, that is, starting from a given map to obtain a flow — this was designated by Smale[50] as getting a *suspension* of a map — would be useful to design various types of chaos as described below. There is no rigorous — analytical — general way to obtain a suspension from a given map, although Gottfried Mayer-Kress succeeded in a special case[51] and I did that by trial and error, that is, in a long journey spent in front of my computer. For instance, the hyperchaotic attractor was found after three months, day and night, spent in front of my digital computer (HP 9845B).

4.3. *Qualitative properties of the expected dynamics*

I came to my first set of equations, starting out with Khaikin's universal circuit[25] and the corresponding S-shaped surface (Fig. 4) which I modified (Fig. 7c) to get the dynamics I wanted to have. The key was to modify (bend-over more and more) the S-shaped surface in order to then have an additional "orientation of flowing" on the other half, that is, eventually a motion with a "twist" would form. The original picture (Fig. 7c) shows a "reversed direction of flow" reinjecting the trajectory (upstair) in a nonlinear way, thereby allowing for an aperiodic trajectory (see Fig. 7 with its original caption).[8] I will not comment here on why I added the Dali's soft watch in the original picture.

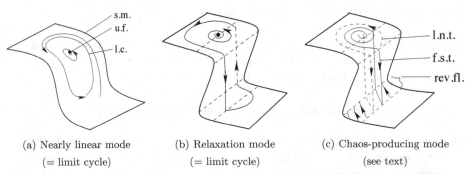

(a) Nearly linear mode (= limit cycle)

(b) Relaxation mode (= limit cycle)

(c) Chaos-producing mode (see text)

Fig. 7. Main trajectorial flow of a universal circuit. s.m.= slow manifold, u.f.= unstable focus, l.c.= limit cycle, the intermediate part of slow manifold in (b) is unstable, f.s.t. "first switched trajectory", l.n.t.= "last non switched trajectory", rev.fl.= reversed direction of flow "downstairs".

Returning to the way in which specific equations were obtained, I introduced them as resulting from an oscillating Turing cell[23] and an Edelstein switch.[52] In a graphical way, this could be expressed as

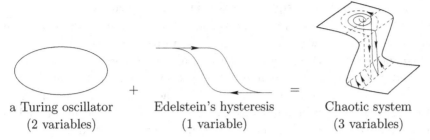

| a Turing oscillator | Edelstein's hysteresis | Chaotic system |
| (2 variables) | (1 variable) | (3 variables) |

Two variables were required to obtain a periodic oscillator and a third one to introduce the switching mechanism. This was equivalent to Andronov and co-workers' description of the universal circuit when they decomposed these behaviors into a two-dimensional slow motion surface F (Fig. 4) that can be described by the two equations in u and z and, "rapid jumps" made in the third dimension. The S-shape surface can only be described in a three-dimensional phase space.

4.4. *The equations and their chaotic solution*

The system I got obeys, under the usual assumptions of wellstirredness and isothermy as well as an appropriate concentration range, the following set of rate equations:

$$\begin{cases} \dot{a} = k_1 + k_2 a - \dfrac{(k_3 b + k_4 c)a}{a + K} \\ \dot{b} = k_5 a - k_6 b \\ \mu \dot{c} = k_7 a + k_8 c - k_9 c^2 - \dfrac{k_{10} c}{c + K'} \end{cases} \quad (3)$$

where a denotes the concentration of substance A, etc., $\cdot = \frac{d}{dt}$, $k_{10} = k'_{10} e_0$, e_0 =constant, and K, K' are Michaelis constants. In fact, once I got the "ideal" reaction scheme, a lot of intuition and time was spent on the analog and/or digital computer varying the coefficients to get the result. The proposed equations were thus slightly altered (some terms were empirically removed). Another system that was proposed in the abstract I sent off on December 1st, 1975 (Fig. 3), here rewritten with the symbols as used in (3) to simplify the comparison

$$\begin{cases} \dot{a} = k_2 a - \dfrac{k_3 ab}{a + K} + c \\ \dot{b} = k_5 a - k_6 b \\ \dot{c} = \mu \left(1 - k_8 c - k_9 c^2 - \dfrac{k'_4 ac}{c + K} \right). \end{cases} \quad (4)$$

Thus, the term $\frac{k_4 ac}{a+K}$ in the first equation was reduced to a linear term $k_4 c$, the term $\frac{k_3 ab}{a+K}$ was removed from the second equation, and the linear term $k_7 a$ was removed from the third equation in which the two nonlinear terms were mixed together.

The S-shaped surface therefore served to design the general structure of the equations and then the parameters were determined by a manual "maieutic" technique. In other words, the "principle for generating chaos" can be summed up into a procedure to pre-define some qualitative properties of the expected behavior and to then use different components introduced as chemical automata[3] to design roughly the structure of the equations (holding exactly in a limit). The final part of the work was just... time and patience in front of a computer (Fig. 8).

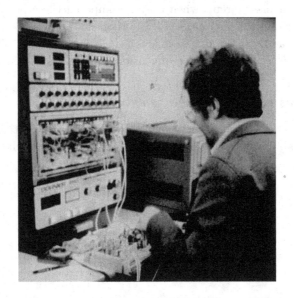

Fig. 8. Otto E. Rössler in front of his computer in 1979.

This means that many parameters were tried and only those leading to the expected dynamical behavior were retained. In other words, during this search for appropriate parameter values, a few terms were set to zero. I must confess that I did not describe this procedure in my original paper. The first chaotic system I thus obtained possessed parameters values as follows: $k_1 = 37.8$, $k_2 = 1.4$, $k_3 = 2.8$, $k_4 = 2.8$, $k_5 = 2$, $k_6 = 1$, $k_7 = 8$, $k_8 = 1.84$, $k_9 = 0.0616$, $k_{10} = 100$, $K = 0.05$, $K' = 0.02$, $\mu = 1/500$; $a_0 = 7$, $b_0 = 12$, $c_0 = 0.2$, $t_0 = 0$ and $t_{\text{end}} = 43.51$. The value of the parameter μ was changed here compared to the one published (1/25) to recover the picture published (Fig. 9).

The simpler system was obtained after nights and days spent in front of my computer. My objective was actually to simplify my set of original equations (3) to obtain a set of simpler equations that "contains just one (second-order) nonlinearity

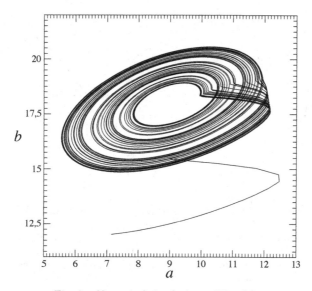

Fig. 9. Numerical simulations of Eq. (3).

in one variable" as mentioned in the abstract in.[14] Once I got a chaotic solution (Fig. 9) to equation (3), I came to the conclusion that "qualitative properties cannot be deduced from simulation alone" and, as Lorenz had done, the dynamics would be better understood by using a Poincaré map. In contrast to what Lorenz had done, I never computed the Poincaré map I provided. I rather drew it after many visual inspections of the dynamics using stereoscopic projections as used in many of my papers to verify the analytical limiting result I had obtained first. Thus, I realized that the map is made up from two branches. One branch is associated with the relaxation process induced by the switching nonlinear mechanism. At a certain threshold value, the linear process is interrupted by the switch, thereby limiting the diverging spiral. I drew the map (Fig. 10) with a qualitatively correct curvature as second author Christophe checked by computing a first-return map to a Poincaré section (Fig. 11).

I showed that the trajectory was bounded in phase space by drawing that the "cut", that is, the threshold at which the relaxation mechanism cuts the linear expansion, induces a "quadratic box" — the term "quadratic box" was replaced with "Li-Yorke box" in[9] — whose edges are bounding the behavior. This results from a common geometric construction for a first-return map. Such a procedure can be considered as a numerical proof for a bounded trajectory although drawn by hand. The so obtained picture was "proving that period-3, and hence chaos, is possible within the box" (Fig. 10).[9] Using the Li and Yorke theorem, I was thus able to deduce the "existence of an uncountable set of repelling periodic trajectories of measure zero".

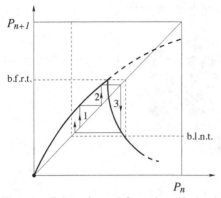
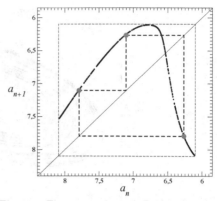

Fig. 10. Poincaré map of a universal circuit in the chaotic mode (see Fig. 7d), provided that $\mu \to 0$. b.f.r.t. = border-line determined by first reinjected trajectory ; b.l.n.t. = borderline determined by last non-reinjected trajectory ; 1, 2, 3 = steps proving chaos.

Fig. 11. First-return map to Poincaré section P for system (3) with the same parameter values as reported in Fig. 9. The box within which the period-3 orbit is observed is drawn with a thin dashed line. The three periodic points are linked by thick dashed lines.

4.5. Topological analysis

To complete the dynamical analysis of the chaotic attractor, I drew a surface by interpreting the stereoscopic representation of the trajectory. The sketch of the flow (Fig. 12a adapted from[8]) had a finite thickness in order to be easily interpreted in terms of a Smale's horseshoe map as the latter is usually represented (Fig. 12b). It contains many details revealing the deep structure of the dynamics I described as

> the "folded pancake" does not display the trajectories themselves, but only an "enveloppe" (made up of surfaces without contact, cf.[20]) which is entered by trajectories (as depicted), but never left. The picture is directly derived from Figs. 7d (turned upside down) and 9, respectively, displaying the principal properties only. The rectangular cross-section on the left-hand side is seen to be mapped diffeomorphically onto a subset of itself, as required from a 2-dimensional map. The "horseshoe" which is formed upon reinjection is also clearly visible.

Today my "folded pancake" could be compared to the "branched manifold" introduced by Williams.[45] Williams explicitly wrote that "the study of the attractor can be reduced to the study of the branched manifold with a semiflow on it" .[45] I used my "folded pancake" for characterizing several different types of chaotic attractor that I observed (see for instance[12]).

The relation to the Horseshoe map was important in order to make a strong connection with periodic orbits and symbolic dynamics, as done by Guckenheimer and his co-workers,[40] although I did not use that too much. I wanted to highlight how the attractor can be split up into two parts as is usually done in the Horseshoe map. Typically, when the latter is investigated, everything is done in the original

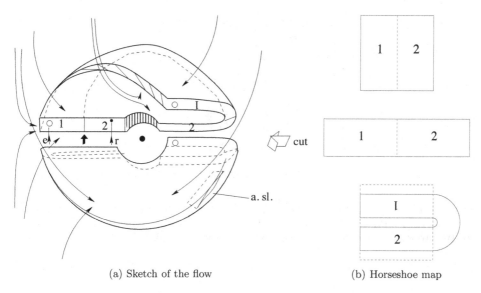

Fig. 12. The "three-dimensional blender". (Cf. Fig. 9a) → = trajectories entering the structure from the outside; 1, 2 = half cross-section (demonstrating the "mixing transformation" that occurs), e = entry point of some arbitrary chosen trajectory, r = reentry point of the same trajectory after one cycle. ↑ = "horseshoe map", a.sl. = allowed slit (see text). In (b), do not forget that the folding is associated with a rotation by π around the central fixed point •, thus orienting the folding to the right and not to the left, as the symbols "1" and "2" would have suggested otherwise. This right-hand sketch of an iteration of the Horseshoe map was not published in.[8]

square (top of Fig. 12b), that is, the curved part is cut out to avoid mathematical complications. Practically speaking, this means that the two parts of the original square can be split and labelled by "1" and "2", respectively. This leads to the "allowed slit" that I introduced to guarantee that "no trajectories are damaged". I thus used a similar cutting process as used by Smale. Such a slit was required to obtain Figure 13 whose purpose was to exhibit the two domains having different topological properties. These two domains are needed to make explicit the mechanism responsible for producing chaos. I thus split my "blender" into two strips (Fig. 13). I thus presented a possible partition of the attractor, based on topological properties. A partition was likewise used, for example, by Guckenheimer and his co-workers to introduce a symbolic dynamics to encode trajectories.[40]

Today, one could compare my "allowed slit" to the "splitting chart" responsible for the stretching as introduced by Robert Gilmore (see chapter 8). And the uppper part of the central core contains the "squeezing mechanism" where the two strips are "glued" together. Stretching and squeezing are the two relevant mechanisms for producing chaos. In Fig. 13, I clearly showed one strip corresponding to a "normal loop" and a second one associated with a "Möbius loop". The only difference between normal and Möbius loop is a "twist".

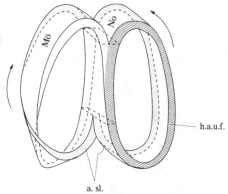

Fig. 13. A structure equivalent to that shown in Fig. 12a. Mö = Möbius loop, No = normal loop; h.a.u.f. = hole around the unstable focus in Fig. 12a; a.sl. = boundaries of the allowed slit in Fig. 12. The two arrows — added by the authors — show the direction of the flow.

5. A Short Walk through Various Topologically Inequivalent Classes of Chaos

In 1975, I was mainly driven by the configuration of the first-return maps. Unimodal maps — made of two monotonic branches — can be classified into three classes (Fig. 14). Various topologies thus differ only by global torsion (π-twists applied to the whole section of the attractor). The smooth unimodal map (Fig. 14a) is known to be associated with a period-doubling cascade as a route to chaos and uses folding as a mixing mechanism. I later identified such a route to chaos in the Lorenz system too.[53]

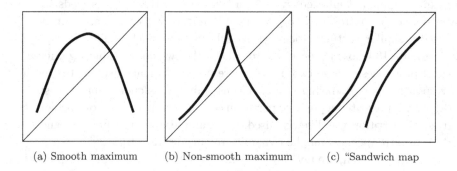

(a) Smooth maximum   (b) Non-smooth maximum   (c) "Sandwich map

Fig. 14. The three different types of unimodal maps.

The non-smooth unimodal map (Fig. 14b) also known as the "Lorenz map" is associated with a tearing as observed near the saddle fixed point of the Lorenz system. With Peter Ortoleva I later proposed a system

$$\begin{cases} \dot{x} = ax + by - cxy - \dfrac{(dz+e)x}{x+K_1} \\ \dot{y} = f + gz - hy - \dfrac{jxy}{y+K_2} \\ \dot{z} = k + lxz - mz \end{cases} \quad (5)$$

producing a chaotic attractor (Fig. 15a) bounded by a genus-one torus, and characterized by a non-smooth unimodal map (Fig. 15b). The tearing mechanism is observed in the bottom-left part of the attractor. The l-value is slightly modified to obtain a Lorenz map without a gap between the two monotonic branches.[41]

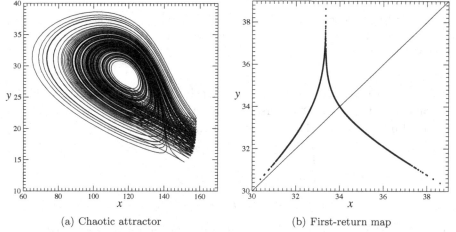

(a) Chaotic attractor (b) First-return map

Fig. 15. Unimodal cut chaotic attractor solution to system (5). Parameter values: $a = 33$, $b = 150$, $c = 1$, $d = 3.5$, $e = 4815$, $f = 410$, $g = 0.59$, $h = 4$, $j = 2.5$, $k = 2.5$, $l = 5.29$, $m = 750$, $K_1 = 0.01$ and $K_2 = 0.01$.

The "sandwich" chaos, or the "unimodal half-cut chaos" so termed since only one branch of the first-return map is inverted, can be viewed as a modified Lorenz system with a broken symmetry if $d \neq 0$. It reads as follows:[10]

$$\begin{cases} \dot{x} = x - xy - z \\ \dot{y} = x^2 - ay \\ \dot{z} = bx - cz + d \end{cases} \quad (6)$$

where the symmetry is broken when $d \neq 0$. An example of a chaotic attractor of the "sandwich" type is shown in Fig. 16.a. At first sight, the attractor should be bounded by a genus-3 torus but there is a path which is forbidden. This attractor is also associated with tearing mechanisms responsible for the discontinuity between the two increasing branches of the first-return map to a Poincaré section. Various avatars of these three types of chaos were discussed in.[54]

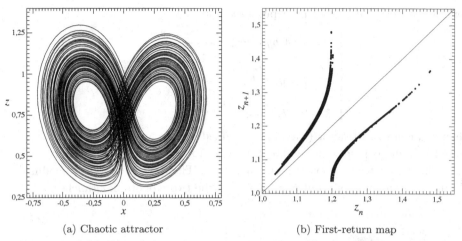

Fig. 16. Unimodal half-inverted cut chaos solution to system (6). Parameter values: $a = 0.1$, $b = 0.09375$, $c = 0.38$ and $d = 0.0015$.

In my attempt to produce the different types of chaos, I proposed a system which has the symmetry of Khaikin's "universal" circuit, that is, there are two symmetric foci connected by fast jumps. The equations for this type of chaos are:[55]

$$\begin{cases} \dot{x} = -ax - y(1 - x^2) \\ \dot{y} = \mu(y + 0.3x - 2z) \\ \dot{z} = \mu(x + 2y - 0.5z) \,. \end{cases} \quad (7)$$

Two plane projections of the chaotic attractor are shown in Figs 17a and 17b. The structure of the attractor is quite complex. A first-return map is computed using two components for the Poincaré section as required for a chaotic attractor bounded by a genus-3 torus. The return map is made up of 16 branches (Fig. 17c). The corresponding template therefore has 16 branches (Fig. 18).

6. Conclusion

Topological chaos is despite great strides made still at its beginning. The hierarchy which it implies — with hyperchaos and fat and superfat attractors — has only barely begun to be mathematically analyzed. The "flatten-and-fold" technique subsequently developed can most probably be applied successfully again. The "playdough method" is not yet exhausted. Besides "stretching" and "folding" and "twisting", a so far undiscovered fourth mechanism can be predicted to exist.

Another still qualitatively under-investigated type of behavior is "flaring" as in the Milnor-attractor derived "flare-attractors". It may have to do with the peculiar unpredictability of the financial chaotic phenomena, also discussed at this conference (Chapter 14). Francisco Doria alerted me to this work and opened up its beauty

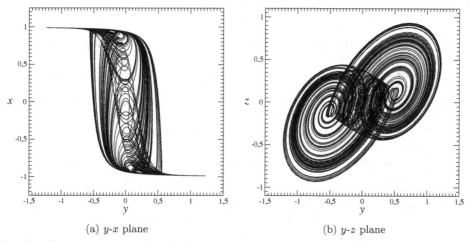

(a) y-x plane   (b) y-z plane

Fig. 17. Phase portrait associated with a chaotic solution to system (7). Parameter values: $a = 0.03$ and $\mu = 0.1$ (and not 10 as reported in Rössler's paper). Initial conditions: $x_0 = -1$, $y_0 = 0.55$ and $z_0 = 0.12$.

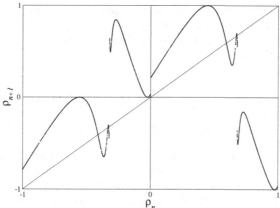

Fig. 18. Multimodal map characterizing the double scroll attractor (Fig. 17) solution to system (7).

in 1994 in the footsteps of a paper by the Yorke school.[56,57] The Wiener-Wald founder's book "From Dish Washer to Millionaire and Back" provided the first sample to the equation found jointly with Georg Hartmann. The topological way of thinking founded by Poincaré and making felt itself in a revigorated manner in this conference dedicated to Bob Gimmore's ongoing activities has only barely begun to raise its head again.

References

1. O. E. Rössler, Ein systemtheoretisches Modell zur Biogenese, *Zeitschrift für Naturforschung B*, **26**, 741-746 (1971).
2. O. E. Rössler & F. F. Seelig, A Rashevsky-Turing system as a two-cellular flip-flop, *Zeitschrift für Naturforsch B*, **27**, 1445–1448 (1972).
3. O. E. Rössler, Basic circuits of fluid automata and relaxation systems, *Zeitschrift für Naturfoschung B*, **27**, 333–343 (1972).
4. O. E. Rössler, Chemical automata in homogeneous and reaction-diffusion kinetics, *Lecture Notes in Biomathematics*, **4**, 399–418 (1974).
5. O. E. Rössler, A principle for chemical multivibration, *Journal of Theoretical Biology*, **36**, 413–417 (1972).
6. O. E. Rössler, A multivibrating switching network in homogeneous kinetics, *Bulletin of Mathematical Biology*, **37**, 181–192 (1975).
7. O. E. Rössler & D. Hoffmann, Repetitive hard bifurcation in a homogeneous reaction system, in *Analysis and Simulations of Biochemical Systems*, H. C. Hemkers & B. Hess (Ed.), Proceedings of the 8th FEBS-Meeting (Amsterdam, North-Holland), pp. 91-102 (1972).
8. O. E. Rössler, Chaotic behavior in simple reaction system, *Zeitschrift für Naturforschung A*, **31**, 259–264 (1976).
9. O. E. Rössler, Chaos in abstract kinetics: two prototypes, *Bulletin of Mathematical Biology*, **39**, 275–289 (1977).
10. O. E. Rössler, Different types of chaos in two simple differential equations, *Zeitschrift für Naturforsch A*, **31**, 1664–1670 (1976).
11. O. E. Rössler, Syncope implies chaos in walking stick maps, *Zeitschrift für Naturforschung A*, **32**, 607–613 (1977).
12. O. E. Rössler, Continuous chaos: four prototype equations, *Annals of the New York Academy of Sciences*, **316**, 376–392 (1979).
13. M. C. Kube, O. E. Rössler & J. L. Hudson, A superfat chaotic attractor, *Chaos, Solitons & Fractals*, **3**, 141–148 (1993).
14. O. E. Rössler, An equation for continuous chaos, *Physics Letters A*, **57**, 397–398 (1976).
15. C. Letellier & V. Messager, Influences on Otto E. Rösslers earliest paper on chaos, *International Journal of Bifurcation & Chaos*, **20** (11), 132 (2010).
16. O. Rössler, Das Ägyptische als semitische Sprache, In: *Christentum am Roten Meer Part I*, Franz Altheim & Ruth Stiehl (Eds), (de Gruyter, Berlin) pp. 263-326 (1971).
17. O. E. Rössler, *Dauerimmunisierung von Albinomäusen mit Rinder-γ-Globulin* [Long-term Immunization of Albino Mice with Bovine-γ-Globulin], University of Tübingen (1966).
18. Seelig F. F., W. Huber & H. Kuhn [1962] Analogiebetrachtungen und Analogrechner zur Behandlung der Korrelation von Elektronen, *Zeitschrift für Naturforschung A*, **17**, 114–121.
19. F. F. Seelig, Numerical solution of 2- and 3-dimensional Schrödinger equations for any molecular potential by iterative variation of numerical test functions with a digital computer — I. Theoretical principles: description of a computer program for solution of 2-dimensional Schrödinger equation, *Zeitschrift für Naturforschung A*, **20**, 416–27 (1965).
20. A. A. Andronov, A. A. Vitt & S. E. Khaikin, *Theory of oscillators*, Pergamon Press (1966).
21. H. Sutaner, *Meßsender Frequenzmesser und Multivibratoren*, Radio-Praktiker Bucherei **128/130**, Franzis Verlag, Munich (1966).

22. N. Rashevsky, An approach to the mathematical biophysics of biological self-organization and a cell polarity, *Bulletin in Mathematical Biophysics*, **2**, 65–67 (1940).
23. A. M. Turing, The chemical basis of morphogenesis, *Philosophical Transactions of the Royal Society B*, **237**, 37–72 (1952).
24. F. F. Seelig & O. E. Rössler, A chemical reaction flip-flop with one unique switching input, *Zeitschrift für Naturforschung B*, **27**, 1441–1444 (1972).
25. S. E. Khaikin, Continuous and discontinuous oscillations, *Zhurnal Prikladnoi Fiziki*, **7**, 21 (1930).
26. K. F. Bonhoeffer, Activation of passive irons as a model for the excitation of nerve, *The Journal of General Physiology*, **32** (1), 69-91 (1948).
27. B. van der Pol, On relaxation oscillations, *Philosophical Magazine*, **2**, 978-992 (1926).
28. K.-P. Zauner, Molecular information technology, *Critical Reviews in Solid State and Material Sciences*, **30**, 33–69 (2005).
29. P. M. Chrilian, Electronic circuits: physical principles, analysis and design, McGraw-Hill, New York (1971).
30. W. H. Eccles & F. Jordan, Sustaining the vibration of a tuning fork by a triode valve, *The Electrician*, **82**, 704 (1919).
31. O. E. Rössler, A synthetic approach to exotic kinetics (with examples), *Lecture Notes in Biomathematics*, **4**, 546–582 (1974).
32. A. T. Winfree, Spiral waves of chemical activity, *Science*, **175**, 624–636 (1972).
33. A. T. Winfree, *The geometry of biological time*, Springer, Berlin (1980).
34. A. T. Winfree, Electrical instability in cardiac muscle: phase singularities and rotors, *Journal of Theoretical Biology*, **138**, 353–405 (1989).
35. May R. & G. F. Oster, Bifurcations and dynamic complexity in simple ecological models, *The American Naturalist*, **110**, 573–599.
36. F. C. Hoppensteadt & J. M. Hyman, Periodic solutions of a logistic difference equation, *SIAM Journal of Applied Mathematics*, **32**, 73–81 (1977).
37. T. Y. Li & J. Yorke, Period-three implies chaos, *American Mathematics Monthly*, **82**, 985-992 (1975).
38. D. Ruelle & F. Takens, On the nature of turbulence, *Communications in Mathematical Physics*, **20**, 167–192 (1971).
39. J. Yorke, Response to David Ruelle, Letter to the editor, *Notices of the American Mathematical Society*, **56**, 1232 (2009).
40. J. Guckenheimer, G. F. Oster & A. Ipaktchi, Periodic solutions of a logistic difference equation, *Journal of Mathematical Biology*, **4**, 101–147 (1976).
41. O. E. Rössler & P. J. Ortoleva, Strange attractors in 3-variable reaction systems, *Lecture Notes in Biomathematics*, **21**, 67–73 (1978).
42. E. N. Lorenz, Deterministic nonperiodic flow, *Journal of the Atmospheric Sciences,*, **20**, 130–141 (1963).
43. Letellier C. & R. Gilmore, Poincaré sections for a new three-dimensional toroidal attractor, *Journal of Physics A*, **42** 015101 (2009).
44. D. Ruelle, The Lorenz attractor and the problem of turbulence, *Lecture Notes in Mathematics*, **565**, 146-158 (1976).
45. R. F. Williams, The structure of lorenz attractors, *Lecture Notes in Mathematics*, **615**, 94–112 (1977).
46. R. F. Williams, The structure of Lorenz attractors, *Publications Mathématiques de l'IHES*, **50**, 73–99 (1979).
47. J. S. Birman & R. F. Williams, Knotted periodic orbits in dynamical systems I: Lorenz's equations, *Topology*, **22**, 47–82 (1983).
48. W. Tucker, The Lorenz attractor exists, *Comptes-Rendus de l'Académie des Sciences*, **328**, 1917–1202 (1999).

49. R. May, Biological populations with nonoverlapping generations; stable points, limit cycles, and chaos, *Science*, **186**, 645–647 (1974).
50. S. Smale, Differentiable dynamical systems, *Bulletin of American Mathematical Society*, **73**, 747–817 (1967).
51. G. Mayer-Kress & H. Haken, An explicit construction of a class of suspension and autonomous differential equations for diffeomorphisms in the plane, *Communications in Mathematical Physics*, **111**, 63-74 (1987).
52. B. B. Edelstein, Biochemical model with multiple steady states and hysteresis, *Journal of Theoretical Biology*, **29**, 57–62 (1970).
53. O. E. Rössler, Horseshoe map in the Lorenz equation, *Physical Letters A*, **60** (5), 392-394 (1977).
54. Letellier C., E. Roulin & O. E. Rössler, Inequivalent topologies of chaos in simple equations, *Chaos, Solitons & Fractals*, **28**, 337–360 (2006).
55. O. E. Rössler, Continuous chaos, In *Synergetics* (H. Haken, ed.), Proceedings of the International Workshop on Synergetics at Schloss Elmau (Bavaria, May 2-7, 1977), Springer-Verlag, pp. 184-197 (1977).
56. J. C. Sommerer & E. Ott, A physical system with qualitatively uncertain dynamics, *Nature*, **365**, 138-140 (1993).
57. E. Ott, J. C. Sommerer, J. C. Alexander, I. Kan & J. A. Yorke, Scaling behavior of chaotic systems with riddled basins, *Physical Review Letters*, **71**, 4134-4137 (1993).

PART 2
Development of the Topology of Chaos

Chapter 6

The mathematics of Lorenz knots

Joan S. Birman

Department of Mathematics, Barnard-Columbia, 2990 Broadway
New York, NY 10027, USA

This is a review article on Lorenz knots. First identified as an interesting class of knots (and links) in 1983, we focus on the progress made by mathematicians in understanding them, up to 2008.

Contents

1. Introduction 127
   1.1. History 128
   1.2. The search for invariants 130
   1.3. Special classes of knots and links 131
   1.4. Some natural questions 133
2. Introducing Lorenz Knots and Links 133
   2.1. A chaotic flow on 3-space 133
   2.2. Every torus link is a Lorenz link 135
3. Parametrizing Lorenz Knots and Links 136
   3.1. The first parametrization, Lorenz braids 136
   3.2. The second parametrization: symbolic dynamics and LR words .. 138
   3.3. The third parametrization: positive twisted torus links ... 139
4. Applications 141
   4.1. The recognition problem for Lorenz knots 141
   4.2. The volume of $S^3 \setminus \mathcal{K}$ and the monodromy of $\mathcal{K}$ ... 143
   4.3. Modular knots and Lorenz knots 144
   4.4. Generalizations of Lorenz knots 147
References 147

1. Introduction

We begin, in §1.1, with a very informal and intuitive review of the history of knot theory. In §1.2 we give precise definitions of knot and link type, and discuss some of the invariants that were discovered over the years. In §1.3 we show how the link classification problem changes when the links in question are limited to special classes, with torus links as an example. In §1.4 this will lead us naturally to Lorenz links, the main subject of this review. They are, in a very precise way, generalizations of torus links.

1.1. History

At the end of the 19th century the well-known Scottish physicist Peter Guthrie Tait (1831-1901) had the idea that the periodic table might be explained by knotting in the 'impenetrable ether'. This lead him to study knots, and while he did not succeed in his original goal, he became fascinated by the intricacies of knotting. This was the beginning of the systematic study of knots, and (in a sense) a beginning of the part of topology that we call *Knot Theory*.

If we widen the discussion to include not just knots but also links, there was even earlier work due to Gauss (1777-1855), who was interested in the question of how the current in a closed knotted wire affected that in another closed knotted wire when the two closed curves were 'linked'. A different but related question also appears in Gauss' work,[2] where he computed the linking number of the earth's orbit with that of certain asteroids.

We've given two reasons why knots and links were of interest in physics, but what about mathematics? Mathematicians first became interested in knots because they are very easily visualized, and their complementary spaces $\mathbb{S}^3 \setminus K$, where K is a knot, are a rich source of examples of 3-manifolds. While knot complements are non-compact, 'Dehn surgery' on knot spaces, defined by associating a rational number to each component of a link,[18] is a way to construct all closed 3-manifolds too. We note that the phenomena of knotting occurs in every dimension (especially in codimension 2) as part of a study of the way that one manifold sits inside another.

Knots also appear in other parts of mathematics where there is no obvious way, at least initially, to visualize knotting. For example, an algebra student who has learned the definitions of a ring and a subring and has worked on some good examples could probably understand that if R' is a subring of a ring R, then the inclusion map $R' \subset R$ might be interesting from a combinatorial viewpoint. Type II$_1$ factors in operator algebras may be thought of analogues of rings. When the Jones polynomial was discovered in 1985, its mathematical origins were in Operator Algebras. It was later understood that it was crucial to the discovery that some sort of knotting-in-disguise was involved in the way that one type-II$_1$ factor sat inside another type-II$_1$ factor. See Birman's article[1] on the work of Jones for a fuller discussion. Data from the early knot tables has survived to this day. Figure 1 shows the first 5 examples from the 1893 knot tables. Those tables have now been extended to 'the first 1,701,936 knots', where the measure of complexity is the number of crossing points in a picture of the knot. See.[20] For example, there is only one knot of crossing number 3, also only one with 4 crossings, but two with 5 crossings and so on. By 'only one' is meant one (up to distinct pictures of the same knot). This is a good moment to mention that, while minimum crossing number, over all possible projections, is by definition independent of the choice of a projection, it has not turned out to be a very meaningful measure of complexity. Indeed, modern knot tables include ones where the measure of complexity reflects more subtle aspects of knot theory. We will have more to say about that in §4.2 below.

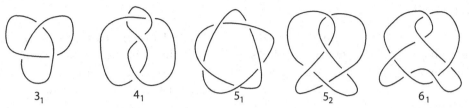

Fig. 1. The first five knots in Tait's tables

A knot is *composite* if you can obtain a representative by tying one knot in a piece of string, and following it by another. It's *prime* if it is not composite. See Figure 2 for examples. Since prime knots are the "building blocks" of all knots,

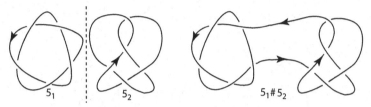

Fig. 2. The knots 5_1 and 5_2 on the left are prime knots from Fig. 1. The right sketch shows the 'composite' knot $5_1 \# 5_2$.

that's why the knots in the tables are all prime. Note also that if O is a picture of the 'unknot', and K is any fixed picture of a knot K, then $K \# O$ just gives us a new picture of K. Caution: there do not exist knots K, K', both different from the unknot, with $K \# K'$ the unknot O. That is, knots form an addition semi-group, using the operation #.

Clearly the pictures in Figures 1 and 2 are not unique, and equally clearly that problem will cause lots of trouble as the number of crossings increases, so it's time for a definition:

- A *knot* K is the image of a circle S^1 under an embedding $e : S^1 \to S^3$ or $\mathbb{R}^3$, that is $e(S^1) = K \subset S^3$. Two knots $K, K^\star$ have the same *type* if there is a diffeomorphism of pairs $(K, S^3) \to (K^\star, S^3)$.

The term 'knot' always means 'knot type'. However, we will be sloppy and speak of $K = e(S^1)$ when we really mean K or any knot $K^\star = e^\star(S^1)$ which has the same knot type as K. Our pictures represent some choice of a projection of the knot onto a plane, with undercrossings and overcrossings distinguished.

The problem of recognizing when two knot diagrams represent the same knot type is a highly non-trivial problem. That's why we have a need for computable invariants that are independent of the projection.

1.2. The search for invariants

To define the earliest known invariant, that of Gauss, we need to generalize the concept of a knot:

- A *link* L is the image under an embedding of $\mu \geq 1$ disjoint circles in $\mathbb{S}^3$ or $\mathbb{R}^3$. If $\mu = 1$ it's a *knot*. Two links L, L^* have the same *type* if there is a diffeomorphism of pairs $(L, \mathbb{S}^3) \to (L^*, \mathbb{S}^3)$ or $(L, \mathbb{R}^3) \to (L^*, \mathbb{R}^3)$. An *invariant* is any computable quantity which takes the same value on all representatives of the link type.

An example of a link-type-invariant is the number of components in a link L, but Figure 3 shows six different 2-component links. The first subtle invariant is due

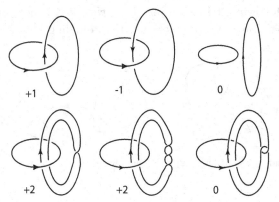

Fig. 3. Six 2-component links, with their linking numbers $\mathcal{L}(K, K')$.

to Gauss, who discovered an iterated line integral over 3-space $\mathbb{R}^3$ that is invariant under isotopy of the pair $(L, \mathbb{R}^3)$, where $L = K \cup K'$:

$$\int_K \int_{K'} \frac{(x'-x)(dydz' - dzdy') + (y'-y)(dzdx' - dxdz') + (z'-z)(dxdy' - dydx')}{((x'-x)^2 + (y'-y)^2 + (z'-z)^2)^{3/2}}$$

The Gauss integral turns out to be an integral multiple of 4π, and the integer is known as the *linking number* $\mathcal{L}(K, K')$. The integers attached to the links in Figure 3 give $\mathcal{L}(K, K')$, but (as one might expect it's a useful but crude measure of the way that the two components link one-another. Note that the linking number can be zero when the components are not visibly disjoint.

Remark: There is, in fact, an easy way to compute the Gauss linking number. See Figure 4 for a way to assign ± 1 to each crossings in a projection. Armed with this definition, color one component in each sketch in Figure 3 red, the other blue. Using the sign convention that's given in Figure 4, count the algebraic sum of the signs at the crossings of red over blue (or blue over red, it will not matter), ignoring all monochromatic crossings. It's the Gauss' linking number.

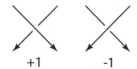

+1   -1

Fig. 4. Sign conventions.

Caution: Our conventions come from Dynamical Systems, and are opposite to those used in Topology! In particular, the value of the Gauss integral is based on conventions in topology. Starting in §2 of this paper, when we begin to discuss Lorenz links, we will adhere, consistently, to the conventions in Figure 4.

Turning to knots, the *minimum crossing number* $c_{min}(K)$ over all possible projections K of a knot is, by definition, a knot or link type invariant, however (except for very low crossing number) it turns out to be astonishingly difficult to compute, and to this day we do not have any real understanding of it in the general case, although if we place some restrictions (see §1.3 below) the situation changes.

The Jones polynomial[13] is a very sophisticated knot type invariant. It's a Laurent polynomial with integer coefficients. Unlike the minimum crossing number, it can be calculated from any projection, even though the calculation may be long and complicated if there are many crossings.

Sadly, it's not a complete invariant. See Figure 5, which shows by example that distinct knots can have the same Jones polynomial.

1.3. *Special classes of knots and links*

If we restrict our attention to special classes of knots or links, then the search for invariants can sometimes be very much simpler than in the general case.

Torus knots and links are a special class. They are defined to be knots and links that can be embedded on a standard torus of revolution in $\mathbb{R}^3$, as illustrated in Figure 6. They are a well-understood class. They are classified by a pair of integers (p,q) (up to the indeterminacy $(p,q) \approx (q,p) \approx (-p,-q) \approx (-q,-p)$). The Jones polynomial, which we have already mentioned above, of a (p,q) torus link is interesting. In the case when p and q are coprime, so that the link is a knot, it's given by the formula:

$$V(t) = \frac{t^{\frac{1}{2}(p-1)(q-1)}}{1-t^2}(1 - t^{p-1} - t^{q-1} - t^{p+q}) \tag{1}$$

The formula is considerably more complicated for links with more than one component, but for both knots and links it has a canonical form that's determined entirely

Fig. 5. 20 distinct knots with the same Jones polynomial (from M. Thistlethwaite).

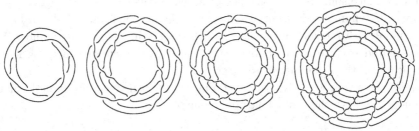

Fig. 6. Four examples of torus knots and links. Their classifying integer pairs, reading left to right, are $(3,5), (5,5), 5, 8), (7,9)$.

by p and q, so as far as the information that it contains it's a fancy way of saying p and q. But the point is that, unlike the integer pair (p, q), the Jones polynomial is defined for *all* knots and links.

The recognition problem for torus links is to decide whether a knot or link is a torus link. It's a hard problem, but if you could solve it, then two integers would suffice to determine its type. In general, the recognition problem for torus links is unsolved.

Positive or negative knots and links are knots and links that admit a projection in which every crossing is consistently signed. Like torus links (which are either positive or negative, depending on the defining parameters p and q), they are a proper subset of all knots and links, and like torus knots and links their study has been very fruitful. It would be a bit of a diversion to discuss their special properties right now, so we ask the reader to be patient (or to look ahead to § 4 below,) where we shall have more to say about them.

1.4. *Some natural questions*

When we encounter a new and interesting family of knots and links, especially one which arises in several different, seemingly unrelated settings (as we shall see, Lorenz knots and links are such a class), some questions we would like to answer are:

- Which knots and links occur?
- Is there a solution to the recognition problem for the class?
- Does the knotting have meaning as regards the setting in which they were discovered?
- Is the mathematics of this class of knots related to phenomena being studied in other contexts?

We will address these questions, as they relate to the class of knots and links that are the main subject of this review, 'Lorenz' knots and links, in the pages that follow.

2. Introducing Lorenz Knots and Links

In this section we introduce Lorenz knots and links and prove that they include all torus knots and links.

2.1. *A chaotic flow on 3-space*

In a famous paper written in 1963 the meteorologist E. N. Lorenz asked the question: "Is weather fundamentally deterministic?" It was clear that there were many many variables that affected the weather, but if we knew all of them, and the precise equations that governed weather patterns, would it really be *predictable* in a precise way? Giving the question a slightly different twist, does *deterministic* $\implies$ *ultimately periodic*? If so, then weather can't be deterministic. He set out to investigate this question. See his very readable article.[14]

His starting point was the Navier-Stokes equations, which he knew described the dynamics of a viscous, incompressible fluid. It had been used to model weather, ocean currents, water flow in a pipe, flow around an airfoil, motion of stars inside a galaxy. All of those are very complex problems. His idea was this: if a problem is too complicated to study it profitably, try to simplify it, preserving its essential features, discarding those which are unimportant. Lorenz was a 'theoretical meteorologist', and in this instance he was thinking like a mathematician, feeling free to modify the problem until it turned into a problem that might be amenable to study. So he discarded variables, modified the equations, testing each time he changed things by using numerical integration. After a very long (and very interesting) search he found a very simple system with the key features that he had wanted to retain. His modified equations turned out to govern fluid convection in a very thin disc that was heated from below and cooled from above. The modified equations were a system of 3 ordinary differential equations in 3-space $\mathbb{R}^3$, and time t:

$$\frac{dx}{dt} = 10(y-x), \quad \frac{dy}{dt} = 28x - y - xz, \quad \frac{dz}{dt} = xy - \frac{8}{3}z \tag{2}$$

There are three space variables: $x, y, z \in \mathbb{R}$ and their meaning is:

x = vertical temperature variation.
y = horizontal temperature variation.
z = rate of convective overturning.

The numbers 10, 20 and 8/3 were particular choices of parameters. The equations are very robust, that is one may vary the parameters 10, 28, and 8/3 in an open set without changing the features that he wanted to study in the solutions.

Of course, one can integrate a system of ODE's in $\mathbb{R}^3$, and if you do the orbits (with time as a parameter) determine a flow. A *Lorenz knot* (resp. *link*) is defined to be a closed periodic orbit (resp. finite collection of orbits) in the flow $\lambda^t : \mathbb{R}^3 \to \mathbb{R}^3$ determined by equations (2). Figure 7 is a picture of what Lorenz saw when he integrated the equations (2).

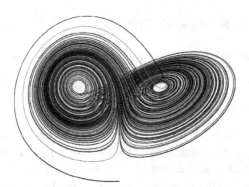

Fig. 7. A beautiful picture of the orbits in the flow on $\mathbb{R}^3$ determined by the Lorenz equations.

Lorenz proved that there exists a bounded region $\mathcal{A} \subset \mathbb{R}^3$ such that forward trajectories ultimately enter $\mathcal{A}$, and once they have entered it they stay there. The region $\mathcal{A}$ is a neighborhood of the butterfly-shaped family of orbits that we see in Figure 7. Notice the layering of orbits. The subset $\mathcal{A} \subset \mathbb{R}^3$ is an *attractor*. Now comes a key point: orbits are extremely sensitive to initial conditions. If you start at two points that are close to one-another, their positions may be very far apart at a later time, even though both are inside $\mathcal{A}$. Indeed, the Lorenz flow has become a prototype for a 'chaotic flow'.

A *template* is a branched 2-manifold with boundary, which is embedded in 3-space $\mathbb{R}^3$. See[9] for an introduction and thorough review of the literature on this

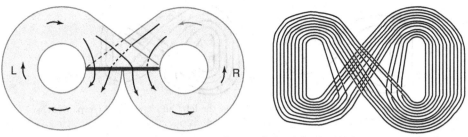

Fig. 8. The left sketch shows the Lorenz template, and the right sketch shows a typical collection of orbits. The orbits are to be compared to those in Fig. 7.

topic. Figure 8 shows the *Lorenz template*. One can almost see the template in Figure 7. The knots that are depicted to the right of the template in Figure 8 are typical orbits supported on the template in the left sketch. All crossings are the same sign, and we will, from now on, adhere to the conventions in Figure 4, i.e. all crossings are *positive*.

The concept of a template first appeared in the early work of Guckenheimer and Williams, in the 1970's, in their studies of the Lorenz flow. They constructed an embedded branched surface in $\mathbb{R}^3$ that supports a 'semi-flow', that is the flow on the branched 2-manifold is oriented and the orientation is not reversible. They reasoned, with very careful estimates and numerical data to back up their ideas, that every finite subset of the closed orbits in the Lorenz flow flow ought to project, simultaneously and disjointly, onto the template. It took over 30 years before indirect mathematical proofs of the existence of the template appeared, in the work of Tucker[21] and of Ghys.[9] Using their results, a Lorenz knot can be defined to be a simple closed curve in 3-space that embeds in the Lorenz template, and a Lorenz link is a finite collection of disjoint simple closed curves carried by the template. From that viewpoint, we can think of them as knots and links in 3-space, deforming them as we wish, as long as the isotopy extends to an isotopy of 3-space. We can also bring all the machinery of knot theory to bear on their study.

2.2. *Every torus link is a Lorenz link*

We observed, earlier, that torus knots and links are a very special class. It's a little bit of a surprise to learn that in fact every torus knot is a Lorenz knot! The proof is surprisingly easy. It begins in Figure 9 with a picture of two knots which are clearly Lorenz, because they can be embedded on the template. These are not the familiar projections of torus knots that we saw earlier, so how can we recognize them? The sketches in Figure 10 shows us how to do it: Since the picture obviously generalizes from (2,3) and (3,5) to every integer pair (p,q) we have learned that the set of all Lorenz knots and links is at least as big as the set of all torus knots and links.

Fig. 9. New images of torus knots of *type (2,3)* (on the left) and *type (3,5)* (resp. right).

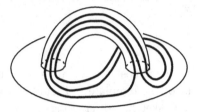 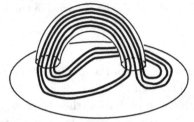

Fig. 10. The projection of Fig. 9 suggests, immediately, a way to embed these particular torus knots on a torus (minus a disc). Cap the single boundary component with a disc to get a closed torus.

3. Parametrizing Lorenz knots and links

We describe three ways to parametrize Lorenz knots. The order in which we choose to present them relates to our own taste, and does not reflect the order in which they were discovered historically.

3.1. *The first parametrization, Lorenz braids*

The first paper[4] in which Lorenz knots and links were studied as a class was written by Birman and Williams, and appeared in 1983, and the parametrization of Lorenz links by Lorenz braids was introduced there. In that paper a Lorenz link was defined to be any finite collection of closed orbits on the Lorenz template, which supports a semiflow. We now introduce a closely related branched 2-manifold, with boundary, which we call the *Lorenz braid template*. It can be seen in the right sketch in Figure 11.

Starting at the top of the braid template, the left and right branches of the template are stretched, and overlap along a horizontal *branch line*, then split apart, with one branch on the left and the other at the right. The four 'bands' are labeled L and R (at the top) and L and R again (at the bottom). A *Lorenz braid* is any finite set of braid strands that embeds on the Lorenz braid template.

An example is given in Figure 12. In the example there are 6,3,3,5 strands of type LL, LR, RL, RR.

We assume that the projection, determined by the ODE's in (2) and illustrated in Figure 12, is onto the region in the xz plane between $z = 0$ (at the top) and $z = -1$

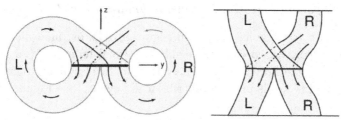

Fig. 11. The left sketch is the Lorenz template. The right sketch shows it cut-open to the Lorenz braid template.

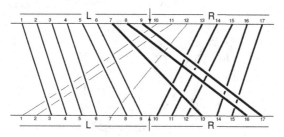

Fig. 12. A typical Lorenz braid.

(at the bottom), and that the initial points of the braid strands have x coordinates $1, 2, 3, \ldots$. Every crossing is therefore between a single overcrossing strand and a single undercrossing strand. Those crossings are always positive, where we follow the sign convention that was given in Figure 4. Note that on each overcrossing strand the final position will always be bigger than the initial position, therefore if the ith overcrossing braid strand starts at (say) $x_i = i$, it will end at $x'_i > x_i$.

Let q_j be the number of strands in the jth group, and let p_j be the difference between the x-coordinates at the start of any one strand in the group at the end. Suppose that there are k such groups in the braid. Then the numbers (p_j, q_j), $j = 1, \ldots, k$ completely determine the braid. For this reason we can parametrize the braids by the $2k$ integers $(p_1, q_1), \ldots, (p_k, q_k)$, where each $q_i \geq 1$, also each $p_i < p_{i+1}$, and finally (a convenient assumptions that eliminates trivial cases) $p_1 \geq 2$ and $q_k \geq 2$. Here (p_i, q_i) means that $p_i, \ldots, p_i$ repeated q_i times. This is our *first parametrization* of Lorenz links. In the example, the parameters are $((2, 4), (3, 2), (6, 1), (8, 2))$.

Observe that in the special case $k = 1$ this reduces to a single integer pair (p, q), and the associated closed braid determines a torus link, as we had already shown in §2.2. Thus Lorenz links are a generalization of torus links! We shall have more to say about this in §3.3.

3.2. *The second parametrization: symbolic dynamics and LR words*

The Lorenz template (see the left sketch in Figure 8) tells us, immediately, that we may associate a word or words in the symbols L (for left) and R (for right) to any a closed orbit in the Lorenz flow. There is no natural starting point for an orbit, therefore the LR words used to describe it are cyclic. See Figure 13. This

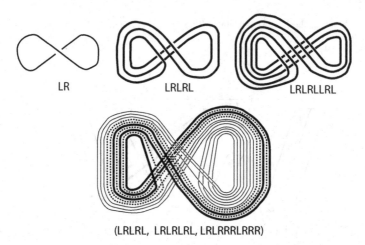

Fig. 13. Examples of cyclic LR words and orbits in the Lorenz template.

is an example of *symbolic dynamics*, a tool introduced to study solutions to ODE's like the Lorenz equations, which can't be integrated in closed form. The bottom sketch in Figure 13 shows a link with 3 components, each described by its word. The thickest line is U = LRLRL (that was our trefoil). The dotted line: is V = LRLRLRL (a type (3,5) torus knot). The thinnest line is W = LRLRRRLRRR (a type (-2,3,7) 'pretzel knot'). It's clear that we don't want our word to be periodic. As for links, we don't want the word for one component to be a power of that for another.

Are there other restrictions? The answer is 'no'. In the 1970's R. Williams proved[4] that the correspondence between left-right words and closed orbits is a very strong one. He proved that a family $\{W_1, \ldots, W_k\}$ of cyclic words represents a Lorenz link if and only if no W_i is periodic, and no W_i is (up to cyclic permutation) a power of any W_j.

We suggest his proof, in Figure 14, by showing how to recover the orbit from the word, in the example U =LRLRRRLRRR. To understand how we constructed the knot, the word U = LRLRRRLRRR determines its 10 cyclic permutations $U = U_1, U_2, \ldots, U_{10}$, in their natural order. Reorder these 10 words lexicographically, using the rule L<R. The new position μ_i is given after each U_i:

U_1 = LRLRRRLRRR   1 U_6 = RLRRRLRLRR   5
U_2 = RLRRRLRRRL   6 U_7 = LRRRLRLRRR   2

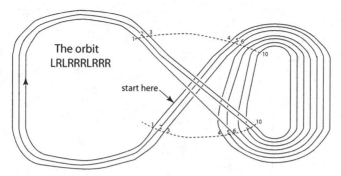

Fig. 14. The Lorenz knot associated to $U = $ LRLRRRLRRR.

$U_3 = $ LRRRLRRRLR   3 $U_8 = $ RRRLRLRRRL   9
$U_4 = $ RRRLRRRLRL   10 $U_9 = $ RRLRLRRRLR   7
$U_5 = $ RRLRRRLRLR   8 $U_{10} = $ RLRLRRRLRR   4

This determines a new cyclic order $(1,6,3,10,8,5,2,9,7,4)$, where strand μ_i begins at μ_i and ends at μ_{i+1}. We have constructed a *permutation braid*. A braid strand i is an *overcrossing* strand if and only if $\mu_i < \mu_{i+1}$, otherwise it's an undercrossing strand. There are 3 (resp. 7) over (resp. under) -crossing strands in the example, hence 3 strands that go around the left ear and 7 around the right ear. Starting with the permutation braid and traversing its associated closed braid, we recover the cyclic word LRLRRRLRRR.

3.3. The third parametrization: positive twisted torus links

To explain the third parametrization, we return to our favorite example, the one in Figure 12. There are 4 types of strands, those that begin at the left (or right) and those that end on the left (or right), and this will always be the case. We label the four types of strands type LL, LR, RR and RL.

Cut open the template along an orbit, as in the leftmost sketch in Figure 15, splitting it into 4 pieces: the first carrying the LL strands, the second the LR strands, the third the RL strands and the fourth the RR strands. Stretch the cut-open template as in sketch (ii) in Figure 15. After the stretching we can see that the LR band can be separated from the LL band and 'uncoiled', as in the passage from sketch (ii) to sketch (iii). Sketch (iv) shows what happens when the strands of the LL band are uncoiled, one strand at a time. (This is illustrated by the example in Figure 16). Finally, in sketch (v) of Figure 15 the RR and RL bands are cut open (as they were in Figure 11). A new braid structure emerges. Observe that the LR braid is always one full twist of the LR band, so it depends only on the number of LR strands in a Lorenz knot.

Let's look at an example. The braids that are supported on the template all have the same LR braid, but their LL-braids differ.

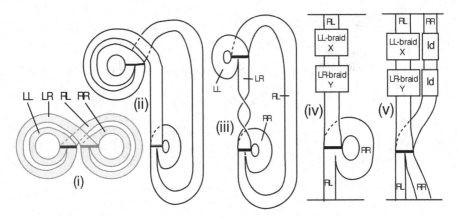

Fig. 15. The steps in cutting open the Lorenz template.

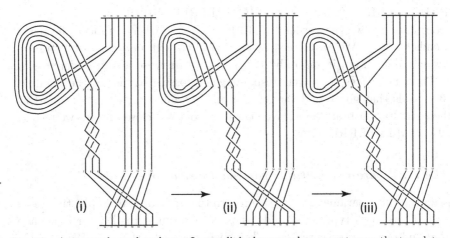

Fig. 16. An example to show how a Lorenz link changes when we cut open the template.

Ilya Kofman and the author have proved[3] that after the cutting and uncoiling processes that is illustrated in Figure 15 we obtain a *T-braid*. Even more, Lorenz links coincide, as a class, with links that we call *T-links*. They are generalizations of torus links. A torus link can be characterized, by a pair of integers (p,q). The most general T-link is defined by a sequence of integers $((p_1,q_1),(p_2,q_2),\ldots,(p_r,q_r))$, where each $p_i < p_{i+1}$. See Figure 17 for an example when the parameters are $((2,3),(4,4),(5,3))$. This braid has 5 strands. The T-link is the closure of a braid on p_r strands. First use the leftmost $p_1 < p_r$ strands ($p_1 = 2$ in the example) to form a torus braid of type (p_1,q_1), then follow it by a torus braid of type (p_2,q_2) on the leftmost p_2 strands, and so forth until, at the last step, all p_r strands are used to form a torus braid of type (p_r,q_r). The braid word (using the standard braid generators $\sigma_1,\ldots,\sigma_4$ for B_5 that describes it is $(\sigma_1)^3(\sigma_1\sigma_2\sigma_3)^4(\sigma_1\sigma_2\sigma_3\sigma_4)^3$ in the

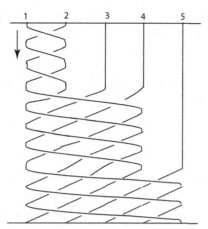

Fig. 17. The T-braid with parameters ((2,3),(4,4),(5,3)).

example. In the general case, when the parameters are $(p_1, q_1), (p_2, q_2), \ldots, (p_k, q_k))$, the braid will be:

$$(\sigma_1 \sigma_2 \ldots, \sigma_{p_1-1})^{q_1} (\sigma_1 \sigma_2 \ldots, \sigma_{p_2-1})^{q_2} \ldots (\sigma_1 \sigma_2 \ldots, \sigma_{p_k-1})^{q_k}$$

Braids with this general form are called *T-braids*, and their closures are called *T-links*. It is proved in[3] that Lorenz links are in one-to-one-correspondence with T-links. This is our third parametrization.

4. Applications

In this section we discuss some of the many consequences of our three viewpoints.

4.1. *The recognition problem for Lorenz knots*

In §1.4 we asked how rich the class of Lorenz knots might be, and whether we could solve the recognition problem. We give partial answers to both questions now. To explain what we know, we need several standard definitions:

- Let c be the number of double points in a projection of a knot K. The *crossing number* $c_{min}(K)$ is the smallest number of double points, over all possible projections of a knot K. As we explained earlier $c_{min}(K)$ is, by definition, a knot type invariant, but it is in general very difficult to compute.
- Every link can be represented as a closed braid, in many ways.[18] The *braid index* $n_{min}(K)$ is the smallest integer such that a knot or link K can be represented as a closed braid on n strands. It too is by definition an invariant, and it too is in general difficult to compute.
- Every knot is the boundary of a compact oriented surface. The *genus* $g(K)$ of a knot K is the genus of the surface Σ of smallest genus such that $K = \partial \Sigma$, and again it's an invariant.

- A knot $K \subset \mathbb{S}^3$ is *fibered* if its complementary space $\mathbb{S}^3 \setminus K$ has the structure of a surface bundle. If so, the fiber is a surface of minimum genus. We note that there are knots that are not fibered, in fact the property of being fibered is quite special.
- A knot is a *positive closed braid* if it has a closed braid projection in which all crossings are positive. We note that every positive closed braid is fibered, and also that there are fibered knots that are not positive closed braids.

We turn to consequences for Lorenz knots. (To simplify things, we do not state the analogues for Lorenz links):

(1) The fact that Lorenz knots are precisely the knots that have natural embeddings in the Lorenz template shows that *Lorenz knots are positive closed braids*. When we began to write this review it was not clear to us whether every knot that is a positive closed braid was a Lorenz knot, so we asked some questions. We learned from Slavik Jablon that the only 10-crossing knot that is Lorenz is the knot 10_{124}. Since the standard projection of the knot 10_{152} exhibits it as a positive closed braid, we conclude that:

- All knots $\supset$ Fibered knots $\supset$ Closed + braids $\supset$ Lorenz knots.

(2) A consequence of positivity is that for any closed positive n-braid representation of a Lorenz knot, the genus g, the crossing number c of the given n-braid representative, and the braid index n of the representative are related by the formula

$$2g(K) = n + 1 - c. \tag{3}$$

Observe that $g(K)$ is a knot type invariant, but n and c are not. From this it follows that the genus $g(K)$ of a Lorenz knot K can be computed from any projection in which all crossing are positive, e.g. the Lorenz template projection, by a simple counting argument. In particular, using the first braid parametrization of a Lorenz knot (see Figure 12), equation (3) shows that:

$$2g(K) = |L| + |R| + 1 - \sum_{i=1}^{i=k} q_i(p_i - 1). \tag{4}$$

(3) The integers $n_{min}(K)$ and $c_{min}(K)$ are (by definition) knot type invariants, but for arbitrary knots they are difficult to compute. However, in the special case of Lorenz knots it is a consequence of the work done by Birman and Williams in [4] and by Franks and Williams in [8] that:

$$n_{min}(K) = \min(|\mathrm{LR}|, |\mathrm{RL}|). \tag{5}$$

Thus we can compute $n_{min}(K)$, too, from the Lorenz braid.

(4) In view of this, one might wonder how the invariant $c_{min}(K)$ is related to $g(K)$ and $n_{min}(K)$? In general, the minimum crossing number of a knot is *not* realized at minimum braid index, however for Lorenz knots, it follows from Proposition 7.4 of[16] that it is. So we have $c_{min}(K) = n_{min}(K) + 1 - 2g(K)$. Thus we can compute $c_{min}(K)$ too from the first parametrization of a Lorenz knot:

$$c_{min}(K) = \sum_{i=1}^{k} q_i(p_i - 1) - |\text{LL}| - |\text{RR}| \qquad (6)$$

Going beyond positivity, we know a little bit more:

(5) Lorenz knots have yet one further property that generalizes the fact that torus knots are types (p, q) and (q, p) have the same knot type, namely Lorenz knots have a very special symmetry of order 2, induced by rotating the template 180 degrees about the z axis in Fig. 11.

(6) Pierre Dehornoy has investigated the zeros of the Alexander polynomial of a Lorenz knots. See his arXiv preprint[7].

It seems possible that, with these two additional properties we already know how to solve the recognition problem for Lorenz knots. To give some evidence, we discuss a remarkable calculation that was done by Pierre Dehornoy and Etienne Ghys, with the help of Slavik Jablon. They proved:

- Among the 1,701,936 prime knots having projections with ≤ 16 crossings, only 19 are Lorenz, and among those only 7 are not torus knots! In particular, they solved the recognition problem, when the crossing number is not ≤ 16.

However this may not say very much about the recognition problem because Lorenz knots generally have high crossing numbers. For example, it was proved in[4] that all algebraic knots are Lorenz, however there is not a single algebraic knot which is not a torus knot and has $c_{min} \leq 16$.

4.2. *The volume of* $S^3 \setminus \mathcal{K}$ *and the monodromy of* $\mathcal{K}$

Up to now, we have been discussing the topology of Lorenz knots. This means that if K is a Lorenz knot, we have been studying properties of the 3-manifold $S^3 \setminus K$ that are independent of any particular choice of a metric on the manifold. But in the early 1980's, William Thurston proved a remarkable theorem about the possibilities for assigning a unique, complete finite-volume metric to the complement of a hyperbolic knot in S^3, also he proves that 'most knots are hyperbolic'. See[19] for a good introduction to this major topic, noting that what he calls the *Geometrization Conjecture* is now a theorem. For hyperbolic knots the hyperbolic volume is a knot type invariant, which becomes an invariant for all knots by defining the volume to be zero when the knot is not hyperbolic. A natural question, then, is to compute the volumes of hyperbolic Lorenz knots.

The study of hyperbolic knot complements has been a focal point for much recent work in 3-manifold topology. The very question changes the focus of knot theory from the properties of diagrams to the geometry of the complementary space. Ideal tetrahedra are the natural building blocks for constructing hyperbolic 3-manifolds, and ideal triangulations can be studied with the help of computer programs. In particular, it has been learned that there are precisely 6,075 noncompact hyperbolic 3-manifolds that can be obtained by gluing the faces of at most seven ideal tetrahedra,[6] with recent extensions to 18,921 examples with eight tetrahedra by Marc Culler and Nathan Dunfield. For a hyperbolic knot, the minimum number of ideal tetrahedra required to construct its complement is a natural measure of its geometric complexity. It is shown in Table 1 of[3] that:

- Of the 201 hyperbolic knots in the[6] census, at least 107 are Lorenz knots.

The number 107 could be too small because, among the remaining 94 knots, it was not possible to decide whether five of them are or are not Lorenz. We remark that the known diagrams for the knots in question did not in any way suggest the Lorenz template.

There is an intuitive reason for this extraordinary data. Torus knots, as observed earlier, have (by definition) hyperbolic volume 0. One expects, then, that one way to construct hyperbolic knots of small volume is to 'tweak' the diagram of a torus knot just a little bit. Now recall that, using the first and third parametrizations, Lorenz knots that are torus knots have parameters (p,q), wherer p and q are coprime integers and $p < q$. Therefore one might expect that Lorenz knots which have parameters $((1,2),(p_2,q_2))$ and are hyperbolic have very small volume, and indeed that is what the data in[3] suggests. This is, as we write, a very active and interesting area of investigation.

There is another aspect of the topology and geometry of Lorenz knots that is related but different from its volume, and we mention it very briefly. As we observed, Lorenz knots are always fibered. We then have the monodromy map to study. By a theorem of Thurston, a fibered knot $\mathcal{K}$ has a hyperbolic complement if and only if its monodromy map and all of its powers, acting on the fundamental group $\mathcal{G} = \pi_1(\Sigma)$ of the fiber, does not fix any conjugacy class in $\pi_1(\mathcal{G})$. The entropy of the monodromy map is then a new invariant of $\mathcal{K}$. This topic is a big one, and space considerations prevent us from saying more, except to remark that just as Lorenz knots appear often in the census of fibered 3-manifolds having small but non-zero volume, so their monodromy maps appear often in the study of surface diffeomorphisms having small positive entropy.

4.3. *Modular knots and Lorenz knots*

Modular knots were introduced by Etienne Ghys in[9] and .[10] As we shall see, they are Lorenz knots in disguise. In this section we explain his work, briefly.

Recall that a model for hyperbolic 2-space $\mathbb{H}$ is $\{z = x + iy \in \mathbb{C} : y > 0\}$ with the hyperbolic metric $ds = (\frac{1}{y})\sqrt{dx^2 + dy^2}$. With this metric, a *geodesic* in $\mathbb{H}$ is either a vertical ray in the upper half plane or a half-circle orthogonal to the x axis. Orientation-preserving isometries of $\mathbb{H}^2$ may be identified with $PSL(2,\mathbb{Z})$, and the quotient space $\mathbb{H}^2/PSL(2,\mathbb{Z})$ is the *modular surface* M. It has a natural metric coming from the metric on its covering space $\mathbb{H}^2$.

The *geodesic flow on* M is defined by choosing a matrix $P = \begin{bmatrix} a & b \\ c & d \end{bmatrix} \in PSL(2,\mathbb{R})$ and sending $P \to H_t P$, where $H_t = \begin{bmatrix} e^t & 0 \\ 0 & e^{-t} \end{bmatrix}$. This is the *geodesic flow* ϕ^t on the modular surface M. An element P defines a periodic orbit in the geodesic flow precisely when $H_t P = PA$ for some hyperbolic matrix $A \in PSL(2,\mathbb{Z})$. Thus closed orbits in ϕ^t are defined by diagonal matrices PAP^{-1}, where $P \in PSL(2,\mathbb{R})$, $A \in PSL(2,\mathbb{Z})$. The condition that A be hyperbolic means that its trace in > 2.

In[9] Etienne Ghys had the idea to lift closed orbits in the geodesic flow ϕ^t on M to a related flow Φ^t on the unit tangent bundle $\tilde{M}$ of M, a 3-manifold. He used the known fact (§ 3.1 of[9]) that $\tilde{M}$ is naturally isomorphic to $PSL(2,\mathbb{R})/PSL(2,\mathbb{Z})$, which is in turn known to be isomorphic to $\mathbb{S}^3 \setminus \mathcal{T}$, where $\mathcal{T}$ is the trefoil knot, i.e. the type $(2,3)$ torus knot, embedded in $\mathbb{S}^3$ in a known canonical way. This enabled him to study the closed orbits in M, a 2-manifold, as knots in $\tilde{M} = \mathbb{S}^3 \setminus \mathcal{T}$.

Definition:[9] A *modular knot* is a closed orbit in the flow Φ^t on the unit tangent bundle $\tilde{M} \cong (\mathbb{S}^3 \setminus \mathcal{T})$ of the modular surface M.

The geodesic flow on M has been studied extensively by number theorists, and is of great interest. However, while the topology and geometry that we just described is well-known, it seems as if nobody had ever really thought of its closed orbits as knots. In particular, Ghys defines and studies the *Rademacher function*, an integer-valued function on closed orbits in the modular flow, via the geometry that we just described, demonstrating that its value on a closed orbit K is in fact the linking number of K with the 'missing trefoil' $\mathcal{T}$.

Lorenz knots have not entered into this picture yet, but we now show that their appearance is very natural and concrete. Modular knots are simple closed curves in $\tilde{M} = \mathbb{S}^3 \setminus \mathcal{T}$. Recall that the fundamental group of $\mathbb{S}^3 \setminus \mathcal{T}$, admits the presentation

$$G = \pi_1(S^3 \setminus \mathcal{T}) :< U, V; \ U^2 = V^3 >.$$

The group G is the free product of the cyclic groups generated by U, V amalgamated along $C = U^2 = V^3$, where C generates the center of G. Thus every free homotopy class in G is represented by a cyclic word W of the form $C^k U V^{\epsilon_1} U V^{\epsilon_r} \cdots U V^{\epsilon_r}$, where $\epsilon_i = \pm 1$.

We have already shown that the choice of a modular knot corresponds to a choice of a hyperbolic matrix A in the group $PSL(2,\mathbb{Z})$, i.e. in the image of G under the homomorphism $G \to PSL(2,\mathbb{Z})$ whose kernel is the central element C. That is, the choice of a cyclic word $\pm UV^{\epsilon_1} UV^{\epsilon_r} \cdots UV^{\epsilon_r}$, where $\epsilon_i = \pm 1$. Let $L = UV$, $R = UV^{-1}$. Then our cyclic word goes over to a cyclic word in L and R. But then, by the parametrization given in §3.2 above, each closed orbit K in the modular flow, lifted to $\tilde{M}$, determines a Lorenz knot!

Theorem (E. Ghys,[9]). *There is a one-to-one correspondence between modular knots and Lorenz knots.*

There is more than this. Ghys proves that the Lorenz template occurs in the setting of modular knots, embedded in a natural way in $\mathbb{S}^3 \setminus \mathcal{T}$, and there is a natural way (using *L-R* words) to embed the family of all modular knots, disjointly and simultaneously, onto the template.

There are two issues about this picture that remain somewhat mysterious. The first is that (a) Ghys does not address the issue that, if they coincide, then there ought to be a 'missing trefoil' in the Lorenz flow. The second is that (b) Ghys does not actually prove that the Lorenz flow in $\mathbb{S}^3$ and the modular flow coincide. Regarding (a), Tali Pinsky[17] has computer evidence that in fact there is a missing trefoil very naturally embedded in $\mathbb{S}^3$ with respect to the Lorenz flow. In[17] she conjectures that the invariant curves connecting the three fixed points of the Lorenz flow (one in the center of each 'ear' of the template, and one on the axis of symmetry) join up to form such a trefoil. See Figure 18. Since the curves in question are invariant curves, closed orbits in the flow would necessarily avoid them. We look forward to her proof that the union of the invariant curves is actually a knot. As

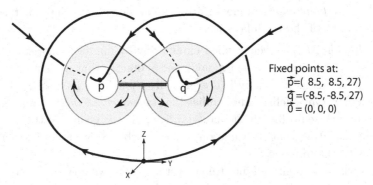

Fig. 18. The missing trefoil.

for (b), while Ghys is a bit vague about how he produced the beautiful pictures in,[10] the picture proofs 'say more than a thousand words'. We recommend them strongly to any reader with a small interest in this topic, as a very non-traditional introduction to Lorenz knots.

In a very different direction, his discoveries reveal a new fact about the Rademacher function: As mentioned very early in this review, the Lorenz flow, which we now reinterpret as the modular flow on $\tilde{M} \cong \mathbb{S}^3 \setminus \mathcal{T}$ is chaotic in the sense that it has an extremely sensitive dependence on initial conditions. This is true for its closed orbits too, so that if two points are very close on the modular surface, and both lie on closed orbits, their linking numbers with $\mathcal{T}$ might be very different. Thus it's a canonical example of what one might call a *chaotic function*.

4.4. *Generalizations of Lorenz knots*

In §2 of this paper we introduced templates, via the example of Lorenz knots. We never even hinted that there were templates for flows on $\mathbb{S}^3$ different from the Lorenz flow. But in fact the Lorenz template is not an isolated tool. We give several examples.

(i) The paper,[4] which initiated the study of Lorenz knots, had a twin[5] in which a different flow on S^3 and its associated template and family of closed orbits was discussed and studied. The second flow arose from a phenomenon which we have already encountered. The knot $\mathcal{T}' = 4_1$ (see Figure 1) is a fibered knot of genus 1, so the fiber is a once-punctured torus. The monodromy map, lifted to the universal cover of the closed torus, i.e. the Euclidean plane $\mathbb{R}^2$, is the linear map $\begin{bmatrix} 2 & 1 \\ 1 & 1 \end{bmatrix}$. This map has a dense subset of periodic points, from which it follows that the closed orbits in the associated flow on $(\mathbb{S}^3 \setminus \mathcal{T}')$, defined by pushing the fiber around the knot, are dense in the flow. Thus, just as Ghys had shown that the modular flow lifted to $(\mathbb{S}^3 \setminus \mathcal{T})$, the flow associated to the fibration is a flow on $(\mathbb{S}^3 \setminus \mathcal{T}')$. In[4] the author and Williams constructed a template for the later. Subsequently, Robert Ghrist proved that it includes *all* knots and links. Its associated template is said to be *universal*, and it seems likely that generic templates are, in fact, universal.

(ii) The research monograph[12] is dedicated to the general topic of knots and links that are defined by templates, through 1997. Referenced in that monograph are papers that deal with the problem of passing from a particular template to an associated flow, a non-trivial question. There are a host of open problems, as we write, about templates that are universal and those that are not.

(iii) In a related but different direction, Pinsky has investigated, in ,[17] templates for geodesic flows on certain hyperbolic manifolds which are different from the geodesic flow on the modular surface, and she has found templates there too, and also analogues of the 'missing trefoil' in the unit tangent bundle of the modular surface. There is a big world out there, and a great deal of structure, waiting to be discovered!

Acknowledgments

Thanks go to to Ilya Kofman for reading a very preliminary draft of this manuscript and commenting on it, and for stimulating discussions; to Morwen Thistlethwaite for the knots in Figure 5; to an unknown colleague for the picture in Figure 7, which we downloaded from the internet.

References

1. J. Birman, On the Work of Vaughan Jones, *Proceedings of the International Congress of Mathematicians 1990*, Springer-Verlag, New York, pp. 9-18 (1992).

2. M. Epple, Orbits of asteroids, a braid and the first link invariant, *The Mathematical Intelligencer* **20** (1), 45-52 (1998).
3. J. Birman & I. Kofman, A new twist on Lorenz knots, *Journal of Topology*, **2**, 227-248 (2009).
4. J. Birman & R. Williams, Knotted periodic orbits in dynamical systems -I: Lorenz's equations, *Topology*, **22**, 47-82 (1983).
5. J. Birman & R. Williams, Knotted periodic orbits in dynamical systems -II: Knot holders for fibered knots, *Contemporary Mathematics*, **20**, 1-60 (1983).
6. P. Callahan, M. Hildebrand, J. Weeks, A census of cusped hyperbolic 3-manifolds, *Mathematics of Computation* **68** (225), 321-332 (1999).
7. P. Dehornoy, On the zeros of the Alexander polynomial of a Lorenz knot, arXiv preprint [Math:GT] 1110.4178v1 (2011)
8. J. Franks & R. Williams, Braids and the Jones polynomial, *Transactions of the American Mathematical Society*, **303**, 97-108 (1987).
9. E. Ghys, Knots and dynamics, *Proceedings of the International Congress of Mathematicians*, **I**, 247-277 (2007).
10. E. Ghys & J. Leys, Lorenz and Modular Flows: A Visual Introduction, *Monthly Essays in Mathematical Topics*, American Mathematical Society (2011).
11. R. Ghrist, Branched 2-manifolds supporting all links, Topology **36**, 423-448 (1997).
12. R. Ghrist, P. Holmes & M. Sullivan, Knots and links in 3-dimensional flows, *Lecture notes in Mathematics* **1654**, Springer-Verlag (1997).
13. V. Jones, Hecke algebra representations of braid groups and link polynomials, *Annals of Mathematics*, **126**, 335-388, (1987).
14. E. N. Lorenz, Deterministic, non-periodic flows, *Journal of Atmospheric Science*, **20**, 130-141 (1963).
15. W. Menasco & M. Thistlethwaite, The classification of alternating knots, *Annals of Mathematics*, **138**, 113-171 (1993).
16. K. Murasugi, On the braid index of alternating links, *Transactions of the American Mathematical Society*, **326**, 237-260 (1991).
17. T. Pinsky, Templates for geodesic flows, *preprint arXiv:1103.4499*
18. D. Rolfsen, Knots and Links, *Publish or Perish, Inc.* Second Printing (1990).
19. P. Scott, The geometries of 3-manifolds, *Bulletin of the London Mathematical Society*, **15**, 401-487 (1983).
20. M. Thistlethwaite, *Knotscape*, http://www.math.utk.edu/ morwen/knotscape.html
21. W. Tucker, A rigorous ODE solver and Smale's 14th problem, *Foundations of Computational Mathematics*, **2**, 53-117 (2002).
22. R. Williams, Lorenz knots are prime, *Ergodic Theory Dynamical Systems* **4**, 147-163 (1983).

Chapter 7

A braided view of a knotty story

Mario Natiello

Matematikcentrum-LTH, Lunds Universitet
Box 118, 221 00 Lund, Sverige

Hernán Solari

Departmento de Física, Facultad de Ciencias Exactas y Naturales
Universidad de Buenos Aires, Argentina

Periodic orbits of 3-d dynamical systems admitting a Poincaré section can be described as braids. This characterisation can be transported to the Poincaré section and Poincaré map, resulting in the braid type. Information from braid types allows to estimate bounds for the topological entropy of the map while revealing detailed orbit information from the original system, such as the orbits that are necessarily present along with the given one(s) and their organisation. We review this characterisation with some examples –from a user-friendly perspective–, focusing on systems whose Poincaré section is homotopic to a disc.

Contents

1. Introduction 149
2. Formal Statement of the Problem 151
3. Elements of the Description 153
4. The Algorithm 156
   4.1. Collapse 156
   4.2. Elimination of cyclically mapped edges 157
   4.3. Description of the algorithm 158
   4.4. Example 159
5. Detailed Enumeration of Implied Orbits 160
   5.1. Orbit forcing 160
   5.2. Orbit trimming 162
6. Concluding Remarks 166
References 167

1. Introduction

The programme that is popularly known as *Topological Methods in Dynamical Systems* was motivated by the interest in understanding the dynamics of low dimensional (3-d ODE's) dynamical systems. Among such systems we count various laser systems and their associated models (modulated losses,[18] saturable absorber,[13] etc.), oscillating chemical reactions[11] and many more.

looseness-1 This programme split naturally in two parts: (a) How to associate experimental data to orbits of a dynamical system and (b) How to extract detailed and precise information from the system given that a set of orbits of a dynamical system is known. Part (b) in its simplest version attempts to decide if a system is (or could be) "chaotic", i.e., to compute the *topological entropy* associated to a given set of periodic orbits and from the topological point of view, to determine a maximal knot holder able to hold the orbits. The detailed version of part (b) is to describe the orbit organisation of the originating system in the most complete way that is possible.

This manuscript will deal exclusively with (b), setting for the detailed version. In particular, the basic assumption is that a set of periodic orbits of a 3-d dynamical system admitting a Poincaré section is known and moreover that this Poincaré section is homotopic to a disc.

It became clear that the Poincaré section imposes strong constraints to the way the orbits look like. Firstly, periodic orbits have a natural period that can be counted by integers, i.e., how many times the orbit crosses the section within one minimal period. A direct consequence is that period-n orbits can be described as elements of the braid group in n strands.[4] Further, relevant braid information can be read directly on the Poincaré section, by studying e.g., how an oriented circle (joining the intersection points of the orbit on the section) is modified by the action of the Poincaré map.[12] Moreover, the oriented circle can be replaced by a line segment.[9,15]

Further, Poincaré maps on the disk are a special case of 2-d homeomorphisms of the disk. Hence, the organising structure given by William Thurston's classification theorem[17] could be used, together with an important result by Dmitri Anosov. It turns out that among irreducible 2-d homeomorphisms of the disk preserving a finite set of points, the one that has the least number of periodic orbits for each period –and hence the lowest topological entropy– is either a rotation or a *Pseudo-Anosov diffeomorphism*. The reducible case can naturally be decomposed into irreducible components before applying the theorem. It is worth to recall at this point the relevant works by Mladen Bestvina and Michael Handel[2] and by John Franks and Michal Misiurewicz[7] on this matter, together with the other references above.

What system information can be extracted in this setup? We list first the tools at hand: (i) Given a 2-d homeomorphism of the disk preserving a finite set of points (in our case the Poincaré map assumed to describe the detected periodic orbits), use one of the available algorithms to decide if this homeomorphism lies in the isotopy class of a Pseudo-Anosov diffeomorphism or, alternatively, in the class of a pure rotation (moving first to the irreducible components if necessary). (ii) In the first case, build a representative of the Pseudo-Anosov map (the *fat representative*). (iii) Compute the topological entropy using the representative map. This will be a lower bound to the topological entropy of the underlying system. If the detected periodic orbits are compatible with a rotation, then this lower bound is zero (in which case the question of being "chaotic" is still undecidable: the given orbits may still arise from a chaotic system, only that they are too "simple" to identify this property). (iv) List all periodic orbits that will necessarily be present in the original system along with the given ones.

Hence, the main question we can address through these methods is: Given a (finite set of) periodic orbit(s), can we decide if the originating system has positive topological entropy? This question has been satisfactorily addressed in many articles[2,4,7,9,15] (among others). Since topological entropy deals with "listing orbits" further questions arise naturally: (a) Which other orbits along with the given one(s) are present in the system? This question is frequently called the *orbit forcing* issue. We will show in this manuscript that the information given by the fat representative map is enough to address the problem. (b) Inspired in the horseshoe map,[14] which is a full-shift in two symbols, we may ask: Given a full shift hosting the known orbit(s), which orbits are **not** present in the system? We indicate this question as the *orbit trimming* issue. These two questions -as far as we know- have not been thoroughly shaped in the form of theorems. A related question, dealing with the relationship between the orbits in the Hénon map and those in the horseshoe, bears the name *orbit pruning*.[5,6]

When this procedure works through, we have a very powerful tool of analysis: Knowing just one orbit, we may say if the originating system is "chaotic" (i.e., when it has positive topological entropy) and even have detailed information about how this chaoticity arises (we can predict an infinite number of orbits that will necessarily be present along with the known one).

The goal of this paper is twofold. On one hand, we want to present a user-recipe of the algorithm developed in a previous paper[15] in order to answer the "chaoticity" question, reshaped as *To which isotopy class belongs a 2-d homeomorphism of the disk preserving the given finite set of points?* and on the other hand we want to advance in the full-answer to the issues of orbit forcing and orbit trimming elaborating the information gained along the process of determination of the fat representative map.

This paper is organised as follows. We continue by stating the problem in a more formal way. Subsequently, we list the relevant part of the previously known results. In Section 4 we describe the algorithm mentioned above. Section 5 describes part of the added value of this manuscript, namely the statement of new results. All sections are illustrated by examples when necessary. The final Section is devoted to some concluding remarks.

2. Formal Statement of the Problem

In the sequel, we consider 3-d Lipschitz-continuous dynamical systems in $\mathbb{R}^3$ (ODE's) admitting a Poincaré section homotopic to a disk $\mathbb{D}$. Let F denote the corresponding Poincaré map. We also assume that a finite set P of periodic orbits of the system is known (we will mostly use one orbit at a time). We label the orbit(s) when necessary according to their (Poincaré) period, i.e., the number of times each orbit passes the Poincaré section during one minimal period.

For the sake of shortness, we assume that the following concepts are known to the reader (when necessary, details can be found in the references[1,12,16]):

(1) To each orbit we can associate an equivalence class of braids under *global torsions* (full rotations around an axis along the flow).
(2) Each equivalence class can be associated to an oriented *circle diagram* on the Poincaré section. The diagram is the image of an oriented circle joining the points of P consecutively, given a choice of ordering. Specifically, the group of circle diagrams with p points is isomorphic with $B_p/Z(B_p)$,[12] the braid group on p strands quotiented with its center.
(3) Circle diagrams can be simplified[15] to *line diagrams*,[9] by deleting the arc joining the last point to the first in the original circle.
(4) Different equivalent choices of the Poincaré section yield different braids and consequently different line diagrams. For the sake of computing the topological entropy, the relevant concept is the *braid-type*,[4] i.e., the equivalence class of braids upon conjugation.
(5) **Thurston's Theorem:**[17] Let $\mathbb{D}$ be the unit disk and P a finite F-invariant set of points. Then F is isotopic to an homeomorphism ϕ on $\mathbb{D}\backslash P$ such that whenever ϕ is irreducible, either ϕ^n is the identity for some positive integer n or ϕ is *pseudo-Anosov*.
(6) A Pseudo-Anosov map has the least number of periodic orbits in its isotopy class.[10]
(7) Pseudo-Anosov maps admit a Markov partition. The largest-modulus eigenvalue of the associated Markov matrix gives the map's topological entropy.[10]

Questions:

- Given a line diagram describing the finite invariant set P (embedded in the unit disk $\mathbb{D}$ onto which F acts), which is the map ϕ isotopic to F on $\mathbb{D}\backslash P$ having the least number of periodic points for each period?
- Forcing: Which periodic orbits are necessarily present along with the set P in any member of the Thurston equivalence class of maps?
- Trimming: Given a full-shift hosting the set P, which periodic orbits of the full-shift are *not* present in ϕ?

Example In Figure 1 we display a hypothetical periodic orbit of a dynamical system defined on $\mathbb{R}^2 \times \mathbb{S}^1$, its braid and the corresponding originating circle (line) and circle diagram (line diagram). A suitably chosen disc on the plane $\{\varphi = c\}$ (constant) serves as a Poincaré section.

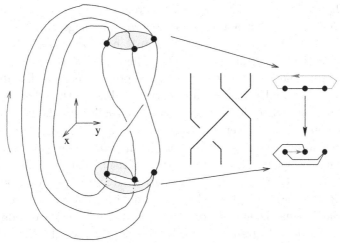

Fig. 1. A period-3 orbit along with its associated braid and circle/line diagram. The blue arrow indicates the direction of the flow while the yellow upper circle represents (a portion of) the Poincaré section.

3. Elements of the Description

We start by devising a model of $\mathbb{D}\setminus P$, and a map $\hat{\theta}$ on this model such that: (a) It lies in the same isotopy class as the Poincaré map of the original system, (b) It provides a natural Markov partition on the disk and — hopefully — (c) It shares with the Pseudo-anosov or pure rotation map in the same isotopy class the property of having the least number of periodic orbits for each period, with the possible exception of a (fully specified and small) finite set of orbits of the same braid-type as the original one.

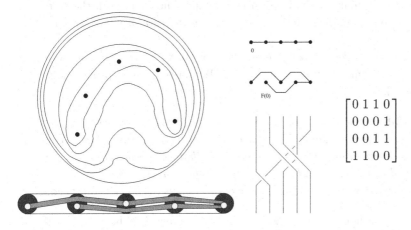

Fig. 2. A period-5 orbit, the map $\hat{\theta}$, line diagram and braid.

Hall[10] established how to decide when this programme can be accomplished using line diagrams (i.e., by schematically computing the image of a line segment joining the points of P through the Poincaré map) and how to proceed for such braids. Figure 2 displays the main ingredients. The unit disk is represented as a topological rectangle, schematically depicted by the pre-image of the line diagram. The rectangle has edges and vertices (blue) corresponding to the schematic ones. The map $\hat{\theta}$ continuously maps the rectangle onto itself following the line diagram, being contractive on the set of vertices and expanding on the edges (green and yellow). The map can be regarded as 1-to-1 on the interior of the rectangle and on part of its boundary, except for a connected portion of the boundary of the rectangle, which is mapped 2-to-1 on the interior. Taking the edges of the rectangle (or on the line diagram) as units, the disk can be provided with a Markov partition. The corresponding Markov matrix indicating where (on which edges) each edge maps can be read from the line diagram or from the image of $\hat{\theta}$ (see the Figure). The topological entropy corresponds to the largest-modulus eigenvalue of the Markov Matrix, in this case $\lambda = 1.722$. Hall establishes also that for other braids the

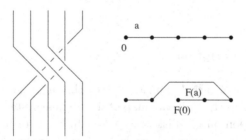

Fig. 3. A more difficult period-5 orbit.

approach via the line diagram does not work, see e.g., Figure 3. In this case, the fat representative map $\hat{\theta}$ differs severely from the pseudo-Anosov map in the same isotopy class. For a given line diagram it can be graphically established whether it belongs to the "doable class" or not.

These ideas have been generalised[15] extending previous works[7,9] in order to provide a tool to understand any periodic orbit. To master this approach some preparation is needed.

Definition 7.1 ((fat) Tree). *A* **tree** *is a connected finite* $1-d$ *CW-complex that does not contain a subset homeomorphic to a circle. In other words, a finite connected set of vertices and edges without loops, such that each edge connects two vertices pairwise and edges do not intersect elsewhere than in the common vertices. When edges are thickened to rectangles and vertices to circles, we have a* **fat tree**.

Note that the fat tree is homotopic to $\mathbb{D}\backslash P$. When necessary (see below) we will assume that the tree lies "centered" on top of the fat tree (each vertex lies at the center of each fat vertex). We will assign a *valence* to each vertex, counting the number of attached edges.

Definition 7.2 (sector). *Adjacent edges on a vertex of a tree generate angles. The corresponding region on the fat tree is called a* **sector***. A fat vertex with $k \geq 1$ edges has k sectors.*

Definition 7.3 (fat representative). *The* **fat representative** *$\hat{\theta}$ is the conjugated map to F acting on the fat tree. This map can be chosen to be 1-to-1 on the interior of the fat tree and on a portion of its boundary, while it may be 2-to-1 from a connected portion of the boundary to the interior of the fat tree. Letting π be the projection from the fat tree to the tree collapsing vertices to points and edges to straight lines, we can define the projected map $\theta = \pi\hat{\theta}$.*

The line diagram corresponds to the image by $\hat{\theta}$ of a tree whose shape is a rectangle.

Definition 7.4 (Fold). *We say that a tree has a* **Fold** *at a vertex V if the image by θ of two adjacent edges locally at V maps 2-to-1 onto one edge at $\theta(V)$. Whenever we have a fold, $\pi\hat{\theta}$ is not locally 1-to-1 and there exists a sector x completely mapped on a portion of an edge.*

Definition 7.5 (Crossing). *A* **Crossing** *occurs when $\hat{\theta}$ maps an edge on a sector at some vertex.*

Definition 7.6 (Bogus Transition). *A* **Bogus Transition** *occurs when the forward image by $\hat{\theta}^k$ ($k \geq 1$) of a crossing maps into a fold.*

We say that the crossing **terminates** in that fold since the next iteration of $\hat{\theta}^k$ followed by projection by π maps fold, sector and crossing onto an edge.

Definition 7.7 (Recurrent Bogus Transition). *A* **Recurrent Bogus Transition** *occurs when the forward image of an edge participating in a fold enters a crossing associated to a bogus transition.*

Definition 7.8 (Minimal Periodic Orbit Structure). *A fat representative has* **Minimal Periodic Orbit Structure** *if it has the least number of periodic orbits for each period in its isotopy class, except for a finite number of border orbits of the same braid-type of the given one.*

After this preparation we can state the following[15]

Theorem 1. *An expansive $\hat{\theta}$ presents no recurrent Bogus transitions if and only if it has minimal periodic orbit structure.*

The first **goal** of our programme is now to modify a (fat) tree in order to eliminate all recurrent Bogus transitions. It proceeds by extending the algorithm by Franks and Misiurewicz[7] to the present setup.

The details of the algorithm have been developed elsewhere,[15] together with a proof that the procedure ends in a finite number of steps. The evolution of the process can be monitored step by step until it comes to an end. An intuitive explanation of how the procedure works has also been given elsewhere.[1] In the procedure of computing the image of the fat tree, the detection of recurrent bogus transitions indicates that some portion of the fat tree (of phase space, indeed) stretches, bends and maps onto itself, in pretty much the way Smale's horseshoe acts. If such a region were collapsed to a point by a homotopy, then all periodic orbits related to it (infinitely many) would collapse as well into at most one new orbit. This indicates two facts in the first place (a) The fat representative was not the map with the least number of periodic orbits for each period and (b) There exists another map on $\mathbb{D}\backslash P$ having a lower number of periodic orbits for each period than the initial representative. The added value of the algorithm is that this latter map can be generated from the previous one by carefully modifying the tree in order to reproduce the collapsing homotopy. After repeating this procedure a finite number of times, we end up with no recurrent bogus transitions and Theorem 1 assures that the fat representative on the final tree indeed is the map we were looking for. In the next Section we describe the detailed actions of the algorithm.

4. The Algorithm

The procedure to obtain a tree such that the fat representative has minimal periodic orbit structure is based in a graphical approach that modifies the fat tree via homotopies. The crucial moves are the *collapse* — which is performed via *collapsing steps* — and the *elimination of cyclically mapping edges*. We recall that the whole detail of definitions, procedure, etc., has been given elsewhere.[15]

4.1. *Collapse*

Let the tree act as a skeleton for the fat tree, i.e., vertices lie at the center of fat vertices and consequently edges lie along the "middle line" of fat edges. The idea of the collapse is that sectors around vertices (delimited by portions of contiguous edges and borders of the corresponding fat edges) that will map into a fold by some forward iterate of $\hat{\theta}$ are identified in such a way that the fold moves (or eventually dissapears). See Figure 4 for a description. Assume that sector x lying along edges a and b around vertex v is involved in a recurrent bogus transition and we want to collapse it. The collapse proceeds by identifying portions of the fat edge a constituting sector x with the corresponding portions of b. We may think of these portions as small rectangles growing away from v along the fat edges. At the end of the identified region, we place a new (fat) vertex, called *added star* having valence 3. The final result is shown in the right picture. Part of the fat edges a and b remain untouched. Other parts dissappear completely (those forming sector x), the parts of the fat edges a and b not forming the sector are joined forming a new edge,

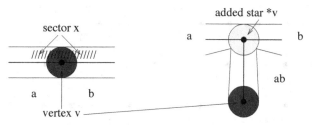

Fig. 4. The process of collapse.

called *ab*, and an added star (called *∗v*) indicates the end of the collapse process. The fold, if it still exists, has moved to *∗v*. The collapse is performed on a vertex having a fold involved in a recurrent bogus transition as well as in all its *pre-images*. The *pre-images* of a fold are the list of sectors mapping locally 1-to-1 by $\hat{\theta}$ onto the sector x involved in the definition of fold.

It is clear that collapsing a smaller or larger portion of a sector is a reversible process as far as homotopies are concerned. In order to produce a substantial change in the tree, the collapse has to be continued until some added star encounters some vertex or other added star. Only there an irreversible change occurs, since the number of vertices+added stars changes by at least one. This action is summarised in the next subsection.

4.1.1. *Collapsing step*

- Collapse until a vertex or added star is encountered.
- If there is no (manageable) fold left: **Stop**.
- If there is no recurrent bogus transition left: **Stop**.
- If two disjoint adjacent collapsing regions meet: **Stop**.
- Otherwise: Keep collapsing until some added star encounters the next vertex or added star.

The need for performing collapses can be monitored by an integer called *complexity*.[15] Each collapsing step reduces complexity, thus leading to a new fat representative having a lower number of periodic orbits. In Figure 5 we illustrate how to perform a collapsing step in the tree of Figure 3. In this case we have just one fold, whose pre-image lies around the middle vertex of the tree. The collapse moves that vertex downwards, adding a star. Gradually increasing the collapse, the image of the added star moves towards another vertex. The step ends when the star is "absorbed" in a preexistent vertex. The fold has moved one vertex to the left.

4.2. *Elimination of cyclically mapped edges*

Collapsing steps may end up adding a new set of vertices to the tree (the added stars) mapping periodically onto themselves, thus building one or more periodic

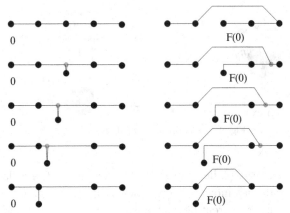

Fig. 5. A collapsing step.

orbits apart from those in P. In this new setup, it may happen that e.g., k edges having an added star on (at least) one endpoint map cyclically among themselves following a period-k orbit among the added stars. As a consequence, the Markov matrix will have an off-diagonal rectangle consisting only of zeros: these edges (let us label them $e_1, \cdots, e_k$) map only among themselves and have no image on the rest of the edges ($e_{k+1}, \cdots, e_m$). Hence, the rectangle constituted by matrix elements M_{sr} where $s \leq k$ and $r > k$ is identically zero.

This sub-cycle among the edges can therefore be simplified by another homotopy, shrinking the k edges to a point. The two vertices serving as endpoints of each edge become one. If only one of the vertices was an added star while the other was an original vertex from P, then this fact is true for the whole sub-cycle. In such a case, the original vertex remains when eliminating each edge. In any case, k added stars and k edges disappear.

After concluding this process, the number of periodic orbits of the fat representative is reduced and also the tree is simplified since the number of edges diminishes. In Figure 6 we illustrate this process on a period-7 horseshoe orbit. In the first line, we mark with dots the sectors in the pre-image of the fold. The second line shows the process halfway along the collapsing step. The arrows show which added vertices will collide at the end of the step. The resulting tree with only four added stars is shown on the third line.

We identify a period-2 orbit among the added stars. Indeed, the edges marked in blue and red in the fourth line map cyclically among themselves and have added stars as (both) endpoints. The edges are collapsed onto an added star on the fifth line. The resulting tree still has a fold but no recurrent bogus transition.

4.3. Description of the algorithm

1. Identify all folds.
2. If there are no folds, **end**.

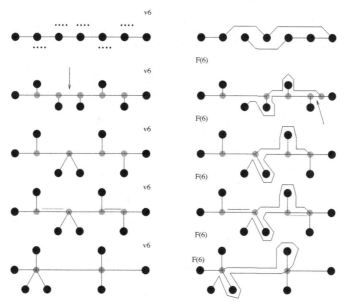

Fig. 6. Elimination of cycles.

3. Select a fold with eventually periodic stars (if any) or otherwise a fold with recurrent bogus transitions. If no fold can be selected **end**; otherwise *Collapse*:
   (a) Mark regions to be collapsed adding valence-3 stars
   (b) Perform one *Collapsing step*.
   (c) Eliminate cycles among edges having at least one added star as endpoint.
   (d) Go to (2)

The *added stars* remaining at the end of the process reveal other periodic orbits along with the starting one(s) not directly evident in the original tree.

Remark 7.1. The edges of the final tree give a natural Markov partition from which a Markov matrix can be computed. The topological entropy associated to the map F and invariant set P can be subsequently computed in the standard way.

4.4. *Example*

We consider in Figure 7 an example from Franks and Misiurewicz[7] since it was incompletely discussed previously.[1,15] We notice two folds in the first line. After performing a collapsing step one fold dissapears (row 2). The light-blue rectangles indicate the pre-images of the fold. After another collapsing step, in row 3, the fold has moved to the only remaining (period-1) added star. The pre-image of the fold is the sector formed with vertices 1 and 4. In the fourth row (missing in previous analysis) the fold has moved to one of the original vertices, revealing a rather simple orbit: The period-6 orbit corresponds to an underlying period-3 that underwent

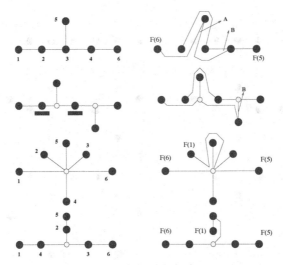

Fig. 7. A period-6 orbit.

period-doubling. The occurrence of this orbit is not enough to guarantee chaotic dynamics, as could be inferred from the fact that the associated Markov matrix has its largest eigenvalue with modulus one. The example is a particular case of a more general result regarding the periodic orbit structure of orientation preserving diffeomorphisms with zero topological entropy.[8]

5. Detailed Enumeration of Implied Orbits

We move now to explore the consequences of having the final tree and the fat representative $\hat{\theta}$ with minimal periodic orbit structure.

5.1. *Orbit forcing*

Lemma 1. *(a) The fat representative $\hat{\theta}$ has a period-n orbit, $n \geq 1$, if and only if a sequence of edges of minimal length n is mapped cyclically by $\hat{\theta}$. (b) The braid associated to the orbit is given by the fat tree and its image.*

Proof. The *only if* part of (a) is immediate. For the reverse statement, consider the action of $\hat{\theta}$ on an arbitrary edge e of the cycle. The map is contracting in one direction (loosely speaking the direction perpendicular to the (thin) edge) while $e \subset \hat{\theta}^n(e)$. Using a standard fixed point theorem on the plane, it follows that $\hat{\theta}^n$ has a fixed point, and consequently the fat representative has a period-n orbit.
(b) Admits a graphical proof: (i) Mark the periodic orbit of (a) on the tree. (ii) Connect the points sequentially following the tree (this construction is homotopic to the pre-image of a line diagram). (iii) The image of the construction in (ii) by

$\hat{\theta}$ gives the line diagram of the orbit, which is associated 1-to-1 to an element of $B_n/Z(B_n)$.[12]

Recall that because of Theorem 1, any diffeomorphism in the same isotopy class of $\hat{\theta}$, in particular the original map F, has a periodic orbit with the braid given by the Lemma.

At this point, the following procedure could be devised: (i) Label all edges in the fat tree. (ii) Construct all possible periodic sequences of edges according to the fat representative. By the previous Lemma, this will give all periodic orbits of the fat representative. Also, the labeling of the edges gives a symbolic name to each orbit, using an alphabet having as many letters as there are edges in the tree.

The construction is immediate, but unfortunately the labeling of orbits is not yet 1-to-1. There can be more than one periodic orbit with the same sequence of edges. See Figure 8, where the edges of the final tree in Figure 6 have been labeled. Edge a maps twice along edge d while edge e maps twice along edge f. However,

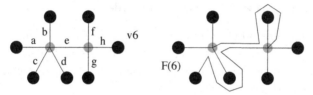

Fig. 8. Labeling of the edges of the final tree of a period-7 orbit.

since the set of vertices is invariant, each edge maps continuously onto a set of edges. Hence, by partitioning each edge into sub-edges when necessary, we can –in a finite number of steps– achieve a situation where each (new) edge maps at most once onto the (new) edges on its image. In such a situation the Markov matrix[a] will have entries that are either 0 or 1 and the symbolic labeling of the orbits in the new alphabet is indeed 1-to-1. We have proved the following

Lemma 2. *There exists a refined Markov partition (or alphabet) obtained by partitioning the edges of the final tree when necessary, that gives a 1-to-1 mapping between symbolic sequences and periodic orbits.*

Example The possible sequences appearing in Figure 8 are: $acgbfdh$, adh, $[e]^\infty$, $a[e]^n fdh$, $n \geq 1$. The first sequence is the one given by the original vertices in P. We identify in Figure 9 the period-5 orbit labeled as $aefdh$.

To refine the Markov partition along the lines of Lemma 2 it is enough to replace edges a and e by edges a_1, a_2, e_1, e_2 such that $\theta(a_1) = c \cup d$, $\theta(a_2) = d \cup e_1 \cup e_2$, $\theta(e_1) = f$ and $\theta(e_2) = f \cup e_2 \cup e_1$.

---
[a]The relabeling does not change the size of the largest-modulus eigenvalue.

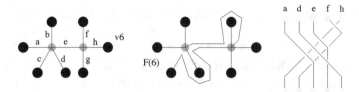

Fig. 9. A period-5 orbit (yellow dots) and its braid.

5.2. Orbit trimming

Lemma 3. *Given a line diagram, there exists a minimal (i.e., with least number of leaves) maximal template holding the orbit(s) given by the F-invariant set P.*

Proof. This follows by inspection of the line diagram. The template will have as many leaves (L) as there are monotonic portions in the diagram. Each leaf is fully expanding to the complete generating interval. The line diagram itself tells if the leaves are twisted or not, as well as their relative ordering. □

The maximal template provides a maximal family of orbits (our "universe") given by the *full shift in L symbols*. The question we denote as *trimming* deals with ruling out the periodic symbolic sequences in the maximal template which are **not** present in the minimal representative.

Note that the generating interval does not need to extend beyond the leftmost and rightmost points of the orbit in the pre-image of the line diagram. We assume in what follows that this is the case. We say that the maximal template is *fitted* to the orbit (we deal hence not with the full shift but with a fitted version). Fitting amounts to delete the leftmost and rightmost intervals in the lexicographic order of the full shift. Of course, those intervals may have infinitely many and even fractally structured interval pre-images, but any periodic orbit having a point on a pre-image will also have a point in one of the deleted intervals and will therefore be automatically eliminated.

We perform collapses at folds because of the presence of (recurrent) bogus transitions. In a bogus transition, an interior portion of some edge will map inside the folding sector (a *crossing* or some of its forward images, see Definitions 7.5, 7.6). When the fold is collapsed, such interior portions dissappear as well (Franks and Misiurewicz[7] named this process *pulling tight*). This process proceeds at various levels as we consider pre-images of higher order. Periodic orbits with points falling in those intervals are absent in the list of periodic orbits of the final map. We say that they are *trimmed*. The edge portions that dissappear in this way correspond to intervals of symbolic sequences of the maximal template. We describe the trimmed regions by enumerating the deleted intervals of symbolic sequences.

To render the trimming rules precise, the key point is to observe that a collapsing step ends (i.e., the region of collapse cannot be further increased without careful re-examination of the situation) when added stars collide either among themselves or with vertices from P. If added stars remain at the end of the process, they reveal a (previously unnoticed) periodic orbit of the fat representative (Lemmas 5 and 6 in Solari and Natiello[15]). In any case, a periodic orbit is involved in the process.

The demarcation of the regions to be collapsed or trimmed is indicated by pairs of points which have different past/present but identical future after mapping into the fold. The collapsing region is then characterised by an interval of symbolic sequences corresponding to those sequences pairwise identified and by indicating the sector to be collapsed at the pre-image of the fold [b]. We denote the symbolic sequences of the regions marked for collapse by $v[A, B]$ where A and B are the endpoints of the collapsing region at the pre-image of the fold (one of them is periodic). v is empty for the pre-image of the fold and it has k letters for the $k+1$-th pre-image of the fold (k-th pre-image of $[A, B]$). Note that the lexicographic order in the full shift implies that whenever v corresponds to an *odd* string, then the interval $v[A, B]$ denotes the (properly ordered) sequences $[vB, vA]$. Otherwise, even v denotes the sequences $[vA, vB]$.

The regions to be trimmed are then interior portions of edges that map onto the regions marked for collapse (i.e., the pre-images of the *crossings*) after one or more iterates. These regions can also be indicated by intervals of symbolic sequences of the form $w[A, B]$, where w is a nonempty string of letters identifying the crossings and their forward images up to mapping on the pre-image $[A, B]$ of the fold.

Lemma 4. *The periodic orbits of the fitted maximal template having substrings belonging to the trimmed regions $w[A, B]$ are not present in the final fat representative.*

Proof. The procedure towards the minimal fat representative proceeds via the deletion of such regions. □

Note that for each collapsing step we will have as many trimmed regions as there are crossings. Note also that the regions to be collapsed in subsequent collapsing steps need not be adjacent at the folding vertex in the original pre-image of the line diagram. Indeed, the folding region in a subsequent collapsing step corresponds to two regions in the original pre-image of the line diagram. One of the regions will have as one endpoint the periodic orbit identified as an added star, while the other region has as corresponding endpoint the pre-image of the added star that was collapsed onto it in the previous collapsing step.

Beside the trimmed regions, that are associated to the bogus transitions, there are a few other sequences that might disappear going from the fitted full shift in L symbols to the final fat representative. These other symbolic sequences correspond to pieces of the collapsing regions that are not associated to periodic orbits. Consequently, they do not belong in the list of trimmed sequences.

---

[b]In the pre-image of the line diagram it is necessary to indicate this sector, as the interval of sequences is not unique to the collapsing region.

Each fat edge has two halves, separated by the thin edge. Whenever the edge is involved in a collapse, each half is associated to different sectors (although they may have the same symbolic sequences). These halves have different fate after a collapsing step: One of them is collapsed away and the other persists as a part of the newly formed edge of the tree. A closer look indicates that the points pairwise identified in the collapse have sequences of the form aV and bV where a, b are nonempty strings that differ in the last symbol. However, the collapse extends up to a periodic orbit (either previously known or previously unnoticed) and by the expansiveness of the map, the sequences $(aV)^n, (bV)^n$, have as limit point this periodic orbit. Hence, of the strings a, b the one that subsists corresponds to that matching the symbolic sequence of the involved periodic orbit. We further notice that by identifying orbits with identical future (i.e., being on the same leaf of the stable manifold) there is at most one periodic orbit per leaf involved in the collapse (an argument that goes back to Birman and Williams[3]). The repetition of this procedure over successive collapses may lead to the complete deletion of some symbolic sequences not associated with periodic orbits. In any case, they are not in the list of trimmed sequences, since they are not related to periodic orbits that are *absent* in the final tree. Rather, they are non-periodic sequences having as limit a periodic orbit that is *present* in the final tree. This is summarised in the following,

Lemma 5. *The only periodic orbits with all of their points belonging to collapsing regions are the orbits identified after the collision of two added stars*

Proof. The sequence of sectors that map into the fold cannot be periodic since it ends at the pre-image of the fold.[15] Hence, for a periodic orbit to be completely included in the collapsing region that begins at the sectors that are pre-image of the fold, it has to belong to the intersection of two collapsing regions. At this intersection the collapse stops and a periodic orbit is identified. □

At this point it is already shown by construction that at the end of the procedure we are left only with the periodic orbits (in fact, braid-types) given by the minimal fat representative. No other periodic orbits are present in the final tree. Using the alphabet of Lemma 2 both the forcing description and the fitted-trimmed description of Lemma 4 are identical.

Theorem 2. *The list of periodic orbits given by the forcing is in 1-to-1 correspondence with the list of periodic orbits given by deleting the trimmed sequences from the fitted maximal template in L symbols.*

Proof. The name of the periodic orbits using the symbols of the maximal template can be found as a translation of the names obtained by the forcing rules (5) after following the orbits mapping for a complete period in the half-edges with their inherited symbol. Orbits not mapping into the interior of the half edge correspond to endpoints of the edges (i.e., vertices –original or added–) and have been identified during the collapse. Trimmed orbits correspond exactly to those orbits trimmed from the tree, being the edges of the new tree precisely what was left of the original edges after trimming. □

The trimming process can be summarised by giving a list of deleted intervals consisting of (a) the two fitting intervals of sequences at both extremes of the maximal template and all the sequences that map into these intervals and (b) the list of trimmed sequences consisting in the silently collapsed pre-images of the pre-image of the fold at each collapsing step. Periodic orbits in the final tree, when regarded under this alphabet, will not have points with sequences falling inside this list of intervals.

Example In Figure 10 we display a period-4 orbit, $(XZXY)^\infty$ named by its leftmost point. The maximal template has three leaves and it is shown in the Figure as well. We have chosen to label the third point of the orbit as $(YXZX)^\infty$. A similar analysis can be done assigning this point to the X-branch.

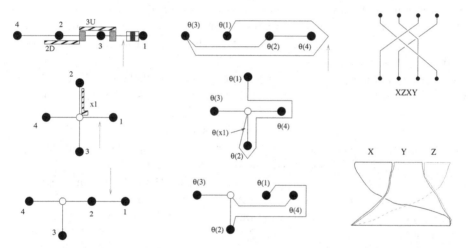

Fig. 10. An example of trimming.

We require that the endpoints in the pre-image of the line diagram coincide with the endpoints of the generating interval for the template. Hence, we fit the template to the orbit by trimming all orbits containing sequences in the semi-closed intervals $[(XZ)^\infty, (XZXY)^\infty) \cup ((ZXYX)^\infty, (ZX)^\infty]$. The first interval arises since the leftmost periodic sequence in the lexicographic order of the full shift is $(XZ)^\infty$, while we want the template to have $(XZXY)^\infty$ as its leftmost point. In the same way, the interval from the rightmost periodic point $(ZXYX)^\infty$ up to the highest periodic sequence in the lexicographic order, $(ZX)^\infty$, is deleted.

At the first collapse, the orbit X^∞ is identified. The upper half of $[X^\infty, YX^\infty]$ and the lower half of $X[X^\infty, YX^\infty]$ are collapsed (dashed regions in Figure 10) while the trimmed regions are: $XX(X^\infty, YX^\infty] \cup YX[X^\infty, YX^\infty]$ indicated in *green* in the same Figure and $Z[X^\infty, YX^\infty]$ indicated in *blue*. It is worth observing that in the initial fitted tree, sequences starting with X are located to the left of the point 3, sequences that start with Y to the right of 3 and up to a point that is a pre-image of 1. From this point up to 1 we have sequences that start with Z.

In the second collapsing step the collapsing region is a sector at the vertex X^∞ in the new tree that corresponds to extending the collapse in the initial tree along one edge up to $X(YXZX)^\infty$ (vertex 2) and along the other edge up to $Y(YXZX)^\infty$ (the dashed region of the second line). This corresponds to the intervals of sequences $X[X^\infty, (YXZX)^\infty]$ and $Y[X^\infty, (YXZX)^\infty]$ (that are adjacent to the new added star X^∞ after the first collapsing step). The trimming regions (yellow) are $ZX[X^\infty, (YXZX)^\infty)$ and $ZY[X^\infty, (YXZX)^\infty]$.

In the second line of Figure 10 we name 0 the period-1 identified as X^∞. Then, points in the fat-edge 4−0 correspond to sequences that start with X in the universal template, and the same is true for points in 0 − 2. Points in 0 − 1 correspond to sequences starting with Y up to the red arrow that signals the pre-image of 1, and from the arrow up to 1 they start with Z. The transition point between X and Y is still 3, hence, in the fat edge 0 − 3 we have sequences starting with X to the left of the thin edge and sequences starting with Y to the right of the thin edge, points on the dividing line start with X.

Going to the last line in Figure 10 points above 4 − 0 − 2 start with X, points below 4 − 0 and to the left of 0 − 3 start with X as well, points to the right of 0 − 3 and below 0 − 2 start with Y as well as points at both sides of 2 − 1 up to the red arrow, from the arrow up to 1 they correspond to symbolic sequences starting with Z. This mapping gives the translation rules for symbolic sequences proper of the last tree into the symbolic sequences of the maximal template.

We note that in the final tree, periodic orbits other than the original one do not have sequences with the substring ZX. This is a consequence of the fitting and trimming, since expanding to standard form all deleted intervals starting with Z we obtain $[ZY(YXZX)^\infty, ZYX^\infty]$, $[ZYX^\infty, ZX^\infty]$, $[ZX^\infty, (ZXYX)^\infty)$ and $((ZXYX)^\infty, (ZX)^\infty]$. Hence, periodic sequences containing the substring ZX (other than the original orbit) as well as part of the sequences containing ZY are not present in the final list.

The list of deleted intervals is: $[(XZ)^\infty, (XZXY)^\infty)$, $((ZXYX)^\infty, (ZX)^\infty]$ (fitting), $XX(X^\infty, YX^\infty)$, $YX(X^\infty, YX^\infty)$, $Z(X^\infty, YX^\infty)$, $ZX[X^\infty, (YXZX)^\infty)$ and $ZY[X^\infty, (YXZX)^\infty]$ (trimming).

6. Concluding Remarks

This manuscript started by reviewing a well established algorithm to understand the orbit structure underlying a 2-d diffeomorphism. Along the road, this understanding brought us to connect this work to other techniques in many levels.

When reconstructing the topological organisation of orbits, maximal templates give a broad sense of the periodic orbit organisation but they lack the precision to distinguish finer details such as those arising from changes of parameters or experimental conditions that do not bring into play new template branches. In contrast, when periodic orbits are dealt with as braid types, ultimately giving the pseudo-Anosov (or perhaps just periodic) minimal fat representative, the detail of the orbit structure displayed by the data will be sharper. Indeed so sharp that even a tiny change of parameters may change the representative tree.

The algorithm to obtain the fat representative from the line diagram, when followed in detail, provides a guide which explains how the orbit forced by the braid type accommodates in the larger (universal) picture of the fitted maximal template (consisting in a full shift on L-symbols fitted to the orbit).

In this work we have unraveled the relation between forcing produced by the braid type viewing it as the complementary action –called trimming– acting on the maximal template. In so doing, we contribute a tool for the analysis of $3 - d$ attractors all of them associated to the same maximal template.

References

1. M. A. Natiello & H. G. Solari, *The User's Approach to Topological Methods in 3d Dynamical Systems*, World Scientific Publishing, New York (2007).
2. M. Bestvina & M. Handel, Train tracks and automorphisms of free groups, *Annals of Mathematics*, **135**, 1–51 (1992).
3. J. S. Birman & R. F. Williams, Knotted periodic orbits in dynamical systems i: Lorenz equations, *Topology*, **22** (1), 47–82 (1983).
4. P. Boyland, Braid types and a topological method of proving positive topological entropy, Preprint, Department of Mathematics, Boston University (1984).
5. A. de Carvallo & T. Hall, Pruning theory and Thurston's classification of surface homeomorphisms, *Journal of the European Mathematical Society*, **3** 287–333 (2001).
6. A. de Carvallo & T. Hall, How to prune a horseshoe, *Nonlinearity*, **15**, R19–R68 (2002).
7. V. Franceschini, C. Giberti & Z. Zheng, Characterization of the Lorenz attractor by unstable periodic orbits, *Nonlinearity*, **6**, 251–258 (1993).
8. J.-M. Gambaudo, S. van Strien & C. Tresser, The periodic orbit structure of orientation preserving diffeomorphisms on D^2 with topological entropy zero, *Annales de l'Institut Henri Poincaré Physique Théorique*, **49**, 335 (1989).
9. T. Hall, The creation of horseshoes, *Nonlinearity*, **7**, 861 (1994).
10. T. Hall, Fat one-dimensional representatives of pseudo-anosov isotopy classes with minimal periodic orbit structure, *Nonlinearity*, **7**, 367–384 (1994).
11. G. B. Mindlin, H. G. Solari, M. Natiello, R. Gilmore & X.-J. Hou, Topological analysis of chaotic time series data from the Belusov-Zhabotinskii reaction, *Journal of Nonlinear Sciences*, **1**, 147–173 (1991).
12. M. A. Natiello & H. G. Solari, Remarks on braid theory and the characterisation of periodic orbits, *Journal of Knot Theory & Ramifications*, **3**, 511 (1994).
13. F. Papoff, A. Fioretti, E. Arimondo, G. B. Mindlin, H. G. Solari & R. Gilmore, Structure of chaos in the laser with saturable absorber, *Physical Review Letters*, **68**, 1128–1131 (1992).

14. S. Smale, Differentiable dynamical systems. *Bulletin of the American Mathematical Society*, **73**, 747 (1967).
15. H. G. Solari & M. A. Natiello, Minimal periodic orbit structure of 2-dimensional diffeomorphisms, *Journal of Nonlinear Science*, **15** (3), 183–222 (2005).
16. H. G. Solari, M. A. Natiello & B. G. Mindlin, *Nonlinear Dynamics: A Two-way Trip from Physics to Math*, Institute of Physics, Bristol (1996).
17. W. P. Thurston, On the geometry and dynamics of diffeomorphisms of surfaces. *Bulletlin of the American Mathematical Society*, **19**, 417 (1988).
18. J. R. Tredicce, F. T. Arecchi, G. P. Puccioni, A. Poggi & W. Gadomski, Dynamic behavior and onset of low dimensional chaos in a modulated homogeneously broadened single mode laser: experiments and theory, *Physical Review A*, **34**, 2073–2081 (1986).

Chapter 8

How topology came to chaos

Robert Gilmore
Physics Department
Drexel University
Philadelphia, PA 19104, USA

We review the steps taken during the development of topological analysis tools for the analysis of chaotic data.

Contents

1. Background 169
2. A Challenge 170
3. Relative Rotation Rates 171
4. Branched Manifolds 174
5. Topological Analysis Program 179
6. Chess Pieces on the Board 182
7. Forcing 183
8. Branched Manifold Perestroikas 185
9. Branched Manifolds Describe Mechanism 186
10. Bounding Tori 187
11. Four Levels of Structure 191
12. Symmetry 192
13. Representation Theory 197
14. Pointers to the Future 200
References 201

1. Background

Feigenbaum's startling and initially unappreciated discovery of the universality of certain scaling ratios[1-3] set off a chain of events that continues to this day. Before this discovery it had been assumed that

> All linear systems are the same.
> Each nonlinear system is nonlinear in its own way.[4,5]

After this discovery it was realized that

> It was a very happy and shocking discovery that there were structures in nonlinear systems that are always the same if you looked at them in the right way.[4,6]

Once a theory has been developed to the point where predictions are possible, there is a rush to experiments to falsify or to confirm the theory. Such was the case with Feigenbaum's scaling predictions. Experiments were carried out on fluids,[7,8] chemical reactions,[9–11] electrical circuits,[12] and lasers.[13–16] The experiments differed in the physics involved and the time scales involved, ranging from months (fluid experiments) to days (chemical reactions) to minutes (electrical circuits) to milleseconds (laser experiments). In the end, all confirmed Feigenbaum's predictions of the universality of scaling relations according to one-dimensional maps of the interval.

On the one hand the return on the investment of time required to do these experiments was magnificent: the experiments had shown that there were at least three previously unknown constants of Nature — scaling ratios in state space ($\alpha = -2.50290...$), in the control parameter space ($\delta = +4.466920...$), and in the space of measured intensity ratios (-8.2 dB). On the other hand the results were disappointing. In the logistic map of the interval, beyond the initial period-doubling bifurcation where the scaling relations are predicted, as well as all subsequent windows, there is a rigid organization of behavior — an organization of periodic windows in a sea of chaotic behavior. Indications of the rigidity of this structure were seen and reported in many of the experiments. But the nature and the structure of this rigidity was yet to be determined.

2. A Challenge

At this point Prof. Jorge Tredicce offered a challenge — and an opportunity. He had a great deal of data "left over" from his "Feigenbaum experiments" carried out on a laser with modulated losses.[16] Since data are always acquired at high cost with pain and difficulty, it would be obscene not to make further use of them. He asked if I could help him to further "understand" these data. This seemed like a wonderful challenge at the time. And with the benefit of hindsight, it was even more than that.

At the time there were only two types of tools available for the study of experimental data from a chaotic source. One type depended on metric measures: fractal dimension estimates of all types .[17–20] The other type depended on dynamical measures: Lyapunov exponents and spectra of Lyapunov exponents .[21–25] Both required very long, very clean data sets, a great deal of computation, resulted in real number estimates with no realistic error estimates ,[26] both were often dependent on where and how to make certain crucial assumptions, and neither was generally reproducible. Further, neither type of tool provided a way to distinguish among the different types of strange attractors that could be seen to be different: for example the mathematical strange attractors of Lorenz[27] and Rössler,[28] not to mention the strange attractors associated with periodically driven nonlinear oscillators such as the Duffing and van der Pol oscillators.

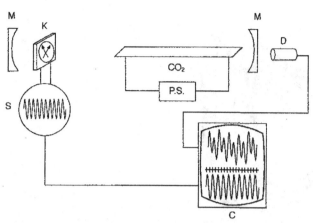

Fig. 1. Schematic diagram of a laser with modulated losses.[16] A carbon dioxide gas tube (CO_2, P.S. is the power source) is inserted between two mirrors (M) that form a confocal resonant cavity. A Kerr cell (K) is inserted in the cavity. This cell is periodically modulated by a signal (S), inducing losses as the polarization of the cell deviates from that produced by the Brewster angle windows. The intensity output is recorded by the detector (D). The input and output signals are recorded in a computer (C).

It was clear that a new type of analysis methodology was called for. It was also clear that this new tool should not depend on metric invariants or dynamical invariants. In fact, looking back to Henri Poincaré,[29] it was clear that this new tool ought to be topological in nature. Listening more closely to Poincaré, it was clear that this new tool ought to involve the periodic orbits "in" a chaotic attractor. A chaotic trajectory winds around in phase space arbitrarily close to any unstable periodic orbit, so it ought to be possible to use segments of a chaotic trajectory as good approximations (surrogates) for unstable periodic orbits.[30,31] The location and identification of such orbits is sometimes simplified because many unstable periodic orbits first appear in their stable avatars when created in saddle-node or period-doubling bifurcations and don't move too far from their original position in phase space as control parameters are changed. It was clear that unstable periodic orbits could not only be extracted from chaotic time series[32] but also serve as the "skeleton" of the strange attractor.[30] This is illustrated in Fig. 2.

3. Relative Rotation Rates

It was our hope that periodic orbits would somehow provide a key to understanding the structure of chaotic attractors. With this idea in mind, Hernan Solari extracted a number of unstable periodic orbits from a mathematical model of Tredicce's laser with modulated losses and undertook to determine the topological properties of these orbits.[33] The simplest and most obvious tool for quantitatively understanding the topological organization of these periodic orbits was to consider them as oriented

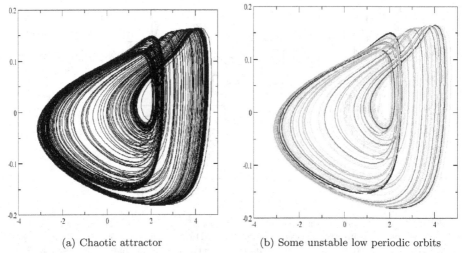

(a) Chaotic attractor (b) Some unstable low periodic orbits

Fig. 2. (Color online) The strange attractor (left) that describes the Belousov-Zhabotinskii reaction is very well outlined by its skeleton, a superposition of unstable periodic orbits of low period (b).

closed loops and compute their Gauss linking numbers. After all, the origin of Gauss linking numbers came from two different branches of physics. The linking number of two oriented closed loops, or periodic orbits A and B, is defined by the Gauss integral

$$LN(A,B) = \frac{1}{4\pi} \oint \oint \frac{(\mathbf{x}(s) - \mathbf{y}(t)) \cdot d\mathbf{x} \times d\mathbf{y}}{|\mathbf{x}(s) - \mathbf{y}(t)|^{3/2}} \tag{1}$$

In this expression $\mathbf{x}(s)$ are the coordinates of the periodic orbit A, the triple of coordinates being parameterized by the scalar s, and similarly for $\mathbf{y}(t)$ and orbit B.

In three dimensions periodic orbits cannot "pass through" each other. The simple reason is that if they could, two different orbits would at some stage have a point in common. This point, being on two different orbits, would not have a unique future. This violates the determinism property characteristic of all sets of ordinary differential equations used by scientists to model physical processes.

The linking numbers of all pairs of periodic orbits (stable or unstable) in the strange attractor are topological invariants so long as the orbits exist, since the flow was in a three-dimensional phase space $D^2 \times S^1 \subset \mathbb{R}^3$. As a result it was possible to use the set of linking numbers for these orbits as a way to characterize/identify the strange attractor of the laser with modulated losses (or at least the mathematical model that described this laser). A table of linking numbers for orbits up to period eight that were extracted from the Belousov-Zhabotinskii strange attractor is provided in Tab. 1

Since the attractor had "a hole in the middle" it was possible to construct another and even more powerful set of topological invariants. Our first impulse

Table 1. Orbits up to period eight were extracted from the experimental strange attractor of the Belousov-Zhabotinskii chemical reaction. The orbits are labeled by a symbol sequence according to where the successive iterations appear on the first return map. (M indicates that iterate occurs at the critical point.) The linking numbers of pairs of orbits are provided in this symmetric table. The linking numbers are orbit pair invariants. Self-linking numbers of the individual orbits are on the diagonal.

Orbit	Symbolics	1	2	3	4	5	6	7	8a	8b
1	1	0	1	1	2	2	2	3	4	3
2	01	1	1	2	3	4	4	5	6	6
3	011	1	2	2	4	5	6	7	8	8
4	0111	2	3	4	5	8	8	11	13	12
5	01 011	2	4	5	8	8	10	13	16	15
6	011 0M1	2	4	6	8	10	9	14	16	16
7	01 01 011	3	5	7	11	13	14	16	21	21
8a	01 01 0111	4	6	8	13	16	16	21	23	24
8b	01 011 011	3	6	8	12	15	16	21	24	21

was to call these "winding numbers" but Tredicce strongly advised us against using that already appropriated name. Instead, we named them *relative rotation rates*, in recognition of their origin. These are fractions that indicate the average rotation of orbits around each other per topological period.

To be explicit, we can construct a Poincaré section for the flow by hinging a half-plane on an axis and passing the axis through the hole in the attractor. Then each time the trajectory "goes around" it will intersect the half-plane, or Poincaré section, once and from the same side. A period-p orbit A will intersect the Poincaré section p times before repeating itself. Similarly, a different orbit B of minimum period q will have q distinct intersections. All intersections must be distinct, otherwise determinism is violated: One point on the Poincaré surface of section would have two distinct futures.

The next step is to connect one of the intersections of one orbit with one of the intersections of the other orbit by a directed line segment (an arrow) in the Poincaré section. As time evolves this arrow will move. We can imagine this process happening in the half plane as we sweep the half plane around its axis which goes through the hole in the attractor. If we rotate the half plane through $p \times q$ full rotations, the arrow will come back to its original orientation. This means that it has rotated through an angle of $2\pi n$ radians, or undergone n full rotations in the moving plane. The *average* number of rotations per period is $n/(p \times q)$. This fraction is the relative rotation rate for the given initial conditions on orbits A and B.[33] If $i = 1, 2, \cdots, p$ and $j = 1, 2, \cdots, q$ specify the intersections of the orbits A and B with the Poincaré section, a relative rotation rate $RRR_{i,j}(A, B)$ can be constructed for each pair (i, j) of initial conditions. There is a simple relation between relative rotation rates and linking numbers:

$$LN(A,B) = \sum_{i=1}^{p}\sum_{j=1}^{q} RRR_{i,j}(A,B) \qquad (2)$$

Relative rotation rates can even be computed for an orbit with itself. The self-relative rotation rates possessed a very attractive feature: They could be used to distinguish between orbits with nonzero topological entropy and "laminar" orbits with zero topological entropy.[33]

This topological index was computed for a number of orbits extracted from the model for a laser with modulated losses. The orbits in this set were identified by a name consisting of a symbol string consisting of 0s and 1s. The symbols were determined from the location of the intersection of the orbit on the first return map which looked basically like a logistic map. Tables of relative rotation rates are useful complements to tables of linking numbers such as the table in Fig. 1: They provide more information.

Periodic orbits in the Smale horseshoe were also located and identified with a symbol name. The relative rotation rates of these orbits in a suspension of the flow with no twisting (zero global torsion) were also computed.[33] Not all of the orbits predicted by Smale horseshoe dynamics were found in the model of the laser with modulated losses. We compared the table of relative rotation rates (also tables of linking numbers) of the orbits found in the laser model with tables for the appropriate subset of orbits in the Smale horseshoe flow suspension and found, to our delight, complete agreement.[33]

With mounting self-confidence we proposed that relative rotation rates could be used as fingerprints to identify strange attractors.[34] In fact, this idea was taken one step further: That the integers associated with relative rotation rates (or linking numbers) could be used to *classify* strange attractors.[35] We had already identified flows that followed a "Smale horseshoe scenario",[33] a "Duffing oscillator scenario",[34] and a "van der Pol scenario", each with distinct sets of linking numbers and relative rotation rates. Why couldn't these indices be used to classify/distinguish one chaotic attractor from another? It seemed reasonable to hope so.

4. Branched Manifolds

At this point my first hope (it was mine: my colleagues Solari and Mindlin weren't nearly as loopy as me) was that we could create a dictionary of scenarios (or processes, or mechanisms) and for each construct a table of linking numbers and/or relative rotation rates for lots of low-period orbits. Then when confronted with experimental data we could pull out the unstable periodic orbits from the data, compute their topological invariants, and then thumb through the dictionary comparing tables until we found an agreement. This program represented a lot of work.

While we were mulling over implementing this program a better solution became available. The beautiful work of Joan Birman and Robert Williams [36,37] gradually seeped into our consciousness. The time scales for implementing the "dictionary program" and for understanding the Birman-Williams Theorem were comparable, but the level of rewards for the latter far outweighed the former. The net result was that we understood the Birman-Williams Theorem at a level sufficient to apply it to our physical needs ("experimentalists' level").

What we understood is this. Suppose we have a dissipative flow in three dimensions whose trajectory "is a strange attractor." There is one positive Lyapunov exponent $\lambda_1 > 0$, one negative Lyapunov exponent $\lambda_3 < 0$, and one zero exponent $\lambda_2 = 0$ "along the direction of the flow". The dissipative nature of the flow requires $\lambda_1 + \lambda_2 + \lambda_3 < 0$. Then it is possible to project points in the phase space "down" along the direction of the stable manifold. This is done by identifying all points with the same future:

$$x \simeq y \quad \text{if} \quad \lim_{t \to \infty} |x(t) - y(t)| = 0 \qquad (3)$$

In this expression $x(t)$ is the future in the phase space of the point $x = x(0)$ under the flow. This Birman-Williams identification effectively projects the three-dimensional phase space down to a two-dimensional set that is a manifold almost everywhere. The two dimensions that remain correspond to the flow direction (with $\lambda_2 = 0$) and the unstable or stretching direction (with $\lambda_1 > 0$). The "almost everywhere" refers to zero- and one-dimensional sets where the "manifold condition" is violated. The points in the projection describe where the flow "splits" and the branch lines describe where flows from two distinct parts of phase space are "squeezed together". These rigorous mathematical structures were prefigured at an intuitive level by Lorenz[27] and Rössler[28] long ago. The Rössler attractor and its branched manifold are shown in Fig. 3 and the Lorenz attractor and its branched manifold are shown in Fig. 4.

Branched manifolds are useful constructions for distinguishing among different mechanisms that generate strange attractors. Four of the most studied strange attractors are those associated with the Lorenz, Rössler, Duffing, and van der Pol dynamical systems. The branched manifolds that describe these strange attractors are shown in Fig. 5.[4] These four branched manifolds are topologically inequivalent. "Equivalence" is by isotopy: Two things are isotopic if it is possible to mold one into the other without tearing or gluing it. As a result, identifying the branched manifold that describes a strange attractor is a powerful tool for distinguishing one (class of) strange attractors from another.

We should point out that branched manifolds can be constructed from 'stretching' and 'squeezing' units. These units are shown in Fig. 6. There are two simple rules for this aufbau construction:

(1) Outputs to inputs;
(2) No free ends.

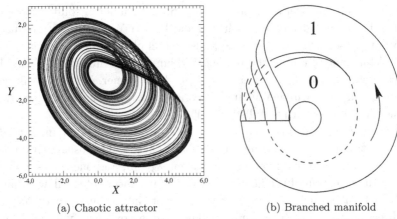

(a) Chaotic attractor (b) Branched manifold

Fig. 3. The Rössler attractor (a) is shown in its projection onto the x-y plane. Its branched manifold is shown on (b). The two branches are labeled 0 and 1. The integer indicates the torsion of the branch. The two branches split at the "splitting point" (near the arrowhead) and join at the branch line.

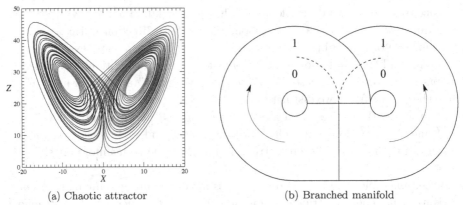

(a) Chaotic attractor (b) Branched manifold

Fig. 4. The Lorenz attractor is shown on the left in its projection onto the x-z plane. Its branched manifold is shown on the right. Neither branch exhibits a twist in this representation of the branched manifold. The two branches split at the "splitting point" which is shown "in" the branch line.

An enjoyable way to construct elegant branched manifolds is to dump a bushel of stretch/squeeze units in front of an enthusiastic class of kindergarteners, along with the instructions above. There is no guarantee that scientists can find a physical system described by the resulting artworks.

There are technical aspects in the statement of the Birman-Williams Theorem that we chose to ignore. There are three in fact. For the proof of the theorem, the flow is assumed to be:

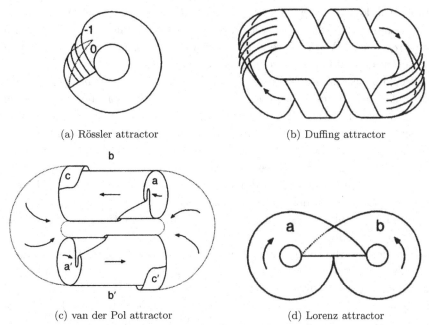

Fig. 5. Branched manifolds for four different attractors. It is clear that no topological move (stretching and folding are OK; tearing and gluing are not) can transform any one of these to any of the others. Therefore it is not possible to find a 1:1 coordinate transformation that converts any of these dynamical systems to any of the others.

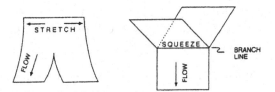

Fig. 6. Stretching and squeezing units serve as the basic building blocks for constructing branched manifolds.

(1) Hyperbolic
(2) In $\mathbb{R}^3$
(3) Dissipative.

We found it useful to ignore the hypothesis on which the theorem is based (at our own peril) for the following reasons:

Hyperbolic: In physics, both in theory and experiment, we have never seen a hyperbolic attractor. Every experimental chaotic attractor that we have seen is continually undergoing bifurcations as the external controls are varied. For example, the logistic map $x' = \lambda x(1-x)$ is hyperbolic only for $\lambda > 4$. It is stable (dynamically but not structurally) only at the knife edge $\lambda = 4$ and a strange *repellor* for $\lambda > 4$. If we were to insist on the assumptions undergirding the theorem it would not be useful to us as physicists.

In $\mathbb{R}^3$: Three dimensional models (e.g., Lorenz[27] and Rössler[28]) generate three dimensional data sets, or scalar data sets (e.g., $x(t)$) that can be embedded in three-dimensional spaces. But physical processes are often described by high dimensional sets of ordinary differential equations, or even by partial differential equations. In short, having collected a scalar data set there is no guarantee at all that it is generated by a three-dimensional set of equations. Even so, if we can find a three dimensional embedding the gods have smiled at us, and there is every anticipation of further smiles allowing the theorem to "work" even though the origin of the data set does not conform to the assumptions underlying the proof of the theorem.

Dissipative: One of the motivating ideas for the theorem comes from classical physics in a conservative manifestation. The magnetic field lines surrounding an infinitely long straight wire carrying a uniform current are closed loops described by two continuous parameters: radius and position along the axial direction of the wire. When the wire is bent most of the closed loops break: typically only a countable measure-zero set remains closed. This is true in particular when the wire is tied into a figure eight knot (carrying a supercurrent). The unbroken magnetic field lines are rigidly organized among themselves. The organization can be discerned by the linking numbers that pairs of closed loops exhibit. Further, these closed magnetic field lines can be isotoped down to the figure eight branched manifold without in any way altering these linking numbers.[4,36,37] In this conservative case the thrust of the Birman-Williams Theorem is valid, even though the proof doesn't cover this case.

I think the Birman-Williams Theorem is more powerful than it appears. Two of the three conditions discussed above are no longer an impediment to important physical applications.

Hyperbolicity: We avoid this constraint by falling back on a pruning argument. This will be explained in Section 7.

In $\mathbb{R}^3$: The theorem is true for *strongly attracting* dynamical systems with one positive Lyapunov exponent. What does this mean? If the spectrum of Lyapunov exponents for a flow in $\mathbb{R}^n$ satisfies $\lambda_1 > \lambda_2 = 0 > \lambda_3 \geq \lambda_4 \cdots \geq \lambda_n$, the system is strongly attracting when $\lambda_1 + \lambda_2 + \lambda_3 < 0$.[a]

As for the third (dissipativity), this is needed for the Birman-Williams projection (Eq. (3)) to work. However, it may not be necessary to project the flow to a branched manifold in order to describe the organization of the unstable periodic

---

[a]R. F. Williams, personal communication to the author.

orbits by the properties of a branched manifold. Indeed, the original inspiration of the unbroken magnetic field lines and the figure-eight branched manifold shows that there is at least one case where the theorem "works" despite the fact that the underlying assumptions are not satisfied. I suspect there are many more such cases (one for every oriented knot); perhaps even a theorem.

5. Topological Analysis Program

Once we understood what information we wanted to extract from the data, it was time to act. We acted first with data from Tredicce's laser. Of course, analyzing experimental stuff is not the same as analyzing computer generated data. Our attempts to extract a topological understanding of the laser data failed. In retrospect the reason was easy to see in the data. The laser acted somewhat like a relaxation oscillator. A large percentage of the time ($\simeq 40\%$) the intensity output (this is the observed variable) was very low. At such times the intensity was recorded in the lowest channel of a multichannel analyzer. When it came time to determine the topological organization of the unstable periodic orbits extracted from the data, we found many crossings occurred in the "lowest channel". Result: we were unable to determine accurately or honestly the crossing properties in low intensity regions. A request to redo the experiment was met with amused incredulity. It had long since been dismantled and pieces sent to new experiments, returned to previous owners, broken, or otherwise no longer available - a typical situation in a working experimental laboratory.

Plan B involved placing an APB for experimental data at each of the Nonlinear Dynamics meetings that we attended during this period. Daniel Lathrop and Eric Kostelich responded to this plea. They had used periodic orbits in order to characterize an experimental strange attractor for the Belousov-Zhabotinskii reaction,[31] and they provided us with a sample of these data. The data had been taken by Harry Swinney's group in Texas .[9-11]

The data were inordinately clean. Even so, we had a few problems attempting to make our favored embedding. Ultimately, these problems were resolved.[38]

We took the following steps to create the first topological analysis of experimental data.

(1) A set of unstable periodic orbits was extracted directly from the scalar time series *before* an embedding was created. This process is indicated in Fig. 7.
(2) A suitable embedding was found. We used an integral-differential embedding. In this embedding each orbit was identified by a symbol string obtained from a suitable return map. Only two symbols (0, 1) were sufficient since the return map had two branches.
(3) A table of linking numbers was made. This involved all orbits found in the first step, and the linking number was determined in the three-dimensional space constructed in the second step.

(4) A branched manifold was identified that "explained" these linking numbers. Our branched manifold had two branches since only two symbols were required to represent the trajectory described by the data set.

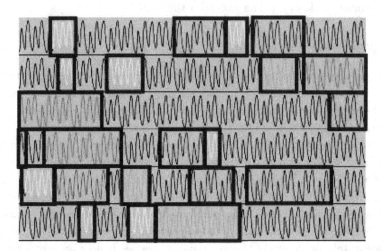

Fig. 7. (Color online) These data were taken by Lefranc, Hennequin, and Glorieux,[40] who redid Tredicce's laser experiment using a logarithmic amplifier-detector to climb out of the lowest-channel bind. The different-period orbits are coded with different colors. The identification preceeds the embedding step. These authors also showed that the fractal dimension computed with and without the data processing differed substantially, 'contradicting' a theorem that fractal dimension is an invariant.

A gloss on the last step is appropriate at this point. The branched manifold supported more orbits of any period p than we actually extracted from the data. We regarded this as a case of the absent orbits having been "pruned" from the hyperbolic limit. We were able to make a 1:1 correspondence between the orbits extracted from the data and an appropriate *subset* of orbits on the branched manifold. The correspondence was by symbol name. A table of linking numbers was constructed for the experimental orbits. This was done in two ways. Two orbits were superposed, the signed number of crossings counted, and the result divided by two.[33] This counting-of-integers method was double checked by computing the Gauss linking integral (Eq(1)). In the three-dimensional embedding each orbit was defined by a vector $(x(s), y(s), z(s))$ of coordinates, where s is a useful parameter. A code for computing the Gauss double integral was developed, extensively tested, and then applied to the embedded experimental data.

On the branched manifold side things took a different turn. We began by determining where each periodic orbit would fall on the branched manifold. This is straightforward but tedious. We continued by superposing the linear crossing segments and counting signed crossings. This is also straightforward but even more tedious up to period $\simeq 4$, after which it became error prone and almost impossible due to blurry eyeballs. In desperation we resorted to doing this by computer (think Geiger counter!). A code was written to do these things faster and better than a human could. The code did two things. It first located periodic orbits on a branched manifold (applied kneading theory[39]). Then it counted crossings of orbit pairs. Two inputs were required. One was the symbol names of the orbits under consideration. The second was a quantitative description of the branched manifold. Part of this description was a square $n \times n$ matrix, where n is the number of distinct symbols required to describe a trajectory in the experimental data as well as the number of different branches in the branched manifold ($n = 2$ in our case). The second was an $n \times 1$ array defining how the two branches are joined at the branch line.[4,41]

We later realized that these quantitative indices — square matrix and array — initially introduced as an *aide comptoir* in our computer code, actually served as an integer description of the branched manifold, which is itself a mathematically rigorous characterization of the experimental chaotic attractor (up to pruning).

Lastly, we realized that the branched manifold could be identified using only a very small number of unstable periodic orbits extracted from the data. This branched manifold could then be used to *predict* the linking numbers or relative rotation rates of all other orbits and orbit pairs extracted from the experimental data. These predictions could then be compared with those constructed from the experimental orbits. Either there was complete agreement, and we could conclude our identification was correct, or else there wasn't — that is, one or more integers in the linking number tables were not the same — in which case we had to *reject* the hypothesis that our identifications, of orbit labels and/or branched manifold characteristics, was correct. The paper in which we reported the results of our analysis of the Belousov-Zhabotinskii data is also the paper in which we announced the Topological Analysis Program.[38]

At last there was a rejection criterion for the analysis of chaotic dynamical systems that wasn't subjective. No error bars are associated with integers!

There were additional useful benefits from this program (Principal of Unexpected Beneficial Consequences). Noise usually degrades the analysis of data. In the case of topological analyses, noise makes it more difficult to extract the longer periodic orbits. The most important orbits for this analysis are the lowest period orbits. The effect of decreasing S/N is to reduce the number of *surplus* orbits, so that the rejection step is carried out with fewer orbits. In spite of this, even with a moderate amount of noise there are more than enough orbits to make a branched manifold identification and then to carry out the rejection tests using the remaining orbits.[41]

Metric analysis methods call for very long, clean, stationary data sets. Any kind of nonstationarity will destroy fractal dimension estimates and seriously degrade dynamical estimates (e.g., Lyapunov exponent spectra). However, it is possible to identify orbits with positive topological entropy in highly nonstationary data. Such

6. Chess Pieces on the Board

It was at this point that Prof. Ennio Arimondo sent us a gift in the form of Francesco Papoff, laded with data. The data had been taken in Arimondo's laboratory using lasers with saturable absorbers.[43] Several different absorbers had been used, and for each saturable absorber the laser had been run under a large number (6 - 10) of operating conditions. This provided us with a serendipitous chance to test another of our favorite hunches.

The idea was this. Suppose you have a physical system operating in a chaotic regime. A strange attractor is produced. The attractor can be investigated as described above, yielding a spectrum of unstable periodic orbits and an underlying branched manifold. This latest hunch was that as 'perturbations' are made, for example changing the operating conditions, the underlying branched manifold remains unchanged but the perturbations "push the flow around" on the branched manifold. We hoped to find that *all* data sets, when analyzed, yielded the same underlying branched manifold but the spectrum of orbits extracted from the various data sets changed from one set to the next. Indeed, this is exactly what we found.[44] Further, the underlying branched manifold was exactly the same as we had previously found for data from the Belousov-Zhabotinskii chemical reaction.

This result, based on the analysis of experimental data, was important in resolving a somewhat philosophical debate on how branched manifolds should be used in physics. This debate is summarized in Fig. 8. On the left in this figure is the Lorenz attractor — its branched manifold is apparent. On the right is a different attractor, the Shimizu-Morioka attractor.[45] Its branched manifold can also be readily inferred. By unwinding the loops on the left and right it bears a close resemblance to the Lorenz branched manifold, but the unwound (writhing) loops now each have a full twist and describe orbit segments of topological period two. Is it more useful to use this as the branched manifold for the Shimizu-Morioka attractor, or rather regard the periodic orbits in this attractor as a subset of the orbits in the Lorenz branched manifold?

Some argued that the branched manifold used to describe a physical system should be the one that contains only and exactly the periodic orbits in the attractor. The implication is that as operating conditions change and the spectrum of orbits in the attractor varies, the associated branched manifold would undergo mind-boggling contortions. Contortions so complicated as to curtail the use of branched manifolds as a nifty tool for understanding chaotic attractors.

On the opposite side of this opinion was our feeling that a branched manifold should be introduced with one branch for each of the symbols required for a unique description of an arbitrarily long trajectory — or at least of the length measured

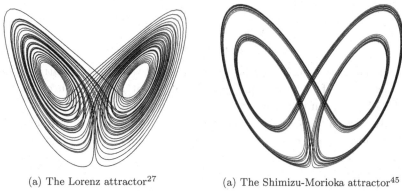

(a) The Lorenz attractor[27]　　　　(a) The Shimizu-Morioka attractor[45]

Fig. 8. It can be argued that these two branched manifolds are different, or that the periodic orbits in the Shimizu-Morioka attractor are simply a subset of those in the Lorenz attractor. In the latter case the Lorenz branched manifold can be used to compute the topological indices of those orbits in the Shimizu-Morioka attractor that are not pruned from the Lorenz attractor.

in an experiment. Such a branched manifold would "contain" all the orbits seen in an experiment as well as a lot more. It would be more useful, we maintained, to regard the missing orbits as having been pruned from the original allowed spectrum (technically, a "full shift on n branches" [46]). With this assumption, all the remaining orbits are organized in exactly the same way as in the branched manifold with unpruned spectrum. And perestroikas — experimentalists love to tweak knobs connected to control parameters — would generally leave the branched manifold untouched while changing the spectrum of orbits in the attractor. This is the source of our idea that changing control parameters serves to push the flow around on an unchanged underlying branched manifold. This simple view is very useful.

7. Forcing

While extracting periodic orbit surrogates from the 25 data sets that Papoff brought to us, Mindlin, Papoff, and Ricardo Lopez-Ruiz noticed that orbits had a social life of their own. In particular, they observed that when one particular orbit was present it was invariably accompanied by a handful of other orbits — always the same handful. Several different orbits possessed this property. The presence of some orbits seemed to "force" the presence of other orbits. This observation cried out for a more careful look.

We approached this problem from a topological perspective, since topology had already been so kind to us. When orbits are created, they are created in saddle-node pairs (neglect period-doubling for the moment). We computed the linking numbers of the pair A_n, A_s (A_n is the node of the saddle node pair of orbits A) with the pair B_n, B_s and arranged the results in a 2×2 matrix (we like matrices):

$$\begin{array}{c|cc} & B_n & B_s \\ \hline A_n & L(A_n, B_n) & L(A_n, B_s) \\ A_s & L(A_s, B_n) & L(A_s, B_s) \end{array} \qquad (4)$$

Depending on the equality or inequality of the four integers in this matrix, it was possible to determine if orbit B could exist before orbit A had been created, or vice versa (see Fig. 9). In this way we were able to piece together an orbit forcing diagram.[47] We carried the calculations out up to period eight. We were also able to develop the idea of a useful subset of orbits, which we called, because of our background, a *basis set of orbits*.

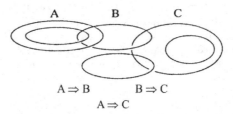

Fig. 9. Orbit pair B must be created before orbit pair A. The two orbits A_n and A_s have different linking numbers with the orbits B_n and B_s. Similarly, orbit pair C must be created before orbit pair B. The existence of the pair A forces the presence of pair B, and $B \Rightarrow C$.

The idea is as follows. Organize all the periodic orbits extracted from an experimental data set. Organize them according to their two-dimensional entropy, using the one-dimensional entropy as a tie-breaker.[4,41,47] For example, we found these orbits in one set of experimental data:

$$\underline{2_1}, \underline{4_1}, \underline{8_1}, \underline{6_1}, 8_2, \underline{7_1}, \underline{5_1}, 8_3, \underline{3_1}, \underline{6_2}, 6_3, 7_5, 4_2, 8_9, \underline{8_4}, 8_7, 7_6, \underline{7_2}, \underline{7_3}, 7_4, \underline{8_5}, 8_6, 8_8, 5_2$$

These orbits are identified by their order of creation in the logistic map. Then, starting with the "highest" orbit (5_2) with the largest entropies, we remove that orbit and "kill off" all the orbits that it forces (these are underlined). If there are any orbits left (yes unless the two-dimensional entropy is equal to the one-dimensional entropy), we continue the process:

$$\underline{6_3}, 7_5, 4_2, 8_9, 8_7, 7_6, 7_4, 8_6, 8_8$$

And again, until no orbits are left. Then the small set of orbits that have been removed (as opposed to "killed off") consists of the basis set of orbits on the branched manifold that describes the chaotic attractor. These are the orbits

$$8_7, 7_6, 7_4, 8_6, 8_8, 5_2 \qquad (5)$$

In truth, this argument works up to whatever finite period forcing information is available. A lower bound on the topological entropy of the flow can be obtained by computing the topological entropy of the braid containing the basis set of orbits.

'Forcing' is a very difficult problem, and in truth our approach is probably the least effective that has been found to work. Other approaches[48-53] are more effective, but much more complicated. The problem of forcing, even on the two branch manifold describing Smale horseshoe dynamics, is still open.

8. Branched Manifold Perestroikas

Normally, small perturbations produce changes in the spectrum of orbits that a branched manifold can support. But sometimes perturbations are not small. Under these conditions the branched manifold will change. What this amounts to is that more symbols are required to uniquely label a chaotic trajectory. Or perhaps fewer symbols are required. Yet more generally, the spectrum of symbols required changes. Since symbols correspond to branches, such perturbations lead to changes in the structure of the branched manifold underlying the description of a chaotic attractor and its perestroikas.

Such changes were studied extensively for the periodically driven Duffing oscillator as a function of changing driving frequency.[54,55] The modifications that occur in the dynamics, the spectrum of stable and unstable periodic orbits, and the number, nature, and organization of the branched manifolds that describe the attractor, and the topological indices that are observable occur with regular predictability. In short, each time the external driving frequency passes through a $1/n$ subharmonic of the natural frequency of the undriven nonlinear oscillator, the underlying branched manifold undergoes another full twist. The global torsion changes systematically by ± 1 in each subharmonic window.

It seems almost as if there is a "branched manifold" that assumes the form of an enormous scroll with branches labelled by successive integers $0, 1, 2, ...$, and the flow is constrained to travel in only a small number of adjacent branches for any value of the control parameters. As the controls are changed the flow is directed over a limited number of contiguous branches, for example $(0,1) \to (0,1,2) \to (1,2) \to (1,2,3) \to (1,2,3,4) \to (2,3,4)\cdots$. This systematic behavior is a consequence of continuity. This behavior has variously been called the "jelly roll scroll" (in the US) and the "gâteau roulé" (in France). One important question is whether the scroll rolls up from outside to inside (yes in the cases we have investigated) or from inside to outside. These two cases can be distinguished by computing linking numbers of appropriate orbits.[4,41,54,55]

Shortly after this study, scrolling behavior was observed in experiments performed on a periodically driven Nd-doped fiber optic laser[57] As the external drive frequency descended through the subharmonics of the natural resonance the branched manifold describing the chaotic behavior became more and more wound up. The systematics of this behavior is indicated in Fig. 10. This winding-up phenomenon is a typical feature when there is a competition between some natural resonance frequency and an externally imposed driving frequency.[56] It has been

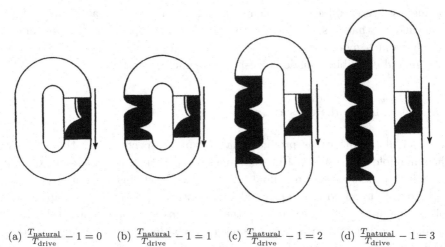

(a) $\frac{T_{\text{natural}}}{T_{\text{drive}}} - 1 = 0$ (b) $\frac{T_{\text{natural}}}{T_{\text{drive}}} - 1 = 1$ (c) $\frac{T_{\text{natural}}}{T_{\text{drive}}} - 1 = 2$ (d) $\frac{T_{\text{natural}}}{T_{\text{drive}}} - 1 = 3$

Fig. 10. As the ratio between the natural frequency (period) of an oscillator and the frequency (period) of an external driver decreases (increases), the strange attractor that may exist becomes increasingly wound up. The global torsion increases systematically as the ratio $\frac{T_{\text{natural}}}{T_{\text{drive}}}$ increases through integer values.

seen in the analysis of chaotic signals from dogfish, catfish, and paddlefish [58,59] and many laser experiments.[60,61]

Instead of winding up in a gateau roulé structure, it is possible that three continuous branches of a branched manifold could fold in an S-shaped structure. This has finally been found in one of the subharmonic domains studied by Javier Used and Juan Carlos Martin in their periodically driven Erbium doped fiber optic laser.[61] This work is summarized in another contribution to these Proceedings.

9. Branched Manifolds Describe Mechanism

Branched manifolds are exactly the right tool for describing the *mechanism* that acts to create a strange attractor and at the same time to organize all the unstable periodic orbits in it. By mechanism we mean the stretching and folding and/or the tearing and squeezing processes that occur repetitively the the phase space.[4,41]

We illustrate two different mechanisms in Figs. 11 (Rössler mechanism) and 12 (Lorenz mechanism). In Fig. 11 a blob of points in the phase space at (a) is stretched along one direction (with $\lambda_1 > 0$) and flattened in another (with $\lambda_3 < 0$) while being displaced in a third direction (with $\lambda_2 = 0$). When the flow exists in a bounded region of the phase space, some mechanism must exist to return this set of points to its initial neighborhood. One mechanism involves a simple fold, shown beginning at (c) in this figure. Eventually the folded-over set of points returns to its initial neighborhood as (c) → (d) → (a). This process repeats indefinitely, building up a flakey structure in the squeezing (λ_3 direction) known as a fractal. In the case depicted in Fig. 11 the mechanism is a simple stretch and fold. This is represented by a simple branched manifold with two branches.

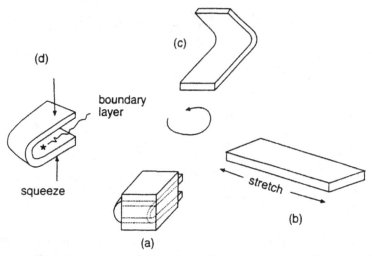

Fig. 11. A set of initial conditions at (a) is deformed by stretching in one direction and squashing in another. As the flow progresses, folding begins to occur at (c) and continues on through (d). This deformed set of initial conditions finally returns to its initial neighborhood (a), where the process is repeated *ad infinitem*.

Another mechanism is illustrated in Fig. 12. In this case a set of initial conditions (the cubes) begin to flow but run into a "buzz saw" that cuts the set into two or more pieces. These pieces move off into different directions in the phase space, where they encounter and are squeezed together into other blobs of points, some with different initial conditions. Eventually these sets of points encounter their initial neighborhoods, and the process continues over and over again. Branched manifolds summarize in a simple and visual way the mechanisms that exist and occur repetitively to build up each strange attractor.[62,63]

10. Bounding Tori

Branched manifolds place constraints on the periodic orbits that can be created or destroyed during a perestroika. At some point we began to wonder if there were larger structures that placed analogous constraints on branched manifolds themselves during a perestroika.

In fact, there are. We began by asking about the topological properties of a "smoothed out" version of a messy fractal chaotic attractor. We could smooth a strange attractor by surrounding each point in the attractor by a ball of ϵ radius (ϵ is "small enough") and constructing the union of all such three-dimensional balls.[64,65] When done correctly, the union was a three dimensional manifold. Since we were

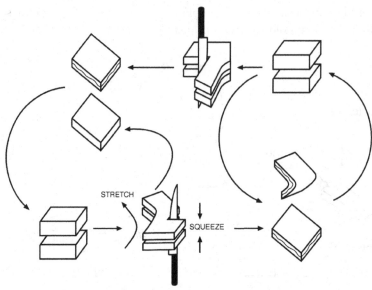

Fig. 12. Sets of initial conditions (cubes) are "sliced", by running into an axis with a stable and an unstable direction (the z-axis for Lorenz-like systems), for example. The different parts flow off in different directions in the phase space, where they may encounter other sliced parts from different regions of phase space. These are squeezed together and eventually return to regions they originated from (recursion).

looking for some structure to enclose or surround the attractor and its branched manifold, we were naturally led to consider the boundary of this manifold. Once again, we wound up talking to topologists. These boundaries were two-dimensional, orientable, and bounded — therefore uniquely tori. A torus is shown in Fig. 13 along with some of the important closed loops on it. A topological torus is characterized by one number, its genus g. Our torus surfaces were "dressed" with the flow that generated the chaotic attractor inside the torus, restricted to the surface. The immediate result was that a strange attractor could be described by an integer, g, the genus of the torus surrounding it, together with another more complicated discrete index that describes the flow on the surface. The second index is not an integer but a transition matrix related to the permutation group P_{g-1}.[64–66]

On the surface the flow has some stagnant points, or fixed points. All fixed points on the surface arise when the tangent vector (recall, $\lambda_2 = 0$) of the flow generating the attractor is perpendicular to the surface. At such points the stability of the restricted vector field is governed by the two remaining eigenvalues, $\lambda_1 > 0$ and $\lambda_3 < 0$. As a result, the index of each fixed point on the surface is -1 and the sum of these indices is related to the genus of the torus by $\sum_{\text{all f.p.}} (-1)^1 = -2(g-1)$. A lot of elegant topology due to Euler and Poincaré goes into this result.[67]

Working by analogy again, we asked if there is an "aufbau principal" for bounding tori the way there is one for branched manifolds. In the latter case a branched

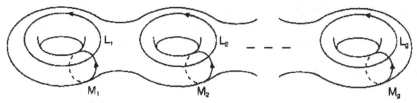

Fig. 13. A torus in $\mathbb{R}^3$ is completely identified by the number of holes in it. This integer index is called its genus. A normal tire tube is a genus-one torus and a sphere has genus zero. A torus of genus g has two useful sets of closed loops on it. These form the generators of its homotopy group. The meridians $(M_1, M_2, \cdots, M_g)$ can be chosen to bound a disk that lies entirely within the torus. The longitudes $(L_1, L_2, \cdots, L_g)$ bound disks that lie entirely outside the torus in the case shown. This isn't always true: to see how this can fail, see Fig. 16. It is always possible to choose g meridians and g longitudes so that they are independent, meridians have no intersections with other meridians and similarly for longitudes, and each meridian intersects only one longitude at one point. The tori bounding strange attractors are "dressed" on their surfaces with the flow that generates the attractor.

manifold can be built up by joining together stretching and squeezing units (c.f., Fig. 6). We didn't have to look too far (not beyond string theory) to find what we needed. The building units we needed were called variously "pairs of pants" or "trinions". These come in two varieties, as do the basic building blocks of branched manifolds. The two building blocks are shown in Fig. 14. Stretching trinions have one input port and two output ports. These contain the branched manifold stretching units (c.f., Fig. 6). Joining trinions have two input ports and one output port. These contain the branched manifold squeezing units.

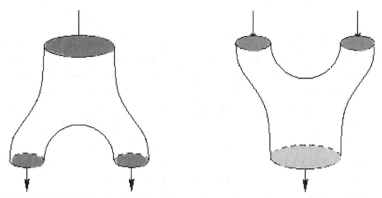

Fig. 14. (Color online) The basic building blocks of bounding tori are stretching (left) and squeezing (right) trinions. A genus-g torus is built up via the aufbau principal using $g - 1$ pairs of stretching/squeezing trinions, $g > 1$. The construction is: Inputs to outputs; with no free ends at the end of the construction. Each joint is colorless. The global Poincaré surface of section is the union of $g - 1$ disks. The disks can be taken as the input ports of the splitting trinions or the output ports of the joining trinions.

We found that it is possible to build up any genus-g torus ($g > 1$) using $g - 1$ *pairs* of these units. Each pair contains one stretching unit and one joining unit. These units must be connected together following the usual rules:

(1) Outputs connect to inputs.
(2) No free ends.

Two pairs of stretching/squeezing units can be used to build up the genus-three torus that bounds the Lorenz attractor, as shown in Fig. 15

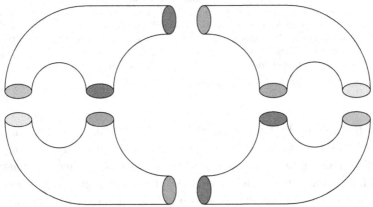

Fig. 15. Two pairs of stretching and squeezing trinions are used to build up the genus-three torus that encloses the Lorenz attractor. Output ports of squeezing trinions flow to the input ports of stretching trinions, etc. All connections are colorless. The global Poincaré surface of section is the union of disks. The disks can be taken as either the input ports of the stretching trinions or the output ports of joining trinions.

An added benefit of this decomposition is that it is now possible to describe the global Poincaré surface of section for a low-dimensional chaotic attractor. Determine its bounding torus. Partition it into $g - 1$ pairs of stretching/joining units. The output port of each joining unit (equivalently, the input port of each splitting unit) is a disk that is one component of the Poincaré surface of section. The section itself is the union of these $g - 1$ disks. A topological period (as opposed to a temporal period, measured in seconds) is a transition from the section to itself: this means from any one of the disks to the next, under the flow. The Poincaré section of the Lorenz attractor is the union of two disks, as seen in Fig. 15.

All of the experimental attractors that we have analyzed that have a "hole in the middle" live in genus-one tori. The mathematical attractors with this feature include the Rössler, Duffing, and van der Pol attractors. The latter is bounded by *two* genus-one tori, one outside and one inside the attractor. The Lorenz attractor lives inside a genus-three torus. A number attractors studied by Aziz-Alaoui[68] live in higher genus tori.

The closed magnetic field lines surrounding a knot tied into the shape of a figure eight knot have the topological organization defined by the figure eight branched manifold. This lives inside a bounding torus of genus $g = 9$. It seems an interesting idea to relate closed magnetic field lines generated by supercurrents in wires tied into the form of various knots with the associated knotholder and relate the genus of the knotholder with the original knot. We point out that tori can be embedded into $\mathbb{R}^3$ in a multitude of bizarre ways, as indicated in Fig. 16.

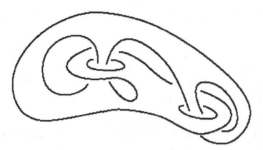

Fig. 16. Tori may be embedded into $\mathbb{R}^3$ in a variety of different ways. For this genus-four torus all meridians bound disks that lie entirely within the torus. However, each of the four simplest longitudes links one other longitude.

11. Four Levels of Structure

We now can address the levels of structure available to describe a strange attractor. We regard attractors that can be deformed into each other smoothly — no cutting or tearing, no creating or annihilating orbits — as isotopic or equivalent. We ask: how do we distinguish (topologically) inequivalent attractors. At present, there are four levels of structure. Each level is discrete. We describe these levels in some semblance of order. At the lowest level there is a basis set of orbits (to any finite period). This set determines all orbits that are present in the attractor, up to some period. At the next level are branched manifolds. These serve as a rigorous caricature for the strange attractor. Branched manifolds constrain the evolution of basis sets of orbits under perestroikas. Above branched manifolds in this hierarchy are bounding tori. Bounding tori constrain the evolution of branched manifolds under perestroikas.

Bounding tori live in $\mathbb{R}^3$. A torus ($g = 1$) can be embedded in $\mathbb{R}^3$ in many inequivalent ways. To determine how many, imagine shrinking the torus surface down to an oriented curve in $\mathbb{R}^3$. Then there as many ways to embed a torus into $\mathbb{R}^3$ as there are oriented knots in $\mathbb{R}^3$. Each such knot defines a torus and within each such torus we can construct a strange attractor. We can do better: we can embed a $g = 1$ strange attractor in each knotted torus in such a way that all the embedded strange attractors are diffeomorphic. They are not isotopic, since the knots cannot be deformed into each other.[69]

For a genus one torus containing a strange attractor there are as many embeddings as there are oriented knot types. Although we cannot yet distinguish inequivalent knots by any known index, the number of knot types is discrete. Similar arguments apply to tori of genus-g. Each genus-g bounding torus can be embedded in many different ways in $\mathbb{R}^3$ (c.f., Fig. 16). This discrete index we call, for want of a better name, the generalized knot index. Constructing a generalized knot index is an even more difficult problem than constructing a knot index for the simpler case of the genus-one torus.[70]

12. Symmetry

We were always on the lookout for new strange attractors. So when Christophe Letellier asked about the relation between symmetry and chaos an immediate bond was formed between us. The first question we asked is: "How does Cartan's theorem about the relation between covering groups and their images play out in dynamical systems theory?"[71] Of course, at that time we had no clue..., and for this reason the chase was exciting.

We were first motivated by the way we could identify symmetric points in the Lorenz attractor and make it "look like" the Rössler attractor. Eventually these visceral feelings gave ground to a more quantitative approach.[72-74]

In one direction (image direction) things are relatively simple. The Lorenz equations are unchanged (equivariant) under the transformation $(X, Y, Z) \to (-X, -Y, +Z)$. By identifying a point with cordinates $(-X, -Y, Z)$ with its partner (X, Y, Z) in the Lorenz attractor we were able to map the original Lorenz attractor into something with one hole in the middle — very much like the Rössler attractor, as shown in Fig. 17. This mapping extends from points to orbits, both closed periodic orbits as well as chaotic trajectories, and eventually to branches of the underlying branched manifold. In the general case, suppose we have a chaotic attractor described by a branched manifold with $2n$ branches and an obvious two-fold symmetry. The Lorenz attractor with four branches springs to mind (c.f., Fig. 4). A two-to-one image is constructed by identifying the symmetrically related branches pairwise. The number of branches is halved, as is the number of branch lines and splitting points. An orbit of minimum period p in the original (cover) attractor is mapped to an orbit of period p in the image system. The image orbit "goes around" either once or twice before closing up. To say this another way, the image orbit has minimum period either p or $p/2$, depending on whether the cover orbit was symmetric or not.

Going in the other direction was yet more exciting. Starting with an image attractor, suppose we wanted to create a two-fold cover with rotational symmetry. Then we have to "lift" the image. The lift can be carried out in many inequivalent ways, depending where we place the rotation axis.[73] For example, suppose we wanted to create a two-fold cover of the Rössler attractor (2 branches) with rotational symmetry. We could put the rotational axis through the hole in the middle.

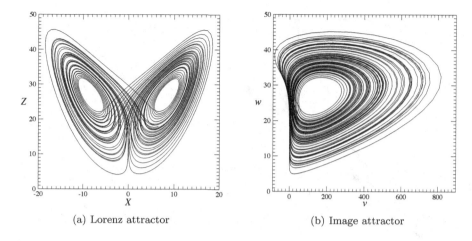

(a) Lorenz attractor   (b) Image attractor

Fig. 17. The Lorenz attractor (a) and its two-to-one image (b). The image is obtained by identifying pairs of points $(X, Y, Z) \leftrightarrow (-X, -Y, +Z)$ in the Lorenz attractor. One convenient way to do this is by introducing a new phase space with coordinates (u, v, w) related to the coordinates (X, Y, Z) through $u = X^2 - Y^2, v = 2XY, w = Z$.

This gives us a cover with a hole in the middle and four branches organized as a double fold. We could put the rotation axis outside the Rössler attractor. The lift would create two disconnected attractors, each with a hole in the middle and two branches, each of the two identical to the original attractor. We could put the rotation axis between branches 0 and 1 in the Rossler attractor: One way to do this results in a four branch lift that is topologically similar to the Lorenz attractor. Another way of placing the rotation axis between branches 0 and 1 results in an attractor quite different from a Lorenz-like attractor.[75]

All these lifts have four branches. They are structurally stable under perturbation in the location/orientation of the rotation axis. However, if the rotation axis is located in such a way that it intersects the attractor, structurally unstable lifts with six-branches result.[75]

We have used one of these structurally stable lifts to relate sunspot number data N (all positive), which has an approximately 11 year cyclic variation, to the underlying magnetic field B, which exhibits an approximately 22 year variation through both positive and negative values.[74] The usual attempts to relate the two involve making a sign change (by hand) at sunspot minima. We made a planar projection of the data onto the $(N, \dot{N})$ plane, constructed a lift with two-fold rotational symmetry, and identified one of the cover variables as a surrogate for the underlying magnetic field. The zero-crossings of the magnetic field appeared in a natural way.[74] The results are shown in Fig. 18. The transformations of the original sunspot data to the time variations of the underlying magnetic field are shown in Fig. 19.

Lifts using larger symmetry groups than the two-fold rotation group can be constructed. The description of many of the exciting things that can happen is a

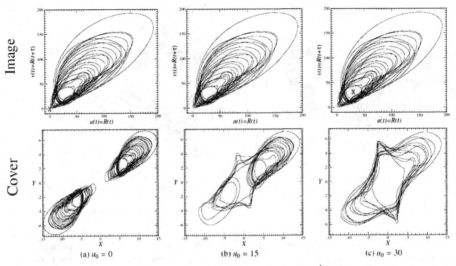

Fig. 18. Top row: Smoothed sunspot data $N(t)$, plotted in the $(N, \dot{N})$ plane. Bottom row: double covers created with different rotation axis. When the axis is outside the attractor (left column) the cover consists of two separate pieces that do not interact. When the axis intersects with the attractor (center) the cover consists of a single attractor, but regions of positive values are not always succeeded by regions of negative values. When the axis is inside the attractor the double cover has a hole in the middle, but there is a deterministic oscillation between positive values and negative values.[74]

long story that is laid out in a work that brought great joy to us.[75] Many of the lifts that we considered leave one or more points fixed (rotations around an axis, inversions in a point, reflections in a plane). These point group symmetries act easily on branched manifolds. If $i = 1, 2, 3, \cdots$ label the branches in the image, the branches in the cover are labeled $i = 1\alpha, 2\beta, 3\gamma, \cdots$, where $\alpha, \beta, \gamma, \cdots$ are the operations in the group. A structurally stable lift of an image with b branches by a point group with g group operations has $b \times g$ branches.

Covers of an image attractor with topological entropy h_T have the same topological entropy. The argument is easy — we provide it for a two-fold cover with rotational symmetry. The lift of a period-p orbit in the image is a symmetric orbit of period $2p$ or two distinct orbits of period p related to each other by symmetry. Assume that the number of orbits of period p grows exponentially in both the cover ($N_C(p) \simeq e^{\gamma p}$) and the image ($N_I(p) \simeq e^{h_T p}$). The number of odd period orbits in the cover is

$$N_C(2p+1) \simeq e^{\gamma(2p+1)} = 2 \times e^{h_T(2p+1)} \simeq 2N_I(2p+1) \qquad (6)$$

It is a simple exercise to conclude that $\gamma = h_T$ in the limit $2p+1 \to \infty$. An argument using orbits of both even and odd period in the cover is slightly less simple but leads to the same place: $h_{\text{Cover}} = h_{\text{Image}} = h_T$. This argument extends without difficulty to covers created using other finite groups.

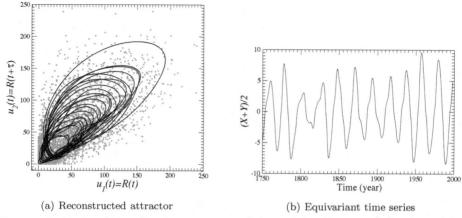

(a) Reconstructed attractor (b) Equivariant time series

Fig. 19. Original sunspot data and processed time series (a), showing an attractor with a hole in the middle. (b) Time series resulting from a projection of a trajectory in the double cover (c.f. Fig. 18, right column) onto an axis that serves as a surrogate for the underlying magnetic field. This result indicates the strength and polarity of the magnetic field underlying sunspot number variability, one with a 22 year cycle, the other (image) with an 11 year cycle.[74]

In addition to symmetries involving point groups, there are symmetries analogous to nonsymmorphic space groups in solid state physics. The simplest such group is the symmetry present when a nonlinear two-dimensional oscillator with inversion symmetry ($\dot{x}_i = f_i(x,y), f_i(-x,-y) = -f_i(x,y)$) is periodically driven: $f_i(x,y) \to f_i(x,y) + a_i \cos(\omega t + \phi_i)$. In this case the dynamics is invariant under the transformation $(x, y, t) \to (-x, -y, t + \frac{1}{2}T)$, where $T = 2\pi/\omega$ is the period of the drive. The periodically driven Duffing and van der Pol oscillators share this symmetry.[4] In these cases it is possible to mod out the order-two symmetry and construct a smaller chaotic attractor that is simpler to analyze.[4,41] This is done by viewing the attractor from a rotating frame of reference. This frame is called the "van der Pol" frame. In fact, there are two such frames: they counterrotate with angular frequencies $\pm\frac{1}{2}\omega$. There is then a natural question: which to use? Since we are physicists we choose the frame in which the kinetic energy is minimum. The concepts of average kinetic energy, average rotational energy, and their variation in a series of rotating frames that satisfy simple boundary conditions, is well-defined and natural. These frames are defined by an integer.[4,77] The results can be seen in Fig. 20.

In the reverse direction, it is possible to lift the image attractor to double covers and p-fold covers, in a very large variety of ways. The multiplicity of possible lifts are distinguished by two integers. The mean energy and mean orbital angular momentum depend on the ratio of these integers, as shown in Fig. 21. Their regularity properties were described in a paper originally entitled "Quantum numbers for strange attractors." But the referee(s) objected to this title, so it was reluctantly changed to something more prosaic.[77] An animated lift of the periodically

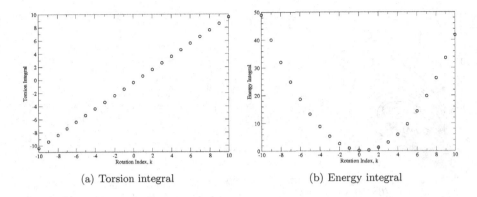

(a) Torsion integral   (b) Energy integral

Fig. 20. It is possible to define an average kinetic energy and an average orbital angular momentum for a dynamical system whose phase space is a torus.[77] These real numbers depend on the representation of the dynamical system: its global torsion n. As the global torsion increases in magnitude, so also do these classical averages. The preferred frame for physicists is the minimum energy frame.

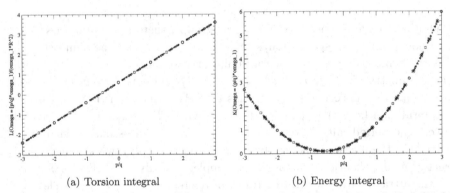

(a) Torsion integral   (b) Energy integral

Fig. 21. If a lift is given a fraction ($\frac{n}{p}$) of a twist per period, then p of these "units" fitted in sequence provide a p-fold cover with global torsion n. The classical integrals depend "smoothly" on the rational fraction $\frac{n}{p}$.

driven van der Pol attractor with three-fold rotational symmetry was constructed by Timothy Jones and can be enjoyed ("eye-candy") at his web site.[b]

We were delighted with the results of our cover-image studies. It *was* possible to answer our original question (about Cartan's theorem) in a way that spoke directly to the close relation between group theory and symmetric dynamical systems. This connection is shown in Fig. 22.

---

[b]http://lagrange.physics.drexel.edu/flash/vanderpol/

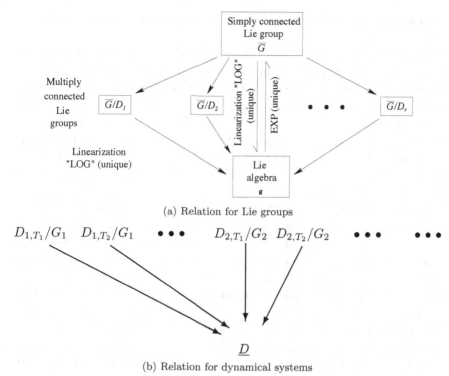

Fig. 22. (a) For Lie groups there is a 1:1 correspondence between Lie algebras and simply connected Lie groups. Each such Lie group is a covering group for *all* Lie groups with the same Lie algebra. These other groups are obtained by "modding out the symmetry". (b) For dynamical systems the relation goes the other way. For each universal image there are many covering dynamical systems. These are distinguished from each other by: the symmetry group G and; the topological index T of the lift. [75]

13. Representation Theory

The first step that needs to be taken for the successful analysis of data taken from a dynamical system behaving chaotically is that a suitable visualization of the system should be created. If the dynamical system is a set of three ordinary differential equations (viz.: Lorenz or Rössler equations) this is a no-brainer. One only has to watch how the trajectory evolves in the natural three dimensional phase space of coordinates: $(x(t), y(t), z(t)) \in \mathbb{R}^3$. If the data are generated in the course of an experiment the situation becomes more exciting. Often the data consists of a single scalar time series $x_i = x(t_i)$. One then hopes to construct a D-dimensional phase space and visualize the trajectory in that phase space. One must construct an *embedding* of the data. Fortunately, this is always possible (in principal) if the original system is finite dimensional ($d < \infty$). This is due to a theorem by Takens[78] that was exploited by Packard, Crutchfield, Farmer and Shaw [79] and that goes back to Whitney.[80] The theorem states that if the data are generated by a dynamical system of dimension d an embedding can always be found (is "generic") in $\mathbb{R}^D$ for $D \geq 2d + 1$.

Many embeddings of a scalar time series are possible. The first choice of many is the time delay embedding. [78,79] This is useful because the signal-to-noise ratio in each component of the embedding is the same. It is less useful because it is not always easy to determine the signs of crossings in two-dimensional projections of time delay embeddings into $\mathbb{R}^3$. My preferred embedding is the differential embedding. In this case $x \to \mathbf{y} = (y_1, y_2, y_3)$, with $y_1 = x, y_2 = \dot{x}, y_3 = \ddot{x}$. It is useful because the signs of crossings in projections to the (y_1, y_2) plane are very simple to determine.[41] It is not useful because the S/N ratio decreases by an order of magnitude (or more) for each higher component of the embedding.

Many inequivalent embeddings of an experimental scalar time series are possible. We emphasize this point by introducing an interesting class of embeddings for the simplest type of dynamical system: a chaotic attractor with a hole in the middle (genus-one strange attractor). Assume we have the scalar time series $x(t)$ and from it we construct the projection into $\mathbb{R}^2$: $(x(t), y(t) = \dot{x}(t))$. This cannot be an embedding but it could have a hole in the middle. Assume it not only has a hole in the middle, but that a straight line segment attached to a fixed point somewhere inside the hole has the property that the projected flow $(x(t), y(t))$ always strikes the segment from the same side as it is rotated through 2π radians around the fixed point inside the hole. It is then possible to reparameterize the projected trajectory in terms of a rotation angle θ in place of time t: $(x(t), y(t)) \to (x(\theta), y(\theta))$.

Now introduce the three coordinates (ξ, η, ζ) for a *harmonic knot*:[81]

$$\mathbf{X}(\theta) = (\xi(\theta), \eta(\theta), \zeta(\theta)) \quad \text{where} \quad \begin{array}{l} \xi(\theta) = \sum_{j=1} A_j \sin(j\theta + \phi_j) \\ \eta(\theta) = \sum_{j=1} B_j \sin(j\theta + \psi_j) \\ \zeta(\theta) = \sum_{j=1} C_j \sin(j\theta + \chi_j) \end{array} \quad (7)$$

Harmonic knots have periodicity 2π under the parameterization given. Any knot can be given a harmonic parameterization.

Introduce a *repère mobile* for the harmonic knot by constructing the unit tangent, normal, and binormal vectors $(\mathbf{t}(\theta), \mathbf{n}(\theta), \mathbf{b}(\theta))$.[82] Under suitable not very restrictive assumptions the data from a chaotic attractor can be embedding using any harmonic knot as a "carrier knot" by

$$x(t) \to x(\theta) \to \mathbf{X}(\theta) + x(\theta)\mathbf{n}(\theta) + x'(\theta)\mathbf{b}(\theta) \quad (8)$$

with $' = d/d\theta$. For each knot type there is an embedding of data generated by genus-one dynamics. The embeddings are topologically inequivalent (nonisotopic) because inequivalent knots, the cores of the tori that "carry" the embedding, are not isotopic.

We now come back to the jackpot question that was raised when the topological analysis program was being formulated:[38]

When you analyze embedded data, what do you learn about the embedding and what do you learn about the dynamical system?

This question struck a chord with me. In Quantum Physics, groups act through their representations. One group can have many inequivalent representations (equivalence is with respect to a change of basis, or similarity transformation). The corresponding question would be: How much can you learn about a group from some/all of its representations.

It seemed reasonable to think that a representation theory of dynamical systems (or their strange attractors) ought to exist[4] which is spiritually similar to the representation theory of groups/algebras that has found so much use in physics. I discussed this problem with Daniel Cross, my graduate student at the time. And presto! After two years of hard work we had a representation theory for strange attractors - at least their low-dimensional varieties.[83–86]

This theory starts from the natural question: What are the labels for inequivalent representations of a (low-dimensional) strange attractor. By representation labels we mean labels that identify distinct, nonisotopic embeddings of a low-dimensional strange attractor into $\mathbb{R}^3$. This isn't exactly an easy question (most questions, asked for the first time, aren't easy). So we began with genus-one attractors as a warm-up exercise. In this case we were able to show that there are only three indices. One is parity. The mirror image of a strange attractor in $\mathbb{R}^3$ is diffeomorphic with the original but not isotopic to the original. A second index is knot type. This has been described above in the context of harmonic knots. The third index is global torsion, known very early on from the initial work with Solari.[33] These three indices serve to distinguish all topologically inequivalent (nonisotopic) embeddings of a genus-one dynamics into $\mathbb{R}^3$.[69]

Now we ask the question: If we raise the embedding dimension by one, mapping $\mathbb{R}^3 \to \mathbb{R}^4$, do some of the formerly nonisotopic embeddings into $\mathbb{R}^3$ become equivalent? After all, the representation labels are in some natural sense *obstructions* to isotopy in $\mathbb{R}^3$. The answer is: Yes, embeddings with different knot types in $\mathbb{R}^3$ all become equivalent in $\mathbb{R}^4$. Embeddings that are reflected images of each other also become isotopic in $\mathbb{R}^4$. Only the global torsion remains an obstruction to isotopy — barely. Embeddings that differ only by a global torsion of 2 become isotopic in $\mathbb{R}^4$. This means that there are only two inequivalent embeddings into $\mathbb{R}^4$: Those with even global torsion and those with odd global torsion. In $\mathbb{R}^5$ every embedding is equivalent. The following table[83,86] shows the progressive extinction of obstructions to isotopy:

Index	$\mathbb{R}^3$	$\mathbb{R}^4$	$\mathbb{R}^5$
Parity	Z_2	–	–
Knot Type	K	–	–
Global Torsion	n	Z_2	–

(9)

In this table K is an index describing knot type (we still haven't a complete handle on this index), Z_2 for parity is ± 1, and Z_2 for global torsion is $(0, 1)$ or n mod 2.

A similar result holds for embeddings of genus-g dynamical systems into $\mathbb{R}^3$. Three indices are required to distinguish among inequivalent embeddings.[70] Raise the embedding dimension by one and only one index remains to describe the residual obstructions to isotopy. In $\mathbb{R}^5$ all embeddings become equivalent.[83,86]

To answer the jackpot question posed above: Anything learned from analyzing the embedding of a "low-dimensional" dynamical system into $\mathbb{R}^5$ is uniquely about the dynamical system, since all embeddings are equivalent in this space.

14. Pointers to the Future

Topological tools have greatly expanded our understanding of low (=3) -dimensional strange attractors. These attractors are on the borderline of our visual comprehension. They live in three-space but are visualized in 2-space: On the screens of computers, for example. A powerful theorem by Birman and Williams allows us to project attractors down to mostly two-dimensional subspaces. Perhaps more can be said about three-dimensional attractors, but I think the most important things have now been said (I hope I'm wrong!). The one remaining piece of information that would be useful has one foot in $\mathbb{R}^3$ and the other in $\mathbb{R}^5$. In $\mathbb{R}^5$ there must be some topological signature that exists that can distinguish inequivalent attractors (as opposed to inequivalent embeddings of a single attractor). What is this index and how does it work? This index identifies *mechanism* (c.f. Sec. 9 and Figs. 11 and 12).

A piece of it identifies genus; another piece identifies components of the global Poincaré surface of section; yet another identifies the transition matrix that describes the flow; and yet another describes the stretching-squeezing process that builds up the strange attractor by infinite repetition of the S & S processes. This topological index is the final piece we need for a complete accounting of three-dimensional strange attractors.

Beyond three there is four, and then five, $\cdots$. All my attempts in the last 15 years to extend the topological analysis program to higher dimensions have foundered on one detail. The detail is that Gauss apparently did not extend his knotting thoughts about closed loops for more than a millesecond to more than three dimensions. The Gauss linking number is at the heart of the topological analysis method. It does not extend beyond $\mathbb{R}^3$. Our analysis methodology therefore does not extend beyond $\mathbb{R}^3$. We are stuck at the starting line!

It almost seems to me that we may be trying too hard for too much. In low dimensions our goal has evolved to one of determining mechanism. Linking numbers have been a tool to this end. Perhaps we should concentrate more on ends and less on means. This means learn how to classify the stretching and squeezing mechanisms that can operator in $\mathbb{R}^D$ ($D > 3$) rather than identifying all the orbits in the attractor, and using them to determine mechanism.

The slides presented at the Birthday Party can be found at.[c]

Acknowledgments

I would like to extend heartfelt thanks to the authors of this celebration for their extreme efforts over a one year period to create this conference in the utmost secrecy. This was quite a surprise party.

References

1. M. J. Feigenbaum, Quantitative universality for a class of nonlinear transformations, *Journal of Statistical Physics*, **19**, 25-52 (1978).
2. M. J. Feigenbaum, The universal metric properties of nonlinear transformations, *Journal of Statistical Physics*, **21**, 669-706 (1979).
3. M. J. Feigenbaum, Universal behavior in nonlinear systems, *Los Alamos Science*, **1**, 4-27 (1980).
4. R. Gilmore & M. Lefranc, *The Topology of Chaos*, NY: Wiley, 2002,
5. L. Tolstoy, *Anna Karenina*, (tr. Constance Garnett), NY: Barnes and Noble (2000).
6. J. Gleick, *Chaos, The Making of a New Science*, Viking, Penguin (1987).
7. M. Giglio, S. Musazzi, & U. Perini, Transition to chaotic behavior via a reproducible sequence of period-doubling bifurcations, *Physical Review*, Lett. **47**, 243-246 (1981).
8. A. Libchaber, C. Laroche, & S. Fauve, Period doubling cascade in Mercury, a quantitative measurement, Le Journal de Physique - Letters **43**, L211-L216 (1982).
9. R. H. Simoyi, A. Wolf, & H. L. Swinney, One-dimensional dynamics in a multicomponent chemical reaction, *Physical Review*, Lett. **49**, 245-248 (1982).
10. J. C. Roux, R. H. Simoyi, & H. L. Swinney, Observation of a strange attractor, Physica **D8**, 257-266 (1983).
11. K. Coffman, W. D. McCormick, Z. Noszticzius, R. H. Simoyi, & H. L. Swinney, Universality, multiplicity and the effect of iron impurities in the Belousov-Zhabotinskii reaction, J. Chem. Phys. **86**, 119-129 (1987).
12. J. Testa, J. Perez, & C. Jeffries, Evidence for universal chaotic behavior of a driven nonlinear oscillator, *Physical Review*, Lett. **48**, 1217-1220 (1982).
13. F. T. Arecchi, R. M. Meucci, G. Puccioni, & J. R. Tredicce, Experimental evidence of subharmonic bifurcations, multistability, and turbulence in a Q-switched gas laser, *Physical Review*, Lett. **49**, 245-248 (1982).
14. R. S. Gioggia & N. B. Abraham, Routes to chaotic output from a single mode, dc-excited laser, *Physical Review*, Lett. **51**, 650-653 (1983).
15. G. P. Puccioni, A. Poggi, W. Gadomski, J. R. Tredicce, & F. T. Arecchi, Measurement of the formation and evolution of a strange attractor in a laser, *Physical Review*, Lett. **55**, 339-342 (1985).
16. J. R. Tredicce, F. T. Arecchi, G. P. Puccioni, A. Poggi, & W. Gadomski, Dynamic behavior and onset of low-dimensional chaos in a modulated homogeneously broadened single-mode laser: Experiments and theory, *Physical Review*, **A34**(3), 2073-2081 (1986).
17. P. Grassberger, On the Hausdorff dimension of fractal attractors, *Journal of Statistical Physics*, **26**(1), 173-179 (1981).

---

[c]http://einstein.drexel.edu/ bob/Presentations/birthday70.pdf

18. P. Grassberger, Generalized dimensions of strange attractors, *Phys. Lett.* **97**(6), 227-230 (1983).
19. P. Grassberger & I. Procaccia, Measuring the strangeness of strange attractors, *Physica* **D9**, 189-208 (1983).
20. P. Grassberger & I. Procaccia, Characterization of strange attractors, *Physical Review, Lett.* **50**(5), 346-349 (1983).
21. J.-P. Eckmann & D. Ruelle, Ergodic theory of chaos and strange attractors, *Rev. Mod. Phys.* 57, 617-656 (1985).
22. A. Wolf, J. B. Swift, H. L. Swinney, & J. A. Vastano, Determining Lyapunov exponents from a time series, *Physica* **D16**, 285-317 (1985).
23. S. Sato, M. Sano, & Y. Sawada, Practical methods of measuring the generalized dimension and largest Lyapunov exponent in high dimensional chaotic systems, *Prog. Theor. Phys.* **77**(1), 1-5 (1987).
24. R. Stoop & J. Parisi, Calculation of Lyapunov exponents avoiding spurious elements, *Physica* **D50**, 89 (1991).
25. M. T. Rosenstein, J. J. Collins, & C. J. De Luca, A practical method for calculating largest Lyapunov exponents from small data sets, *Physica* **D65**, 117-134 (1993).
26. D. Ruelle, *Proc. Roy. Soc. London, Ser. A* **427**, 241 (1990).
27. E. N. Lorenz, Deterministic nonperiodic flow, *Journal of Atmospheric Science*, **20**, 130-147 (1963).
28. O. E. Rössler, An equation for continuous chaos, *Physics Letters A*, **57**(5), 397-398 (1976).
29. H. Poincaré, *Les Methodes nouvelle de la mécanique céleste,*, Paris: Gauthier-Villars (1892).
30. D. Auerbach, P. Cvitanović, J.-P. Eckmann, G. Gunaratne, & I. Procaccia, Exploring chaotic motion through periodic orbits, *Physical Review Letters*, **58**, 2387-2390 (1987).
31. D. J. Lathrop & E. J. Kostelich, Characterization of an experimental strange attractor by periodic orbits, *Physical Review A*, **40**, 4028-4031 (1989).
32. J.-P. Eckmann, S. O. Kamphorst, & D. Ruelle, Recurrence plots of dynamical systems, *Europhysics Letters*, **5**, 973-977 (1987).
33. H. G. Solari & R. Gilmore, Relative rotation rates for driven dynamical systems, *Physical Review A*, **37**, 3096-3109 (1988).
34. N. B. Tufillaro, H. G. Solari, & R. Gilmore, Relative rotation rates: Fingerprints for strange attractors, *Physical Review A*, **41**, 5717-5720 (1990).
35. G. B. Mindlin, X.-J. Hou, H. G. Solari, R. Gilmore, & N. B. Tufillaro, Classification of strange attractors by integers, *Physical Review Letters*, **64**, 2350-2353 (1990).
36. J. S. Birman & R. F. Williams, Knotted periodic orbits in dynamical systems I: Lorenz's equations, *Topology*, **22**(1), 47-82 (1983).
37. J. S. Birman & R. F. Williams, Knotted periodic orbits in dynamical systems II: Knot holders for fibered knots, *Contemporary Mathematics*, **20**, 1-60 (1983).
38. G. B. Mindlin, H. G. Solari, M. A. Natiello, R. Gilmore, & X.-J. Hou, Topological analysis of chaotic time series data from the Belousov-Zhabotinskii reaction, *Journal of Nonlinear Science*, **1**, 147-173 (1991).
39. J. Milnor & W. Thurston, On iterated maps of the interval, *Lecture Notes in Mathematics*, **1342**, 465-563 (1988).
40. M. Lefranc, D. Hennequin, & P. Glorieux, Homoclinic chaos in a laser containing a saturable absorber, *Physics Letters A*, **163**, 239-249 (1992).
41. R. Gilmore, Topological analysis of chaotic dynamical systems, *Reviews of Modern Physics*, **70**, 1455-1530 (1998).

42. A. Amon & M. Lefranc, Topological signature of deterministic chaos in short nonstationary signals from an optical parametric oscillator, *Physical Review Letters*, **92** (9), 094101 (2004).
43. D. Hennequin, F. de Tomasi, B. Zambon, & E. Arimondo, Homoclinic orbits and cycles in the instabilities of a laser with saturable absorber, *Physical Review A*, **37**, 2243-2246 (1988).
44. F. A. Papoff, A. Fioretti, E. Arimondo, G. B. Mindlin, H. G. Solari, & R. Gilmore, Structure of chaos in the laser with saturable absorber, *Physical Review Letters*, **68**, 1128-1131 (1992).
45. T. Shimizu & N. Morioka, On the bifurcation of a symmetric limit cycle to an asymmetric one in a simple model, *Physics Letters A*, **76**, 201-204 (1980).
46. N. B. Tufillaro, T. A. Abbott, & J. P. Reilly, *An Experimental Approach to Nonlinear Dynamics and Chaos*, Reading, MA: Addison-Wesley (1992).
47. G. B. Mindlin, R. Lopez-Ruiz, H. G. Solari, & R. Gilmore, Horseshoe implications, *Physical Review E*, **48**, 4297-4304 (1993).
48. T. Hall, Weak Universality in two-dimensional transitions to chaos, *Physical Review Letters*, **71**, 58-61 (1992).
49. T. Hall & A. de Carvalho, How to prune a horseshoe, *Nonlinearity*, **15** R19-R68 (2002).
50. T. Hall & A. de Carvalho, The forcing relation for horseshoe braid types, *Experimental Mathematics*, **11** 271-288 (2002).
51. M. A. Natiello & H. G. Solari, Remarks on braid theory and the characterisation of periodic orbits, *Journal of Knot Theory and its Ramifications*, **3**(4), 511-529, (1994).
52. M. Lefranc, Alternative determinism principle for topological analysis of chaos, *Physical Review E*, **74**, 035202(R) (2006).
53. P. Collins, Forcing relations for homoclinic orbits of the Smale horseshoe map, *Experimental Mathematics*, **14**(1), 75-86 (2005).
54. J. L. W. McCallum & R. Gilmore, A geometric model of the driven Duffing oscillator, *International Journal of Bifurcation & Chaos*, **3**, 685-691 (1993).
55. R. Gilmore & J. W. L. McCallum, Superstructure in the bifurcation diagram of the Duffing oscillator, *Physical Review E*, **51**, 935-956 (1995).
56. R. Gilmore, C. Letellier & M. Lefranc, Chaos Topology, *Scholarpedia*, **3** (7), 4592 (2007).
57. G. Boulant, M. Lefranc, S. Bielawski & D. Derozier, Horseshoe templates with global torsion in a driven laser, *Physical Review E*, **55** (5), 5082-5091 (1997).
58. R. Gilmore, X. Pei, & F. Moss, Topological analysis of chaos in neural spike train bursts, *Chaos*, **9** (3), 812-817 (1999).
59. R. Gilmore & X. Pei, The topology and organization of unstable periodic orbits in Hodgkin-Huxley models of receptors with subthreshold oscillations, *Handbook of Biological Physics*, **4**, *Neuro-informatics, Neural Modeling*, (F. Moss and S. Gielen, Eds.), Amsterdam: North Holland, pp. 155-203 (2001).
60. J. Used & J. C. Martin, Reverse horseshoe and spiral templates in an Erbium doped fiber laser, *Physical Review E*, **79**, 046213 (2009).
61. J. Used & J. C. Martin, Multiple topological structures of chaotic attractors ruling the emission of a driven laser, *Physical Review E*, **82**, 016218 (2010).
62. G. Byrne, R. Gilmore, & C. Letellier, Distinguishing between folding and tearing mechanisms in strange attractors, *Physical Review E*, **70**, 056214 (2004).
63. C. Letellier, T. D. Tsankov, G. Byrne, & R. Gilmore, Large scale structural reorganization of strange attractors, *Physical Review E*, **72**, 026212 (2005).
64. T. D. Tsankov & R. Gilmore, Strange attractors are classified by bounding tori, *Physical Review Letters*, **91** (13), 134104 (2003).

65. T. D. Tsankov & R. Gilmore, Topological aspects of the structure of chaotic attractors in $\mathbb{R}^3$, *Physical Review E*, **69**, 056206 (2004).
66. J. Katriel & R. Gilmore, Entropy of bounding tori, *Entropy Online*, **12**, 066220 (2010).
67. V. I. Arnold, *Ordinary Differential Equations*, Cambridge, MA: MIT Press (1980).
68. M. A. Aziz-Alaoui, Differential equations with multispiral attractors, *Internation Journal of Bifurcation & Chaos*, **9**, 1009-1039 (1999).
69. N. Romanazzi, M. Lefranc, & R. Gilmore, Embeddings of low-dimensional strange attractors: Topological invariants and degrees of freedom, *Physical Review E*, **75**, 066214 (2007).
70. R. Gilmore, C. Letellier, & N. Romanazzi, Global topology from an embedding, *Journal of Physics A*, **40**, 13291-13297 (2007).
71. R. Gilmore, *Lie Groups, Lie Algebras, and Some of Their Applications*, NY: Wiley (1974).
72. C. Letellier & R. Gilmore, Covering dynamical systems: Two-fold covers, *Physical Review E*, **63**, 016206 (2000).
73. C. Letellier & R. Gilmore, Dressed symbolic dynamics, *Physical ReviewE*, **67**, 036205 (2003).
74. C. Letellier, L. A. Aguirre, J. Maquet & R. Gilmore, Evidence for low-dimensional chaos in sunspot cycles, *Astronomy & Astrophysics*, **449**, 379-387 (2006).
75. R. Gilmore & C. Letellier, *The Symmetry of Chaos*, Oxford: University Press (2008).
76. C. Letellier & R. Gilmore, Symmetry groups for 3D dynamical systems, *Journal of Physics*, **A40**, 5597-5620 (2007).
77. R. Gilmore, Two-parameter families of strange attractors, *Chaos*, **17**, 013104 (2007).
78. F. Takens, Detecting strange attractors in turbulence, *Lecture Notes in Mathematics*, **898**, 366-381 (1981).
79. N. H. Packard, J. P. Crutchfield, J. D. Farmer, & R. S. Shaw, Geometry from a time series, *Physical Review Letters*, **45**, 712-716 (1980).
80. H. Whitney, Differentiable manifolds, Annals of Mathematics **37**(3), 645-680 (1936).
81. http://www.math.uiowa.edu/ jsimon/aaron.html
82. D. J. Struik, *Lectures on Classical Differential Geometry*, Reading, MA: Addison-Wesley, 1950.
83. D. J. Cross & R. Gilmore, Representation theory for strange attractors, *Physical Review E*, **80**(1), 056207 (2009).
84. D. J. Cross & R. Gilmore, Differential embedding of the Lorenz attractor, Physical Review **E81**, 066220 (2010).
85. D. J. Cross & R. Gilmore, Equivariant differential embeddings, Journal of Mathematical Physics **51**, 092706 (2010).
86. D. J. Cross & R. Gilmore, Complete set of representations for dissipative chaotic three-dimensional dynamical systems, Physical Review **E82**, 056211 (2010).

Chapter 9

Reflections from the fourth dimension

Marc Lefranc

Laboratoire de Physique des Lasers, Atomes, Molécules, UFR de Physique, Université Lille 1 Sciences et Technologies, 59655 Villeneuve d'Ascq cedex, France

The knot-theoretic characterization of three-dimensional strange attractors has proved an invaluable tool for comparing models to experiments, understanding the structure of bifurcation diagrams, constructing symbolic encodings or obtaining signatures of chaos.[1–4] In four dimensions and above, however, all closed curves can be deformed into each other without crossing themselves. Therefore, the fundamental idea of topological analysis, namely that the topological structure of a strange attractor provides signatures of the stretching and folding mechanisms which organize it, must be translated into a different formalism. Here, we discuss two modest attempts to make progress in this direction. First, we illustrate the relevance of catastrophe theory in the analysis of higher-dimensional systems by describing experimental signatures of cusps in weakly coupled chaotic systems.[5] Second, we note that determinism not only precludes intersection of two trajectories but also orientation reversal of phase space volume elements. Enforcing this principle on dynamical triangulations of periodic points advected by the flow leads to higher-dimensional analogues of braids, which in three dimensions appear to provide the same information as usual approaches.[6]

Contents

1. Introduction 206
   1.1. Signatures of determinism in chaotic systems 206
   1.2. Topological analysis of three-dimensional systems 207
   1.3. Towards a topological analysis in higher dimensions 209
2. Cusps in Weakly Coupled Chaotic Systems 210
   2.1. Coupled logistic maps 211
   2.2. Coupled diode resonators 213
3. Characterizing Periodic Orbits with Dynamical Triangulations 214
   3.1. Orientation preservation: a topological Liouville's theorem ... 215
   3.2. Representing surfaces by dynamical triangulations 216
   3.3. Enforcing orientation preservation on triangulations 217
   3.4. Induced dynamics in the itinerary space 218
   3.5. Path growth and entropy 219
   3.6. Combinatorial representation of the invariant manifold 221
   3.7. Triangulations in higher dimensions 223
4. Conclusion 224
References 226

1. Introduction

1.1. *Signatures of determinism in chaotic systems*

The fact that dynamical systems obeying deterministic equations of motion can evolve irregularly in time, as is illustrated by the turbulent motion of a flow, has been since long a fascinating subject of investigation and remains so today. When one examines a time series delivered by a low-dimensional chaotic dynamical system, such as the output intensity of a modulated CO_2 laser (Fig. 1), the first impression is that it "does not every time quite change, nor yet quite dwell".[7] The time series never repeats itself exactly, but comparing different parts of the signal inevitably creates a strong feeling of "déjà vu".

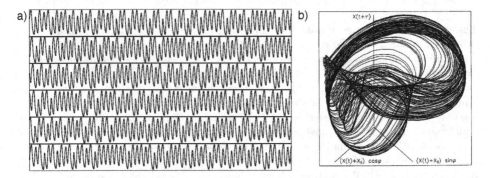

Fig. 1. (a) Output intensity of CO_2 laser with modulated losses as a function of time. (b) The trajectory associated with this signal in a reconstructed state space with cylindrical coordinates $\{X(t), \varphi, X(t+\tau)\}$ where τ is a time delay and φ is the modulation phase.

A geometric representation of the dynamics allows us to disentangle what changes and what does not change. At each time, the state of the system can be described by a set $\{X(t), X(t+\tau), X(t+2\tau), \ldots\}$ of time-delayed values of the time series. The assumption is that a given sequence, if the time delay τ and the number of coordinates are suitably chosen, can be generated by one and only one dynamical state. If we plot the locations visited by the representative point of the system in this reconstructed state space (Fig. 1(b)), it appears clearly that the trajectory of the system does not evolve at random but is confined to a highly organized geometric object, a "strange attractor".[8]

As Fig. 2 illustrates, what varies continuously is the state of system. What does not change is the set of equations governing the dynamics. Each time the system returns close to a point of phase space previously visited, it behaves almost as during the last visit, because location specifies state and state determines the future. Consequently, trajectories going through neighboring states are essentially parallel (Fig. 2). This is a signature of the uniqueness theorem, which states that there is only one solution of a system of differential equations for each set of initial conditions. This precludes that two trajectories intersect, except if the velocity vector field vanishes at that point. This simple but fundamental property is precisely what gives pictures of strange attractors their unmistakable appearance.

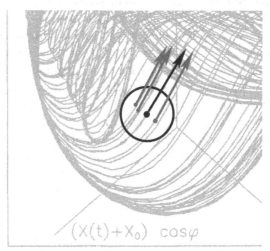

Fig. 2. If the state space is correctly reconstructed, a point and its neighbors have almost parallel trajectories, because they have close short-term futures.

Amazingly, the uniqueness theorem, and more generally the deterministic nature of dynamics, constrain the structure of strange attractors so deeply that they can be harnessed to design robust and powerful methods for characterizing chaotic signals. To this end, a crucial ingredient is the presence in any chaotic attractor of infinitely many unstable periodic orbits. These periodic orbits are intimately linked to recurrent behavior. Since a system can return arbitrarily close to any point in the attractor, and since these recurrences can be deformed to an arbitrarily close closed orbit, it results that these periodic orbits densely fill the attractor and have arbitrarily large periods.[8] Being unstable, their neighborhood is transiently visited by the system at irregular time intervals, which translates into bursts of almost periodic behavior in the time series. This allows one to extract low-period orbits from chaotic signals by searching systematically for such close-return events. Periodic orbits are precious because they can be fully characterized in a finite time.

1.2. *Topological analysis of three-dimensional systems*

In dynamical systems with three state variables, unstable periodic orbits are associated with closed curves in a three-dimensional state space. Because these curves cannot intersect without violating the uniqueness theorem, their intertwining can be characterized using knot theory. Indeed, the latter provides us with topological characterizations of closed curves which do not vary when curves are deformed continuously without inducing crossings. In particular, the knot type of a periodic orbit remains the same on its whole domain of existence, its topological invariants providing genuine fingerprints of the orbit. Similarly, how many turns one periodic orbit winds around another is a dynamical invariant.[1]

Computing knot and link invariants for all orbits detected in a chaotic system would only result in large tables of numbers if there was not a global organization in the linking of periodic orbits. This organization results from the action of the stretching and squeezing mechanisms which generate chaotic behavior, taking apart neighboring trajectories exponentially fast while maintaining all trajectories within a bounded region of phase space. Stretching and squeezing act repeatedly on periodic orbits as they do on any trajectory and braid them in a systematic way. This is reflected in the existence for any attractor of a branched manifold, named a template (or knot-holder), such that any finite link of periodic orbits can be deformed to the surface of the manifold without modifying its topological invariants, as shown by Birman and Williams[9] (Fig. 3). The template describes concisely how stretching and squeezing shape the global topological organization of the strange attractor.

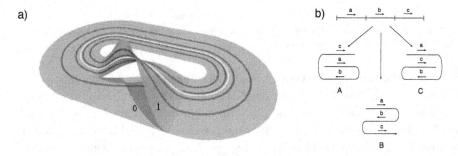

Fig. 3. (a) The Smale's horsehoe template, carrying a period-1 orbit and a period-4 orbit. Their linking number is 2, as in the original attractor from which they have been projected. (b) The template indicates in which way the unstable manifold folds onto itself. After Gilmore and Lefranc.[1]

The key idea in the Birman and Williams proof is to identify points of the original phase space which have identical futures, their orbits converging to each other for time going to positive infinity. Thus, the template may be seen as resulting from squeezing the original phase space along the local stable direction. This operation does not modify topological invariants of periodic orbits because their stable manifolds cannot intersect (no trajectory converges simultaneously to two different periodic orbits) : the squeezing therefore amounts to an isotopy.

After squeezing a three-dimensional flow along the stable direction, only the flow (neutral) and unstable direction survive, so that the template is two-dimensional. The flow direction is the one followed by the semi-flow as one rotates around the central hole, while the unstable direction is transverse to it. Fig. 3(a) shows how the two branches are stretched in the transverse (unstable) direction before joining together on the branch line. The order in which branches stack together provides an important piece of information (Fig. 3(b)).

Thus, the template can be viewed as specifying how the intersection of the unstable manifold with a Poincaré section evolves as one sweeps the Poincaré section around the attractor, before being eventually folded onto itself (Figs. 3(b) and 4(a)). The template and the semi-flow carried by it describe the action of the flow restricted to the unstable manifold. When mapping a one-dimensional object into itself, only the simplest type of singularity, namely fold singularities, can occur. As we shall see below, this interpretation of a template is precious in that it is does not depend on how we characterize periodic orbits, with knot theory or something else, and can therefore be generalized to higher dimensions.

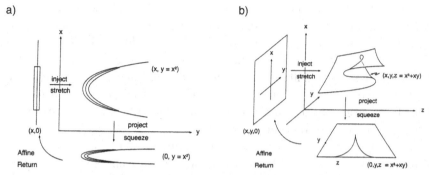

Fig. 4. (a) As one follows the semi-flow on a template (Fig. 3(a)), the section of the template, representing the unstable manifold, evolves in the plane before being folded onto itself. (b) In a four-dimensional system with two unstable directions, the intersection of the unstable manifold with a Poincaré section is a two-dimensional surface evolving in a three-dimensional space before being mapped onto itself, generating cusp singularities in the process. After Gilmore and Lefranc.[1]

1.3. Towards a topological analysis in higher dimensions

Unfortunately, part of what has been discussed in the previous section breaks down in higher dimensions. For one thing, two closed curves can be always be deformed into each other without inducing crossings. In principle, one could imagine to dress periodic orbits with their stable manifolds and to study how dressed periodic orbits are organized relative to each other. Indeed, k-dimensional spheres (periodic orbits+invariant manifolds) embedded in $k+2$-dimensional spheres (ambient space) can have distinct topological configurations.[10] More generally, knot theory can be viewed as the study of codimension 2 embeddings.

As a first step towards topological characterization of chaos in higher-dimensional systems, Mindlin and Solari studied how invariant manifolds rotate as one proceeds along a periodic orbit.[11] In doing so, they found beautiful signatures of tori and Klein bottles. This shows that the distinction made in three-dimensional systems between saddle and flip orbits, whose invariant manifolds rotate by integer and half-integer numbers of turns, respectively, still holds in higher dimensions.

However, extracting periodic orbits from experimental times series is already difficult in higher-dimensional systems, identifying their invariant manifolds is a formidable task, so that a method to characterize periodic orbits using only their trajectory in phase space would be highly desirable. In section 3, we will discuss an attempt to design such a method.

To analyze the topological structure of higher-dimensional systems, we not only need analogues of knots but also analogues of templates. The latter are probably easier to imagine, based on our three-dimensional experience. For the sake of simplicity, let us restrict to the case where a four-dimensional strange attractor has almost everywhere two unstable directions and one strong stable direction. Then, the intersection of the unstable manifold with a Poincaré section is two-dimensional. As the Poincaré section is swept around the attractor, this two-dimensional object evolves in a three-dimensional space before being eventually mapped into itself (Fig. 4(b)). There are now two types of singularities which occur generically in the mapping from the unstable manifold onto itself : folds, which occur along lines, and cusps, isolated points which are found at the joining of two fold lines. There is a hierarchical classification of singularities in the sense that cusps are singularities of fold lines, which indeed meet tangentially at a cusp.

Just as templates for three-dimensional flows can be viewed as specifying a collection of folds and how they interact, templates for higher-dimensional flows should describe the relative organization of singularities specific to a given chaotic system. This is illustrated in Fig. 5 where different ways in which a surface is mapped to itself are shown, resulting in different crystals of folds and cusps. The higher the dimension of the unstable manifold, the more complicated the singularity crystal, with higher and higher order singularities appearing, such as swallowtails, butterflies or umbilics.[12] The presence of such singularities and of the tangencies between invariant manifolds associated to these singularities implies that high-dimensional systems are less and less hyperbolic as their dimension increases. In particular, they are most likely responsible for the phenomenon of unstable dimension variability which has been described in the literature.[13]

2. Cusps in Weakly Coupled Chaotic Systems

From the above discussion, cusp singularities should generically underlie the dynamics of hyperchaotic systems. However, such singularities may be be difficult to detect, because they only exist in an abstract description of the dynamics, after the stable directions had been squeezed. Alternatively, they may be seen as singularities of the infinitely iterated Poincaré return map, an extremely complicated object.

In order to provide simple evidence of the relevance of cusps in hyperchaotic systems and of their genericity, we consider here systems obtained by coupling two strongly dissipative systems with one unstable direction each, weakly enough that synchronization does not occur.[5] The two unstable directions then survive and

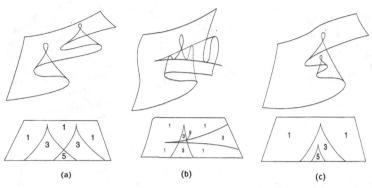

Fig. 5. Different ways of folding a surface into itself are shown, resulting in different singularity structures. The numbers indicate the number of preimages of points in a given region of the plane. After Gilmore and Lefranc.[1]

generate together a two-dimensional unstable manifold. If the original systems are sufficiently dissipative that their Poincaré maps are close to one-dimensional maps, their locations in their respective Poincaré sections can be specified with a single coordinate. The dynamics of the coupled system is then naturally represented in the plane made of the two subsystem coordinates. Below, we consider both a mathematical system (coupled logistic maps) and an experimental device (coupled diode resonators), which display cusp singularities in the very same way.

2.1. *Coupled logistic maps*

We consider here two logistic maps acting on the x and y variables and coupled to each other by a small term of amplitude ϵ:

$$x_{n+1} = a - x_n^2 - \epsilon y_n = f(x_n, y_n) \quad (1a)$$
$$y_{n+1} = a - y_n^2 - \epsilon x_n = g(x_n, y_n) \quad (1b)$$

Figs. 6a and 6b show iterates of system (1) in the plane, respectively without ($\epsilon = 0$) and with coupling ($\epsilon \neq 0$).

The dynamics of the logistic map $x_{n+1} = a - x_n^2$ is very much controlled by the critical point where the map $x \to a - x^2$ is non invertible (i.e., the singular point $x = 0$ where $\partial x_{n+1}/\partial x_n = 0$). In particular, a partition of the interval in subintervals located left and right of the critical point provides a faithful symbolic dynamical coding of the dynamics.[1,8] As for the mapping (1), it loses invertibility at points where its Jacobian

$$d = \det J = \det \begin{bmatrix} \dfrac{\partial x_{n+1}}{\partial x_n} & \dfrac{\partial x_{n+1}}{\partial y_n} \\ \dfrac{\partial y_{n+1}}{\partial x_n} & \dfrac{\partial y_{n+1}}{\partial y_n} \end{bmatrix} = \det \begin{bmatrix} -2x_n & -\epsilon \\ -\epsilon & -2y_n \end{bmatrix} = 4x_n y_n - \epsilon^2 \quad (2)$$

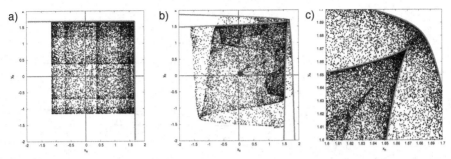

Fig. 6. Iterations of (1) for $a = 1.68$ and (a) $\epsilon = 0$, (b),(c) $\epsilon = 0.1$. Figure (c) shows a closer view of the image of the cusp located in (b). The thin lines in the middle are the singularity lines, the thick lines on the top and right are their images. The cusp singularity is located at the dot indicated on the upper fold line in (b), its image is the turning point clearly visible in (b) and (c). After Jérémy Oden et al.[5]

becomes zero. The set of singular points where $\det J = 0$ is a called a fold line, and separates points which have the same image under the mapping (1). The structure of fold lines is quite different depending on whether the logistic maps are coupled or not. When $\epsilon = 0$, folds are located along the lines $x = 0$ and $y = 0$ intersecting at the origin. These two lines, which correspond to singularities of the individual logistic maps, divide the plane into four quadrants (Fig. 6(a)). As soon as the coupling constant ϵ becomes non zero, the singular set consists of two curves of equation $y = \epsilon^2/4x$, which no longer intersect and thus divide the plane in three regions (Fig. 6(b)).

This abrupt change stems from the appearance at $\epsilon \neq 0$ of a higher-order singularity, a cusp. The manifestation of this singularity is the turning point appearing on the image of the fold line, clearly visible in Fig. 6(c). This indicates that the restriction of (1) to the upper fold line is itself singular, this singularity inside a singularity being the signature of a cusp.

More generally, the singularities of a differential map $f : M \to N$ are organized in a hierarchical way.[12] There is typically a set of points of M where the differential df has not maximal rank. The restriction of f to this singular set can itself display higher-order singularities. By iteratively considering restrictions to singular sets of restrictions, one finally arrives at point singularities, which are the highest-order singularities of the system. Considering a mapping of the plane into itself, its differential has rank one (Jacobian is zero) on fold lines. Images of fold lines can further be folded over themselves, corresponding to isolated singularities, cusps. A cusp is generic and structurally stable: it cannot be removed by perturbing slightly the two-dimensional mapping, exactly like a critical point of a mapping of the interval into itself is robust to perturbations.

The cusp location is determined by requiring that the differential of (1) is zero for a displacement along the fold line $y = \epsilon^2/4x$. Thus, the tangent vector $t =$

$(x, -\epsilon^2/4x)$ must be in the kernel of the Jacobian (2), evaluated on the fold line:

$$J.t = \begin{bmatrix} -2x & -\epsilon \\ -\epsilon & -\epsilon^2/2x \end{bmatrix} \cdot \begin{bmatrix} x \\ -\epsilon^2/4x \end{bmatrix} = \begin{bmatrix} 0 \\ 0 \end{bmatrix} \quad (3)$$

The solution of this equation is $x = \epsilon/2$, which indicates a cusp in $(\epsilon/2, \epsilon/2)$, with its image in $(a - 3\epsilon^2/4, a - 3\epsilon^2/4)$. We can therefore conclude that the cusp singularity is generic and present whenever $\epsilon \neq 0$, thus for arbitrarily weak couplings. This supports the idea presented above that the dynamics of any hyperchaotic system with two unstable directions is governed by underlying cusp singularities associated with the dynamics in the unstable space.

2.2. Coupled diode resonators

To show that the conclusions obtained for a mathematical model persist in an experimental system, Oden, Bielawski and Lefranc[5] have studied a hyperchaotic system consisting of two capacitively coupled diode resonators[14](Fig. 7). This simple device displays a rich dynamics.[15] A common driving sinusoidal signal is simultaneously applied to the two resonators. Depending on coupling strength, one can observe uncorrelated outputs (no coupling), completely synchronized outputs (strong coupling), or partially correlated outputs, corresponding to a hyperchaotic regime, which is obviously the case of interest here. Each diode resonator in Fig. 7 is a strongly dissipative system, whose dynamics is well approached by a one-dimensional first return map (Fig. 7(b)).

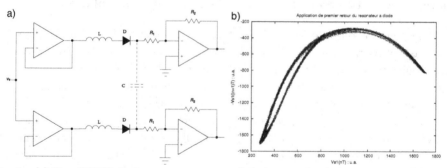

Fig. 7. (a) Two 1N4007 diode resonators are coupled via a small capacitive coupling (a fraction of a picoFarad). $R_2 = 1k\Omega$, $R_1 = 100\Omega$, $L = 2.2mH$. Operational amplifiers LF356N only serve to decouple inputs and to convert currents into voltages at outputs. (b) Stroboscopic first return map for an isolated resonator, obtained by sampling output voltages once in a modulation period, and plotting sample value as a function of the previous one. After Oden et al.[5]

Accordingly, this system displays the same singularity structure as the coupled logistic map system, as can be seen in the stroboscopic Poincaré sections of Fig. 8, whose coordinates are the two output voltages sampled once per modulation period. Fig. 8(a) shows that without coupling, the probability density functions of the two voltages are totally independent (images of fold lines are vertical or horizontal).

As soon as the two circuits are coupled, the joint density function is significantly modified and displays a global structure where multiple signatures of cusps are clearly visible (Fig. 8(b)). Just as light in optics can be focused to create caustics, images of cusps and fold lines in Fig. 8(b) clearly stand out against their background due to their higher intensity. Indeed, these singularities concentrate orbits in their neighborhood around their images, creating peaks of high density, just as in a rainbow, allowing easy visualization of these singularities on an oscilloscope screen.

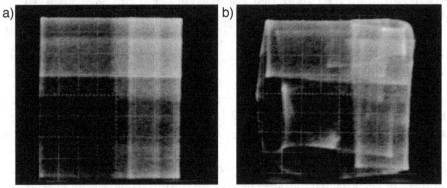

Fig. 8. Stroboscopic Poincaré sections, with the two output voltages sampled once per modulation period as coordinates (a) Uncoupled case; (b) Coupled case. The image of the primary cusp can be seen top right, as well as its image bottom left. After Oden et al.[5]

3. Characterizing Periodic Orbits with Dynamical Triangulations

Classical mechanics is often formally divided into kinematics and dynamics. Kinematics consists of everything needed to describe motion (velocity, acceleration) while dynamics concentrates on what actually causes motion (forces, etc.). In the topological analysis of three-dimensional chaotic systems, there is a similar distinction between knots and links on one side, and templates on the other side. Knots and links are useful to specify the topological organization of periodic orbits, but they do not give us any information other than how periodic orbits are intertwined. Templates, on the other hand, make the connection between the stretching and squeezing mechanisms which organize the strange attractor and knot and link invariants.

As discussed in previous sections, we have some ideas about how to describe higher-dimensional analogues of templates, using tools of singularity theory. However, this does not lead us anywhere unless we can describe the topological organization of higher-dimensional periodic orbits and distinguish topologically inequivalent configurations. As mentioned previously, this can be achieved if we dress periodic orbits with their invariant manifolds.[11] However, reconstructing invariant manifolds from short and noisy experimental data seems difficult. It it thus desirable to construct an approach to characterize periodic orbits using only their trajectory, which can be extracted from chaotic signals using the close-return technique, and which works in any dimension.

In this section, we describe a formalism which is reminiscent of braids and which adapts naturally to phase spaces of any dimension. It is still unknown whether a nontrivial higher-dimensional analysis method can be built on it, however it appears that at least in three dimensions, it is equivalent to the traditional approach.

3.1. *Orientation preservation: a topological Liouville's theorem*

Our starting point is that classical knot theory is not fundamental to template analysis but is simply a convenient tool to characterize how determinism constrains trajectories in three-dimensional spaces. That the knot type of a three-dimensional orbit is well-defined results from the fact that two trajectories cannot intersect. However, avoiding intersections of two one-dimensional closed curves only creates topological obstructions in three dimensions, because then $3-1-1$ is exactly 1. This no longer holds in higher dimensions, and thus another formulation of determinism must be sought.

There is no geometric object whose dimension tracks that of phase space more naturally than a phase space volume element. As time elapses, such a volume element is advected by the flow and can take arbitrarily complicated shapes under the action of stretching and squeezing. Like with a droplet in flow, however, its interior and exterior remain disjoint at all times. Indeed, they are robustly separated by the boundary of the volume element, a closed hypersurface which cannot undergo self-intersections without violating determinism (Fig. 9(a)). A requirement of non-intersection is clearly more constraining for an hypersurface than for two curves.

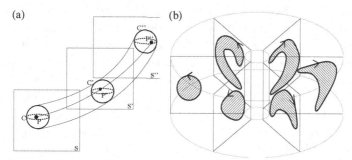

Fig. 9. (a) The interior and exterior of a phase space volume element remain separated as the volume element is advected by the flow. In particular, the image of a point located inside the volume element remains inside. (b) Volume elements advected by a flow maintain their orientation as do their intersections with Poincaré sections. This is illustrated here for two-dimensional sections of a three-dimensional flow. Oriented curves bounding the area elements all rotate in the same sense in the plane regardless of how they are stretched and squeezed. After Lefranc.[6]

Mathematically, the non-intersection of the volume element boundary is equivalent to the preservation of the volume element orientation[a] along trajectories. Indeed, orientation can only change if the volume element goes to zero, which in turn implies that infinitely many trajectories are intersecting themselves. In fact, orientation preservation can be seen as an integral version of the requirement of non-intersection of two curves. In classical mechanics, the Liouville's theorem states that for a Hamiltonian system, the volume element measure is preserved under action of the flow.[16] Requiring that orientation, and only orientation, is preserved may thus be viewed as a topological version of Liouville's theorem, suitable for dissipative systems.

If we restrict ourselves to attractors embedded in $\mathbb{R}^n \times S^1$, and which can thus be sliced in n-dimensional Poincaré sections parameterized by φ, a similar orientation preservation theorem can be stated for n-dimensional volume elements of the Poincaré section (Fig. 9b). This is the point of view we will adopt in the remainder of this chapter, in order to stay close to what we know in three dimensions.

3.2. *Representing surfaces by dynamical triangulations*

Characterizing how volume elements keep their orientation or how surfaces do not self-intersect is not easier than studying the arrangement of invariant manifolds, unless we can do it using only trajectories of periodic orbits, as with knots and braids. If only the motion of periodic points is known, one approach to reconstruct the time evolution of surfaces is to represent the dynamics in a triangulation whose nodes are periodic points. More precisely, consider a set of periodic points P_i in a Poincaré section, belonging to one or several periodic orbits, which are exchanged among themselves by the Poincaré return map. In the simplicial space built on the triangulation, periodic points P_i are 0-cells, segments $\langle P_i, P_j \rangle \equiv \langle ij \rangle$ between two periodic points are 1-cells, triangles $\langle P_i, P_j, P_k \rangle \equiv \langle ijk \rangle$ are 2-cells, etc (Fig. 10a). Similar ideas have been used by Sciamarella and Mindlin[17] to analyze the static structure of a chaotic attractor, but we focus here on the dynamics. When we follow the flow from Poincaré section to Poincaré section, periodic points move in the Poincaré section and the m-cells attached to them change accordingly (Fig. 10b).

We can now construct analogues of m-dimensional surfaces from concatenations of contiguous m-cells, whose set will be denoted by S_m. The dynamics of the periodic points comprising S_0 induces a dynamics in S_m, which will be represented by maps F_m taking concatenations of m-cells to other such concatenations. These maps cannot be simple restrictions of the nonlinear return map F however they should reflect its action on m-dimensional surfaces as closely as possible. Specifically, we require i) that the F_k are invertible, ii) that they satisfy determinism as expressed by orientation preservation and iii) that their action can be obtained

---

[a]If the volume element is defined using local basis vectors, its orientation is related to the sign of the determinant of the basis vectors, and can only change if the determinant goes to zero, indicating degeneracy among the vectors.

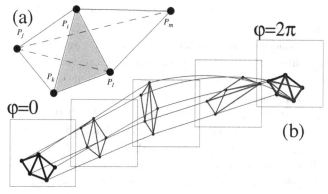

Fig. 10. (a) Triangulation of periodic points in a Poincaré section. The two-cell $\langle ikl \rangle$ is shown in gray. (b) As the Poincaré section is swept across S^1, the triangulation is advected by the flow and evolves in time, as illustrated here for a period-5 orbit. After Lefranc.[6]

from a continuous deformation of k-cells, just as F is a continuous deformation of identity. As we shall see below, enforcing orientation preservation leads necessarily to a dynamics that differs from trivial advection.

3.3. *Enforcing orientation preservation on triangulations*

Let us restrict ourselves to the simple case of a 3D flow. For a triangulated set of periodic points in a two-dimensional Poincaré section, the volume element is a triangle $T = \langle P_i, P_j, P_k \rangle$ based on three periodic points. If this 2-cell were valid at all times and remained firmly attached to the three points, then the induced 2-cell return map would be trivially given by $F_2(\langle P_i, P_j, P_k \rangle) = \langle F(P_i), F(P_j), F(P_k) \rangle$.

However, it typically happens that one of the three points passes between the two others at some time, so that the triangle becomes degenerate. As Fig. 11 illustrates, the orientation of the candidate 2-cell is thereby modified: the curve going from P_i to P_j and then to P_k rotates in opposite directions in the two extreme Poincaré sections, which clearly violates determinism. To work around this problem, we must remember that only node motion is constrained by experimental data, and realize that the reconstructed simplicial dynamics can be adjusted in a very simple way to enforce orientation preservation, as we discuss now.

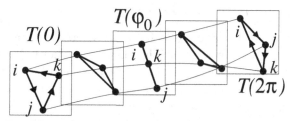

Fig. 11. Triangles based on triplets of periodic points typically become degenerate at some times, changing their orientation. After Lefranc.[6]

At degeneracy, the boundary of T decomposes into two sides, $\langle ij \rangle$ and $\langle jk \rangle + \langle ki \rangle$, which are facing each other with opposite outer normals. Determinism is violated not only because the two sides seem to go through each other, but also because interior and exterior, as defined with respect to outer normals, are exchanged. However, we can prevent all this to happen, and cancel the orientation reversal, simply by swapping the two opposing sides at degeneracy.

This operation is illustrated in Fig. 12 where the two 1-cell chains facing each other at triangle degeneracy are drawn as a solid and a dashed line. The key point is that we have to construct the edge dynamics so that the left (solid line) and right (dashed line) sides do not go through each other and remain at the left and right, respectively. Now, the left and right sides are respectively associated with itineraries $\langle ik \rangle + \langle kj \rangle$ and $\langle ij \rangle$ before degeneracy and with itineraries $\langle ij \rangle$ and $\langle ik \rangle + \langle kj \rangle$ after degeneracy. Thus, everything can be made consistent if we apply the following substitutions when a triangle reversal occurs:

$$\langle ij \rangle \to \langle ik \rangle + \langle kj \rangle \tag{4a}$$

$$\langle ik \rangle + \langle kj \rangle \to \langle ij \rangle \tag{4b}$$

This dynamical rule takes into account the fact that the left and right sides are associated with different periodic points before and after degeneracy (more precisely, the periodic point P_k is transferred from one side to each other). Note that $\partial T = \partial \langle ikj \rangle = (\langle ik \rangle + \langle kj \rangle) + \langle ji \rangle$ is mapped by (4) to $\langle ij \rangle + (\langle jk \rangle + \langle ki \rangle) = \partial \langle ijk \rangle$. Thus, the substitution (4) compensates for triangle inversion: in the end, orientations of T and ∂T are preserved, and so is determinism. Note that the expanding transformation (4a) and the contracting transformation (4b) are discrete analogues of the stretching and squeezing processes, respectively.

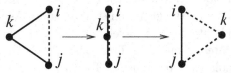

Fig. 12. A triangle is inverted as P_k passes between P_i and P_j. Identifying the solid (resp., dashed) paths in the initial and end configurations leads to substitution (4). After Lefranc.[6]

3.4. Induced dynamics in the itinerary space

Each substitution (4) can be viewed as defining an operator σ_{ij}^k acting on concatenations of edges (i.e., on itineraries on the 2D triangulation). Note that these operators verify $(\sigma_{ij}^k)^2 = 1$ and thus are invertible. By listing all triangle inversions which occur as one rotates along the attractor between two traversals of a Poincaré section, and composing the corresponding substitutions σ_{ij}^k accordingly, one obtains a map F_1 which maps triangulation paths to other triangulation paths. This map should reflect how the original return map acts on smooth one-dimensional

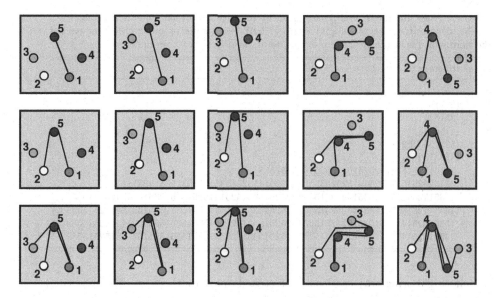

Fig. 13. The motion in section plane of periodic points of orbit 00111 of a suspension of the horseshoe map is displayed for three consecutive periods (one row corresponds to one period). Also shown is the evolution of path $\langle 15 \rangle$ under the dynamics specified by (4). The length of this itinerary grows rapidly with time.

curves in the Poincaré section. Note the similarity of the substitution operators σ_{ij}^{k}, associated with the orientation reversal of a triangle, with the braid generators σ_i, associated with the crossing of two adjacent strands (and thus the apparent orientation reversal of the interval between the two strands).

To illustrate this point, we now consider a suspension of the standard horseshoe map and its periodic orbit whose name is 00111 in the usual symbolic coding (i.e., the orbit passes three times in a row through the folding part). The braid of this orbit is the one shown in Fig. 10(b). As periodic points move from their initial location in the section plane to that of their image under the return map, triangle inversions occur when point 4 successively crosses the four edges $\langle 15 \rangle$, $\langle 13 \rangle$, $\langle 25 \rangle$ and $\langle 23 \rangle$ (Fig. 13). The induced return map is thus given by $F_1 = N \circ \sigma_{23}^{4} \circ \sigma_{25}^{4} \circ \sigma_{13}^{4} \circ \sigma_{15}^{4}$, where N corresponds to a trivial action of the return map (e.g., $N(\langle 12 \rangle) = \langle 23 \rangle$ expressing that $f(P_i) = P_{i+1}$), and the action of the σ_{ij}^{k} given by (4). The action of F_1 on its invariant set is completely specified by the following rules:

$$\langle 14 \rangle \to \langle 25 \rangle, \quad \langle 15 \rangle \to \langle 25 \rangle + \langle 51 \rangle, \quad \langle 25 \rangle \to \langle 35 \rangle + \langle 51 \rangle, \quad \langle 35 \rangle \to \langle 41 \rangle \quad (5)$$

3.5. Path growth and entropy

In parallel with Fig. 13, which shows how an itinerary starting from $\langle 15 \rangle$ grows with time in the triangulation, Table 1 displays some iterates $F_1^m(\langle 15 \rangle)$ computed using (5). Their length is seen to diverge exponentially as $m \to \infty$, which reflects the

Table 1. A few iterates $F_1^m(\langle 15 \rangle)$ are given by their itineraries along periodic points [e.g., (35152) denotes the path $\langle 35 \rangle + \langle 51 \rangle + \langle 15 \rangle + \langle 52 \rangle$]. For iterates of lowest order, the part of the word which is identical to the previous iterate in reverse has been underlined.

m	Itinerary of $F_1^m(\langle 15 \rangle)$
0	(15)
1	(2<u>51</u>)
2	(35<u>152</u>)
3	(415<u>25153</u>)
4	(5251<u>535152514</u>)
5	(15351525<u>14152515351525</u>)
6	(2514152515351525<u>251535152514152515351</u>)
7	(35152525153515251415251535<u>1535152514152515351525251535152514152</u>)
10	(15351525141525153515252515351525251535152514152515351535152514152515351535152514152515351525141525 1535...)
15	(15351525141525153515252515351525251535152514152515351535152514152515351535152514152515351525141525 1535...)
20	(15351525141525153515252515351525251535152514152515351535152514152515351535152514152515351525141525 1535...)
100	(15351525141525153515252515351525251535152514152515351535152514152515351535152514152515351525141525 1535...)

fact that neighboring trajectories are stretched apart by the flow. The asymptotic growth rate:

$$h(P) = \lim_{m \to \infty} \frac{\ln |F_1^m(\langle 15 \rangle)|}{m} \quad (6)$$

is expected to be the entropy $h_T(P)$ of orbit P, defined by Philip Boyland as the minimal topological entropy of a map containing an orbit with the same braid type as P.[18] The growth rate can be directly obtained as the logarithm of the leading eigenvalue of matrix $(M_{ee'})$, whose entries count how many times edge e' or its reverse appears in $F_1(e)$ given by (5). In the present case, we find $h(00111) \sim 0.5435$. For most periodic orbits, however, the computation of the growth rate is more difficult because some edges in the F_1-invariant set trigger a "squeezing" rule (4b), in which case an adjacency matrix $(M_{ee'})$ cannot be directly defined.

Let us for example consider the case of the horseshoe orbit 0010111. In this case, the dynamics in the F_1 invariant set involves the squeezing substitution $\langle 16 \rangle \langle 67 \rangle \to \langle 17 \rangle$. Then $F_1(\langle 16 \rangle + \langle 67 \rangle) \neq F_1(\langle 16 \rangle) + F_1(\langle 67 \rangle)$, which precludes using a transition matrix for entropy computations, although estimates can still be obtained by direct iteration. However, it appears that by considering contracting paths such as $\langle 16 \rangle + \langle 67 \rangle$ as new indecomposable basis "edges", and applying other recodings then required for consistency, one can always rewrite F_1 as an ordinary substitution similar to (5) so that the topological entropy of the orbit can be computed from the associated transition matrix. For example, the induced return map for horseshoe orbit 0010111 can be rewritten as (noting $e_{ij} \equiv \langle ij \rangle$ and $e_{ijk} \equiv \langle ij \rangle + \langle jk \rangle$):

$$e_{14} \to e_{25}, \; e_{15} \to e_{257} \, e_{76}, \; e_{17} \to e_{257} \, e_{71}, \; e_{25} \to e_{37} \, e_{76},$$
$$e_{37} \to e_{41}, \; e_{67} \to e_{71}, e_{167} \to e_{25} \, e_{51}, \; e_{257} \to e_{37} \, e_{761}$$

Besides e_{167}, a new basis path e_{257} has to be introduced so that its image covers e_{167}. A transition matrix can then be obtained, with entropy $h(0010111) \sim 0.4768$.

The conjecture that the growth rate (6) is equal to the topological entropy given by the train-track algorithm[18–20] [b], was tested numerically for all 746 periodic orbits of the horseshoe map up to period 12.[6] Agreement to machine precision was obtained in each case. This result is important because the topological entropy of a braid can be used as a powerful indicator of chaos: finding a single positive-entropy orbit in a system is sufficient to show that it is chaotic, at least in some regions of parameter space. In particular, this has been used to provide evidence of chaos in a nonstationary optical parametric oscillator.[21] It would be extremely useful to be able to do the same for higher-dimensional systems, because time series in this case are generally too short for conventional characterization techniques, so that it is very difficult to distinguish chaos from noise.

3.6. *Combinatorial representation of the invariant manifold*

A closer examination of Table 1 reveals that the dynamics constructed in Secs. 3.3 and 3.4 shadows that of the original flow much more closely than just by predicting correct values for entropies. Indeed, it can be seen that $F_1^{5k}(\langle 15 \rangle)$ converges to an infinitely long path p_∞ satisfying $F_1^5(p_\infty) = p_\infty$. As we shall see below, the invariant path p_∞ is in fact the exact analog of the infinitely folded unstable manifold of the periodic orbit.

In Table 1, the itineraries of order 1 to 7 have been divided in two parts, with the trailing one underlined. Quite amazingly, this trailing part is a full copy of the previous iterate, written in reverse! Furthermore, starting with the fourth iterate, the leading (non-underlined) part can be obtained from the previous iterate just by removing a few symbols at the head. Since each iterate is in principle obtained from the previous one by applying a complex set of substitution rules (4), it is highly remarkable that there is such a simple relation between successive iterates.

For example, consider $p_3 = F^3(\langle 15 \rangle) = (41525153)$. It is the shortest subpath of p_∞ that visits all four edges in the invariant set of (5), shown in bold in Fig. 14(a). As mentioned above, the image $F_1(p_3) = (5251535152514) = (525153) + (35152514)$ consists of a subpath of p_3, obtained by removing from p_3 its two leading symbols, concatenated with a full reverse copy of p_3. As Fig. 14(b) shows, this property can be given a very clear geometric interpretation if we consider the path p_3 as an embedding of an interval of the real line in the plane, even though it is convoluted and visits periodic points several times. The dashed line in Fig. 14(a) shows a possible embedding of p_3. Indeed, it is seen that *the path p_3 is folded onto itself by a singular one-dimensional map*. This results from the fact that $F_1(p_3)$ smoothly connects direct or reversed subpaths of p_3, which are naturally embedded in p_3.

But this is not the only surprise. First, the one-dimensional map previously identified has by construction a period-5 orbit whose periodic points are the P_i, and which is shown in Fig. 14b together with the graph of the map. Now, there is

---

[b]T. Hall, TRAINS, Software available from
http://www.liv.ac.uk/maths/PURE/MIN_SET/CONTENT/members/T_Hall.html.

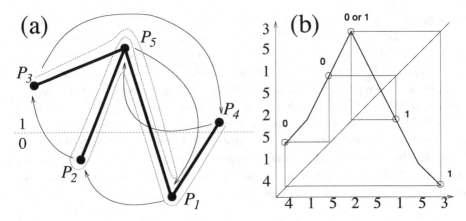

Fig. 14. (a) Periodic points of horseshoe orbit 00111 and their schematic trajectory in section plane between two crossings of the Poincaré section. Bold lines indicate edges belonging to the invariant set of (5). The dashed line shows a possible embedding of the path $F_1(\langle 15 \rangle)$ (the third iterate of Table 1); (b) Path $P_4 P_1 P_5 P_2 P_5 P_1 P_5 P_3$ folds onto itself under action of induced return map F_1. The unimodal map obtained has 00111 as a periodic orbit. After Lefranc.[6]

a natural symbolic coding of periodic orbits of maps of an interval into itself, with points at the left (resp., right) of the critical point being labeled 0 (resp., 1). As can be seen in Fig. 14b, the 1D coding of the orbit is $001^0_1 1$ (the fourth symbol is ambiguous because the corresponding point is located at the critical point). This recovers the exact 2D symbolic coding of the horseshoe orbit 00111 without having to construct a generating partition in the plane! The minor ambiguity is in fact natural as it reflects that the characterization performed cannot tell between the 00111 and 00101 horseshoe orbits, which are twin orbits born in the same saddle-node bifurcation. This is consistent with the conjecture that the approach developed here provides us with genuine topological invariants. In fact, the one-dimensional map in Fig. 14b is the simplest one that permutes the periodic points as for the 00111 orbit of the logistic map, the permutation being the one-dimensional analogue of the three-dimensional knot type).

Moreover, it turns out that what has been described above for the itinerary $p_3 = F^3(\langle 15 \rangle)$ also holds for all subsequent iterates $F^m(\langle 15 \rangle)$, hence for the infinite invariant path p_∞! At each step, it is possible to construct a one-dimensional map describing the folding along the path described by the iterate. In the limit, we get an infinitely detailed first return map which describes precisely how the unstable manifold is folded onto itself by the flow. The average slope of this 1D map is nothing but the growth rate computed above.

Importantly, our numerical simulations indicate that regardless of the initial triangulation, it converges rapidly to an invariant triangulation which reproduces itself at each iteration of the Poincaré return map. Indeed, it be seen in Fig. 14a that there is only one triangulation incorporating the edges in the invariant set of the substitution dynamical system.

The fundamental result here is that the asymptotic dynamics on p_∞ is *singular* even though transformations (4) are *invertible*. This reflects a key property which results from the identifying points whose orbits converge to each other asymptotically: any chaotic invertible return map (e.g., the Hénon map) can be associated to an underlying lower-dimensional noninvertible map (e.g., the logistic map), which describes the dynamics restricted to the unstable manifold (after the identification). What is remarkable here is that the combinatorial description allows one to extract the singular map in a finite number of steps. In particular, the one-dimensional map of Fig. 14b contains all the relevant information about the topological structure of the orbit. As with three-dimensional templates, this is because the future of a periodic orbit is fully predictable from its behavior on a finite time interval, both in forward and backward time. That the symbolic name of an orbit can be recovered using the underlying one-dimensional map is also promising for using topological analysis to construct global symbolic codings.[22]

Together, our results suggest that the simplicial space approach is equivalent to the train track algorithm in three dimensions. The fundamental properties of chaos are also reproduced: the dynamics is deterministic (by construction), invertible, and the stretching and squeezing processes are described in a symmetrical way. But the most seducing aspect of the triangulation-based approach is that the idea illustrated in Fig. 12 can in principle be generalized to higher-dimensional phase spaces.

3.7. *Triangulations in higher dimensions*

Let us now sketch with examples how these ideas could be extended to higher-dimensional systems. Four-dimensional chaotic flows embedded in $\mathbb{R}^3 \times S^1$ have three-dimensional Poincaré sections. Inside a Poincaré section, four periodic points define a tetrahedron, which can be viewed as a 3-cell of a simplicial space. As one sweeps the Poincaré section across S^1, nodes move in the three-dimensional section and tetrahedra can become degenerate and reverse their orientation (Fig. 15). Exchanging opposing sets of facets can then restore orientation, as in Fig. 12.

In four dimensions, there are two types of degeneracies, sketched in Figs. 15a and 15b. In the first case, we exchange a single 2-cell with a set of three 2-cells. In the second case, we swap two sets of two 2-cells. Let us consider the example in Fig. 15a, where a single 2-cell $\langle 132 \rangle$ collides with a set of three opposing 2-cells $\langle 124 \rangle + \langle 143 \rangle + \langle 234 \rangle$. Orientation can be preserved across degeneracy if the following substitutions are applied:

$$\langle 132 \rangle \to \langle 134 \rangle + \langle 243 \rangle + \langle 142 \rangle \tag{7a}$$

$$\langle 124 \rangle + \langle 143 \rangle + \langle 234 \rangle \to \langle 123 \rangle \tag{7b}$$

We have then $\partial \langle 1234 \rangle \to \partial \langle 1243 \rangle$, which compensates the orientation reversal of $\langle 1234 \rangle$, so that the outer normals are preserved. By composing such substitutions, we should be able to build a substitution dynamical system mapping concatenations of triangles into concatenations of triangles. If a nontrivial dynamics is generated

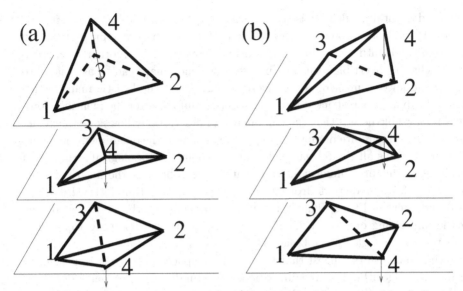

Fig. 15. As time flows (from top to bottom), tetrahedra defined by quartets of periodic points in three-dimensional Poincaré sections can become degenerate and change orientation. Two configurations can be distinguished according to how 2-cells are grouped into opposing sets. In (a), a single 2-cell ⟨132⟩ opposes a set of three 2-cells ⟨124⟩ + ⟨143⟩ + ⟨234⟩. In (b), a complex with two 2-cells ⟨124⟩ + ⟨143⟩ opposes another complex with two 2-cells ⟨132⟩ + ⟨234⟩.

by this system, this would allow us to compute lower bounds on topological entropy, as obtained from computing growth rates, but also to extract signatures of higher-order singularities.

Indeed, a remarkable property of the simplicial approach is how easily a signature of the fold singularity can be obtained in a three-dimensional flow. The substitution dynamical system describing the stretching and squeezing mechanisms has an infinitely long invariant path, which is mapped onto itself by one-dimensional map. This map typically displays one or several fold singularities (Fig. 14b).

Similarly, it is expected that for four-dimensional flows, the induced return map mapping chains of two-cells to other chains will be singular and display fold lines separating triangles folded onto each other. At the intersection of such fold lines, we should find the combinatorial equivalent of a cusp singularity. The chaotic dynamical systems will then be characterized by the relative organization of these discrete cusps, as discussed in Sec. 1.3.

4. Conclusion

Obviously, it is highly desirable to extend the concepts and tools of topological analysis to higher-dimensional spaces, so that we can classify attractors and understand their structure, explain bifurcation structures, construct symbolic codings and obtain signatures of chaos.

Higher-dimensional analogues of templates, which describe how stretching and squeezing knead phase space, can more or less be imagined. In systems which can be embedded in a torus $\mathbb{R}^n \times S^1$, and where only one contracting direction is involved in the dynamics, the template should describe how the $(n-1)$-dimensional intersection of the unstable manifold with an n-dimensional Poincaré section evolves as the latter is swept along the torus. Because the unstable manifold is stretched and folded before being mapped onto itself, this mapping is typically singular, with higher and higher order singularities appearing generically as the dimension of the unstable dimension increases. For example, folds and cusps should be ubiquitous in templates for four-dimensional flows, whose templates will be associated with a singular map of a two-dimensional surface into itself.

In support to this idea, we have discussed how cusps can be evidenced in weakly coupled strongly dissipative systems, both in a mathematical system (two coupled logistic maps) and in an experimental device (two coupled diode resonator systems). In both examples, cusps appear for extremely small couplings, indicating that they are generic in hyperchaotic systems. In weakly dissipative systems, cusp singularities will not be directly visible in return maps because, strictly speaking, they only exist in the infinitely iterated return map. In numerical simulations, an interesting idea would be to study the local arrangement of invariant manifolds at different points of the attractor and to search for higher-order tangencies.

To characterize the structure of periodic orbits, we also need higher-dimensional analogues of braids that can reconstructed from experimental signals. This requires to express determinism other than by precluding the intersection of two trajectories, as in three dimensions. A promising approach is to construct a dynamics on triangulations of periodic points advected by the flow, and to enforce orientation preservation. For three-dimensional flows, a simple formalism is obtained, which not only reproduces the correct values for the topological entropies of periodic orbits but also describes accurately the folding of the unstable manifold onto itself. In fact, preserving orientation in a simplicial space whose nodes are moving is a general and natural problem, with important applications in fluid mixing, for example.

A consistent triangulation-based formalism has not yet been built for higher-dimensional systems, unfortunately. Such a formalism would allow us to distinguish deterministic chaos from random noise, obtain lower bounds on topological entropy, and determine the organization of the singularities underlying the dynamics. This would greatly advance our understanding of high-dimensional chaos.

Last but not least, I would like to deeply thank Bob Gilmore for many things, in particular for laying out an extraordinarily beautiful programme for analyzing chaotic dynamical systems, for giving stimulating challenges and also for all the nice discussions we had, which somehow changed me. May this contribution represent a step in the right direction. Bob, I wish you again a very happy 70th birthday!

References

1. R. Gilmore & M. Lefranc, *The Topology of Chaos*, second revised and enlarged edition, Wiley-VCH (2011).
2. R. Gilmore & C. Letellier, *The Symmetry of Chaos*, Oxford University Press (2007).
3. N. Tufillaro, T. Abbott & J. Reilly, *An Experimental Approach to Nonlinear Dynamics and Chaos/Book and Disk*, Addison-Wesley (1992).
4. H.G. Solari, M.A. Natiello & G.B. Mindlin, *Nonlinear Dynamics: A Two-Way Trip from Physics to Math*, Taylor & Francis (1996).
5. J. Oden, S. Bielawski & M. Lefranc, Les singularités du chaos : doubles plis et fronces dans des systèmes couplés simples, *Comptes-rendus des Rencontres du Non-Linéaire*, **12**, 155-160 (2009) [c] — S. Bielawski, J. Oden, R. Gilmore & M. Lefranc, in preparation.
6. M. Lefranc, Alternative determinism principle for topological analysis of chaos, *Physical Review E*, **74**, 035202(R) (2006).
7. P. Verlaine (author), G. Hall (translator), *Poems of Paul Verlaine*, Duffield and Company (1906).
8. E. Ott, *Chaos in Dynamical Systems* Cambridge University Press, Cambridge (2002).
9. J. S. Birman and R. F. Williams, Knotted periodic orbits in dynamical systems I: Lorenz's equations, *Topology* **22**, 47–82 (1983).
10. D. Rolfsen, *Knots and Links*, AMS Chelsea Publishing (2003).
11. G. B. Mindlin & H. G. Solari, Tori and Klein bottles in four-dimensional chaotic flows, *Physica D* **102**, 177 (1997).
12. V. Arnold, A. Varchenko, & S. M Gusein-Zade, *Singularités des applications différentiables*, Edition Mir, Moscou (1986). — R. Gilmore, *Catastrophe theory for scientists and engineers*, Dover (1993).
13. E. J. Kostelich, I. Kan, C. Grebogi, E. Ott, & J. A. Yorke, Unstable dimension variability: A source of nonhyperbolicity in chaotic systems, *Physica D* **109**, 81 (1997).
14. P. Linsay, Period doubling and chaotic behavior in a driven anharmonic oscillator, *Physical Review Letters*, **47**, 1349 (1981) — J. Testa, J. Perez & C. Jeffries, Evidence for universal chaotic behavior of a driven nonlinear oscillator, *Physical Review Letters*, **48**, 714 (1982).
15. R. Van Buskirk & C. Jeffries, Observation of chaotic dynamics of coupled nonlinear oscillators, *Physical Review A*, **31**, 3332–3357 (1985).
16. L. Landau & E. M. Lifschitz, *Mechanics*, Butterworth-Heinemann (1976).
17. D. Sciamarella & G. B. Mindlin, Topological structure of chaotic flows from human speech chaotic data, *Physical Review Letters*, **82**, 1450 (1999).
18. P. Boyland, Topological methods in surface dynamics, *Topology and its Applications*, **58**, 223 (1994).
19. M. Bestvina & M. Handel, Train-tracks for surface homeomorphisms, *Topology*, **34**, 109–140 (1995).
20. T. Hall. The creation of horseshoes, *Nonlinearity*, **7**, 861–924 (1994).
21. A. Amon & M. Lefranc, Topological signature of deterministic chaos in short nonstationary signals from an optical parametric oscillator, *Physical Review Letters*, **92**, 094101 (2004).
22. J. Plumecoq & M. Lefranc, From template analysis to generating partitions I: Periodic orbits, knots and symbolic encodings, *Physica D* **144**, 231–258 (2000).
23. J. S. Birman, *Braids, Links and Mapping Class Groups*, Annals of Mathematical Studies **82**, Princeton University Press (1975).

---

[c] http://nonlineaire.univ-lille1.fr/SNL/media/2009/CR/Oden.pdf.

Chapter 10

The symmetry of chaos

Christophe Letellier

CORIA UMR 6614 - Rouen University
F-76801 Saint-Etienne du Rouvray cedex
France

When a dynamical system possesses a symmetry property, it can be seen as a cover of an image which does not have any residual symmetry. In the original phase space of such an equivariant dynamical system, a phase portrait of an n-order symmetry can be viewed as a fundamental domain copied n times. For instance, the Lorenz system has an order-2 symmetry, namely a rotation by π around the z-axis: the Lorenz attractor has thus two wings, one being the symmetric of the other. As a direct consequence of this feature, the Poincaré section is no longer defined by a simple surface of section but rather a set of n components, each of them being a section of the attractor in one copy of the fundamental domain in the phase space.

It is possible to modd out the symmetry property to obtain an image whose analysis is much easier than working in the original phase space. Contrary to this, some symmetry can be introduced in any dynamical system by using some coordinate transformation: a cover is thus obtained. In this chapter, the very special case of image for which the rotation axis crosses the attractor will be discussed and the analysis of a toroidal attractor will be detailed and applied to the sunspot numbers.

During the last ten years, with Robert Gilmore, we developed a full theory which describes the properties of systems with symmetry. These years of work (fights) lead us to the monograph entitled *The symmetry of Chaos*. This presentation will be completed by some problems we are still fighting about, ensuring us a lot of fun for next years.

Contents

1. Introduction 228
2. Earliest Works on the Lorenz System 228
3. How I Started with Topology 229
4. Injecting Symmetry into the Sunspot Data 234
5. Lorenz-like Attractors 237
6. Symmetry Group for Three-dimensional Systems 239
7. Symmetric Toroidal Chaos 243
8. Conclusion and Perspectives 245
References 245

1. Introduction

While I was student I was questioning myself about the deep nature of waves, particles, ... I was reading Louis de Broglie's books.[1,2] I wanted to explore in a deep way the so-called 'wave-corpuscule duality' and spent my fifth year at the University studying particle physics at Orsay. I thus learnt, among many other things, that "physical properties are invariant under symmetries". During that year, it was the DEA *Fields, particles and matter* then managed by Luc Valentin who pushed me to follow the good feeling I had about chaos, after some seminars about chaos theory given by Pierre Bergé, Monique Dubois and Vincent Croquette. Very attracted by the fascinating picture of chaotic attractors, I decided to spend a PhD on that topic. Unfortunately, Bergé and Dubois did not propose a subject on chaos that year, and I returned to Rouen where Gérard Gouesbet proposed me a thesis on global modeling.[3]

2. Earliest Works on the Lorenz System

I thus started to learn about chaos with the two well-known systems, namely the Lorenz[4] and the Rössler[5] systems. While investigating the Lorenz system, I was considering how were reconstructed the phase portraits from each of the variable, and not only from variable x as everybody was doing by that time. I remarked that the phase portraits reconstructed from variable x (or y) and from variable z were not equivalent (Fig. 1). As recognized by Greg King and Ian Stewart,[6] the origin of the problem was the rotation symmetry the Lorenz system has. I then looked to find a way to apply the principle according which "physical properties are invariant under symmetry properties," that is, to recover a phase portrait reconstructed from variable x or y, but equivalent to the phase portrait reconstructed from variable z. I was convinced that the latter, although without any residual symmetry, should have all the relevant dynamical properies of the Lorenz dynamics.

I was pushed in that direction by the power spectrum computed from variable z which exhibits a main frequency related to the pseudo-period associated with the revolutions surrounding the saddle-foci (Fig. 2) responsible for the two leaves of the Lorenz attractor. Contrary to this, the power spectrum computed from variable x or y presents no particular frequency and corresponds to a continuously decreasing broad band.[7] I was thus convinced that variable x does not contain the right information. I thus started to work with the absolute value $|x|$ which provides a spectrum equivalent to those computed from variable z. Moreover, the phase portrait reconstructed from $|x|$ was topologically equivalent (modulo a sign) to the portrait reconstructed from variable z: it thus provided what I named the fundamental domain, as Cvitanovic and Eckhard [8] developed as I later discovered. This was a part of my PhD thesis performed in 1992 (although only published in 1996) and this matter lead to my first paper submitted.[9]

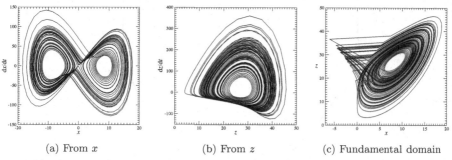

(a) From x (b) From z (c) Fundamental domain

Fig. 1. Phase portrait reconstructed using derivative coordinates from variable x (a), variable z (b), and the fundamental domain (c). The two latter are topologically equivalent and are different representations of a 'fundamental domain'.

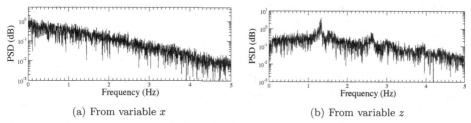

(a) From variable x (b) From variable z

Fig. 2. Power spectrum computed from variable x (a) and z (b) of the Lorenz system. Parameter values: $R = 28$, $\sigma = 10$, and $b = 8/3$.

Then comes the problem to correctly recover the 'Lorenz map' as a first-return map to a Poincaré section. Lorenz used a first-return map to maxima of variable z. I showed that this was equivalent to compute a first-return map to a two-component Poincaré section and using an 'invariant' variable, that is,

$$P \equiv \left\{ (x_n, z_n) \in \mathbb{R}^2 \mid y_n = \pm\sqrt{b(R-1)}, \dot{y}_n \lessgtr 0 \right\}.$$

A unimodal map with a cusp is thus recovered when built on $|x_n|$ or z_n (Figs. 3). This was a first use of multi-component Poincaré section. A complete theory for such Poincaré section was later developed by Tsvetelin Tsankov and Robert Gilmore.[10]

3. How I Started with Topology

During my PhD thesis, I was fascinated by the 'paper sheet model' or the 'blender' drawn by Rössler[11] (see Chapter 5). I thus started to redraw all these pictures to understand them. I then used a more formal approach as introduced by Bob's 'mafia'[12–14] as I named its group involving Gabriel Mindlin, Hernan Solari, Ricardo Lopez-Ruiz, and Nicholas Tufillaro. I also used a lot Nick's book.[15] Using the

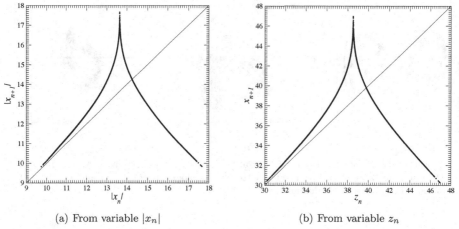

(a) From variable $|x_n|$ (b) From variable z_n

Fig. 3. First-return map to a two component Poincaré section to recover the Lorenz map, computed by Lorenz using the maxima of variable z. Parameter values: $R = 28$, $\sigma = 10$, and $b = 8/3$.

fundamental domain, I got a template for the Lorenz attractor (Fig. 4) which was contradicting the template Mindlin and co-workers published.[12]

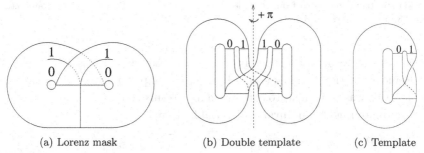

(a) Lorenz mask   (b) Double template   (c) Template

Fig. 4. Transformation of the Lorenz mask (a) — as introduced by Lorenz using isopleths,[4] and later used by Rössler[16] and Williams[17] — to a double template (b) and, once the symmetry was modded out, into a simple template (c).

From a paper published by Rick Miranda and Emily Stone,[18] I discovered that the 'fundamental domain' was nothing else than an image (or the proto-Lorenz system as named by Miranda and Stone). From such an image system, these two authors constructed n-fold cover of the proto-Lorenz system as exemplified by the quartic cover ($n = 4$) as shown in Fig. 5a. The phase space is tesselated by four copies of the 'fundamental domain D' under the order-four symmetry $\mathcal{R}_z(\frac{\pi}{2})$, that is, the group

$$\mathcal{G} = \left\{ \mathbb{I}, \Gamma_{\frac{\pi}{2}}, \Gamma_{\pi}, \Gamma_{\frac{3\pi}{2}} \right\} \text{ where } \Gamma_\theta = \begin{bmatrix} \sin\theta & \cos\theta & 0 \\ -\cos\theta & \sin\theta & 0 \\ 0 & 0 & 1 \end{bmatrix}.$$

The chaotic attractor (Fig. 5a) was analyzed in terms of a template (or a mask) showing the four copies of the fundamental domain. The four-component Poincaré section was clearly shown (Fig. 5b). This was formalized using the bounding tori as introduced by Tsankov and Gilmore.[19]

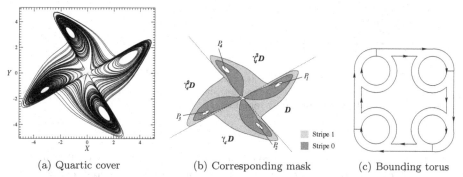

(a) Quartic cover   (b) Corresponding mask   (c) Bounding torus

Fig. 5. Quartic cover of the proto-Lorenz system introduced by Miranda and Stone.[18] The corresponding mask[20] showed how a four-component Poincaré section is constructed, to compare to the bounding torus proposed by Tsankov and Gilmore.[19]

While reading a paper by Robert Shaw,[21] I noted the system

$$\begin{cases} \dot{x} = -S(x+y) \\ \dot{y} = -y - Sxz \\ \dot{z} = Sxy + V \end{cases} \qquad (1)$$

which produces an attractor (Fig. 6a) presenting a topology differing from those of the Lorenz and the Rössler attractor. I thus proposed a four-branches template (Fig. 6b).[22]

Once remarked that this system has a rotation symmetry $\mathcal{R}_z(\pi)$, it was possible to construct the first-return map to a Poincaré section of the fundamental domain (Fig. 7a). The smooth unimodal map thus obtained allowed to invoke the unimodal order to predict the bifurcation diagram. With Pascal Dutertre,[23] we thus introduced the map

$$\Phi = \begin{vmatrix} \Phi^{-1}(10) = \overline{1} \\ \Phi^{-1}(11) = 0 \\ \Phi^{-1}(01) = 1 \\ \Phi^{-1}(00) = 2 \end{vmatrix} \qquad (2)$$

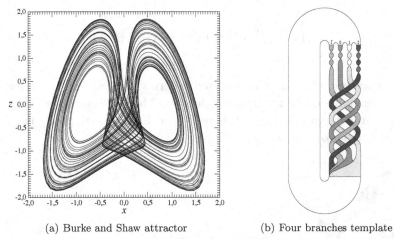

(a) Burke and Shaw attractor (b) Four branches template

Fig. 6. Chaotic attractor solution to the Burke and Shaw system (1) with parameter values: $S = 10$ and $V = 4.271$, and the corresponding template.

to switch from a four-symbol to a two-symbol symbolic dynamics. The bifurcation diagram of the quartic map (Fig. 7b) was thus completely predicted using this transformation applied to the unimodal order of orbital sequences.[22]

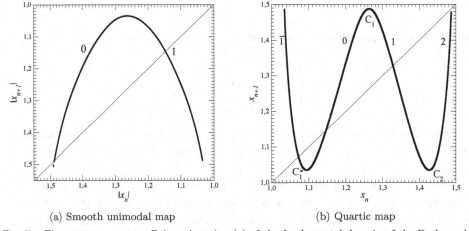

(a) Smooth unimodal map (b) Quartic map

Fig. 7. First-return map to a Poincaré section (a) of the fundamental domain of the Burke and Shaw attractor. The original map to a single component Poincaré section — since the bounding torus has a genus equal to one — is a quartic (three-modal) map (b).

This is from a question asked twice by Bob about the preceeding work that we started to work together. Starting from Miranda and Stone's paper,[18] we spent a whole week of hard work to construct a general procedure for modding out symmetry properties. One of our first results was to construct a two-fold cover of the Rössler system (Fig. 8) by inverting the coordinate transformation

$$\phi \equiv \begin{vmatrix} x = \text{Re } (X+iY)^2 \\ y = \text{Im } (X+iY)^2 \\ z = Z \end{vmatrix} \qquad (3)$$

where (x, y, z) are the coordinates of the original Rössler system.[24]

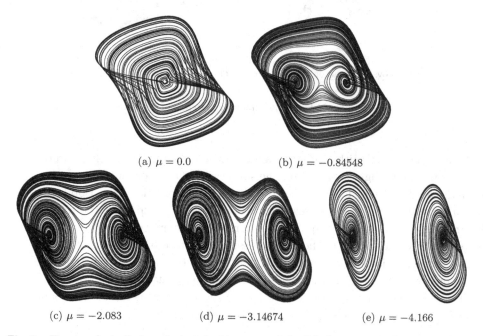

(a) $\mu = 0.0$ (b) $\mu = -0.84548$

(c) $\mu = -2.083$ (d) $\mu = -3.14674$ (e) $\mu = -4.166$

Fig. 8. Five topologically inequivalent double covers of the Rössler attractor depending on the location of the rotation axis along the x axis.

Using mapping

$$\phi_I = \begin{vmatrix} x = \text{Re}(X+iY)^2 \\ y = \text{Im}(X+iY)^2 \\ z = 2XZ, \end{vmatrix} \qquad (4)$$

we computed the image of the Kremliovsky system (Fig. 9)

$$\begin{cases} \dot{x} = -y - z \\ \dot{y} = x + ay \\ \dot{z} = bx + z(x^2 - c) \end{cases} \qquad (5)$$

which is equivariant under an inversion symmetry. We also investigated the embeddings which can be constructed from the different variables of the Lorenz system. All this matter lead to our first joint paper.[24]

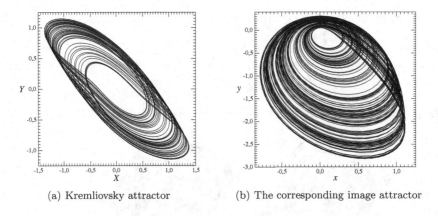

Fig. 9. Chaotic attractor produced by the Kremliovsky system and its corresponding image.

(a) Kremliovsky attractor (b) The corresponding image attractor

The first version submitted of that paper[24] did not contain the peeling bifurcation. This was done slightly later after the submission. One morning I came telling Bob that I produced new nice pictures by displacing the rotation axis. Bob looked at them and told me: "this is a new bifurcation". Indeed, displacing the rotation axis can lead to topologically inequivalent covers of a given image. This is illustrated in Figs. 8 where one can follow how it is possible to deform the unique symmetric attractor into two disconnected asymmetric attractors through successive 'peeling' bifurcations when the rotation axis intersects the image attractor.[24] We quickly submitted a second paper but the referee — in fact, the same as for the first paper — asked to include this peeling bifurcation to the first paper.

4. Injecting Symmetry into the Sunspot Data

The sunspots are known to be dark spots (due to a temperature up to 2000 K below the 'normal' superficial temperature which is about 6000 K). They are due a to strong magnetic field[25] which blocks the convection, preventing heat exhaust. These dark spots were intensively investigated by Thomas Harriott, Galilei Galileo,[26] and Christoph Scheiner.[27] For a review on the debate which occurred between Galileo and Scheiner, see William Shea[28] and Albert van Helden.[29] More than one century later, Heinrich Schwabe investigated the sunspots to draw a map of them (his first task was to look for a planet between Mercury and the sun): this was the first attack on a 'dark matter' problem. 25 years later remarked that the sunspot number was fluctuating with a period about 11 years.[30] This period was later confirmed by Rudolf Wolf.[31] Today, the sunspot numbers are investigated using the Wolf index $R = k(10g + f)$ where g counts the number of sunspot groups and f counts the individual sunspots. The factor k was introduced as a 'normalization' factor among the different observers who contributed. The data now available are shown in Fig. 10ª. The 11 year period is also associated with an inversion of the magnetic field as discovered by Hale and co-workers.[32]

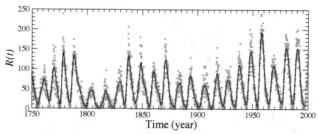

Fig. 10. Monthly averaged sunspot numbers using Wolf's index. The smoothed data ($w_s = 30$) — thick line — are superimposed to the monthly averaged sunspot numbers.

After an additional smoothing to remove short term fluctuations (we were only interested in cycle-to-cycle fluctuations),[33] a phase portrait was reconstructed using delay coordinates (top of Fig. 11). An embedding dimension is found to be around 3.[33] There is then a need for introducing the symmetry required by the inversion of the magnetic field. In order to do that, Ronald Bracewell suggested to invert each even cycle.[34] Unfortunately, according to the peeling bifurcation point of view, this is equivalent to locate the rotation axis at the origin of the phase space and two disconnected attractors should be obtained (Fig. 11a) and not a single symmetric one as expected.[33] In fact, the 'Bracewell trick' forces a connection between the two disconnected attractors; the discontinuity induced by such an 'artificial' connection prevents any successful global modeling.[35] This could be one of the reasons that, only a noisy limit cycle was found by Pablo Mininni and co-workers.[36]

If the rotation axis crosses the phase portrait reconstructed from the sunspot number, irregular reversals are obtained (Fig. 11b), contrary to what is observed. It is only when the rotation axis is located in the neighborhood of the 'fixed point' around which the trajectory is assumed to be structured (Fig. 11c) that reversals at each cycle are observed. This was therefore this cover we retained for a global modeling.

Starting from the transformed data set (Fig. 12) available at the ATOMOSYD website[b], Luis Aguirre obtained the difference equation[37]

$$X_k = 4.1011\, X_{k-1} - 7.5055\, X_{k-2} + 8.3094\, X_{k-3} - 6.1966\, X_{k-4} \\ + 2.9831\, X_{k-5} - 0.69788\, X_{k-6} + 0.84207 \cdot 10^{-4}\, X_{k-11} X_{k-10}^2 \,. \quad (6)$$

This equation is a Nonlinear Auto-Regressive Moving Average with eXogeneous input model (NARMAX) as introduced by Stephen Billings and Chen.[38] The model

---

[a]The time series used here is available on the web-site of the National Geophysical Data Center (NGDC) in Boulder, Colorado, USA at www.ngdc.noaa.gov.
[b]http://www.atomosyd.net/spip.php?article38.

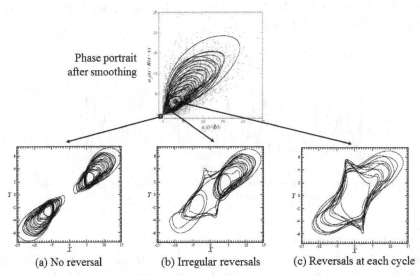

Fig. 11. Different topologically inequivalent covers of the phase portrait reconstructed from the sunspot number using delay coordinates ($\tau = 16$ months). The 23 cycles recorded since 1749 are shown here.

(6) was good for predicting the maxima but not for the minima of the sunspot number (a side effect of the two-to-one mapping).[37]

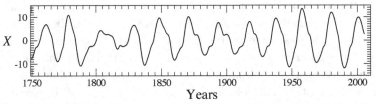

Fig. 12. Time series of the double cover of the phase portrait reconstructed from the sunspot numbers. These data were obtained by applying Φ^{-1} to the original time series (yielding $\{X(t)\}$) and subsequently decimating by a factor 4.

This model produces a symmetric attractor (Fig. 13a) whose image (Fig. 13b) is characterized by a three-branches template (Fig. 13c). It can be thus considered as the two-fold cover of a Rössler-like attractor with a global π-twist.[37] With this model, an evidence for a deterministic component driving the solar dynamics (observed through the sunspots) is obtained. The chaotic nature of the solar dynamics is therefore very likely.

This model was used for predicting the next two solar cycles (Fig. 14). The confidence bands widen as the prediction horizon increases. Up to 2025, the confidence bands are reasonably narrow, but after they become very wide indicating that there is great diversity among individual realizations. Hence, based on the

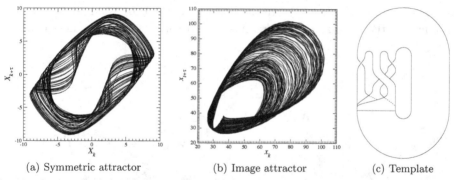

(a) Symmetric attractor   (b) Image attractor   (c) Template

Fig. 13. Symmetric chaotic attractor (a) solution to the NARMAX model (6) obtained from the transformed data set (Fig. 12), only using data after 1850 (see[33] and[37] for explanations). Its image (a) and the corresponding template (c) are also shown.

"reasonable" consistency among the different realizations up to 2025, this is taken as the prediction horizon. As for any other predictions made for the next solar cycle, results should be considered with caution. When the data and the confidence bands are transformed back to Wolf's sunspot numbers (Fig. 14), the forecasted peaks are 65 ± 16 (September-October 2012) and 100 ± 34 (January-February 2021) units of sunspot numbers, respectively for the 24th and 25th cycles.

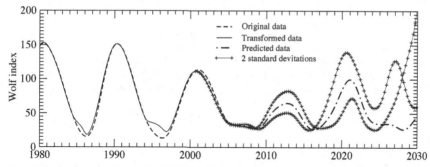

Fig. 14. The continuous line up to 2005 is the time series produced by model (6) transformed back to be compared to the sunspot numbers. The dashed line is the ensemble average of 100 realizations predicted using model (6). The crosses indicate $\pm 2\sigma$. The continuous line from 2005 is the single forecast obtained by using model (6).

5. Lorenz-like Attractors

With Tsvetelin Tsankov and Greg Byrne we investigated the role the symmetry axis plays in the structure of two topologically inequivalent attractors Lorenz-like systems produce.[39] We called Lorenz-like systems any set of differential equations equivariant under a rotation symmetry $\mathcal{R}_z(\pi)$ and producing a Lorenz attractor. Most of these systems also produce a Burke and Shaw attractor for different parameter values. In Fig. 15 are shown the Lorenz and the Burke and Shaw attractors produced by the Lorenz system. In Table 1 are reported some of the Lorenz-like systems.

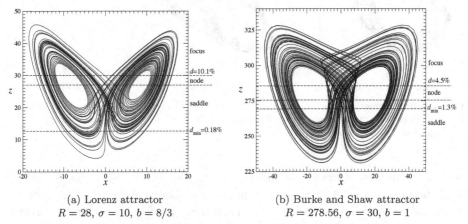

(a) Lorenz attractor
$R = 28$, $\sigma = 10$, $b = 8/3$

(b) Burke and Shaw attractor
$R = 278.56$, $\sigma = 30$, $b = 1$

Fig. 15. Projection in the x-z plane of the two types of chaotic attractors solution to the Lorenz system. The transverse stability of the z-axis is reported versus the z-value. The minimum distance $d_{\min}$, between the attractor and the z-axis is also indicated as well as the distance at $z = z_c$.

Table 1. Cœfficients of the nine quadratic systems with a rotation symmetry around the z-axis. The first five systems have three fixed points, one located at the origin of the phase space and two which are related by the rotation symmetry. The last four systems have only the two symmetry-related fixed points. L_1=Lorenz system, L_6=Burke and Shaw system.

System	x a_1	y a_2	yz a_3	x b_1	y b_2	xz b_3	— c_0	z c_1	x^2 c_2	xy c_3	y^2 c_4	Ref.
(L_1)	$-\sigma$	$+\sigma$	0	R	-1	-1	0	$-b$	0	+1	0	4
(L_2)	$-\sigma$	$+\sigma$	0	$R-\sigma$	R	-1	0	$-b$	0	+1	0	40
(L_3)	$-\sigma$	σ	0	0	-1	-1	Ra	-1	0	+1	0	41
(L_4)	0	+1	0	+1	$-\mu$	-1	0	$-\alpha$	+1	0	0	42
(L_5)	0	+1	0	$-\lambda$	$+\kappa$	-1	0	-1	+1	0	0	43
(L_6)	$-S$	$-S$	0	0	-1	$-S$	$\mathcal{V}$	0	0	$+S$	0	21
(L_7)	$-\sigma$	$+\sigma$	0	0	0	+1	b	0	0	-1	0	44
(L_8)	$-\sigma$	$+\sigma$	0	0	0	+1	b	0	0	0	-1	44
(L_9)	$-\mu$	0	+1	$-\alpha$	$-\mu$	+1	+1	0	0	-1	0	45

All these Lorenz-like systems have two-symmetry-related saddle-foci but some of them (L_8 and L_9) do not present a saddle as a third fixed point. Nevertheless, for all of them, the rotation axis is a one-dimensional singular set which eventually ejects the trajectory to infinity. By investigating the transverse stability to the z-axis, with Tsvetelin and Greg, we showed that the axis can be split in two parts: one with foci and one with saddle. Lorenz attractors are located in the neighborhood of the axis where the transverse stability of the z-axis is associated with saddles (thus responsible for a tearing mechanism), and Burke and Shaw attractors are thus located in the neighborhood of the axis where foci are responsible for a folding.[39] When the attractor only surrounds saddles, there is only a tearing in the attractor, when it only surrounds foci, there is a folding mechanism. When the attractor is located in the neighborhood of the transition between saddles and foci along the z-axis, tearing and folding (multimodal map) are observed in the attractors (Fig. 16). If obviously, fixed points — zero-dimensional singular sets — structure the phase portraits, some one-dimensional singular set can also structure the phase portrait. Moreover, such higher-dimensional singular set can help to understand some characteristics these Lorenz-like attractors share.

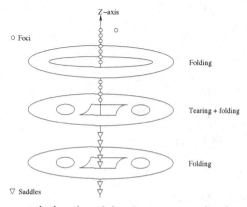

Fig. 16. As parameters vary the location of chaotic attractors related to the rotation axis — and the transverse stability — change, and so the corresponding mechanism responsible for chaos. Folding occurs when the transverse stability is that of a focus and tearing occurs when the trasnverse stability is that of a saddle. Both occur in the transition region.

6. Symmetry Group for Three-dimensional Systems

With Robert Gilmore, we then extended our investigations of dynamical systems in $\mathbb{R}^3$ with symmetry. The map ϕ between the n-fold cover and its image system is one-to-one almost everywhere, that is, the Jacobian of map ϕ is singular only on zero- and one-dimensional sub-space of the phase space $\mathbb{R}^3$. Such mappings can be conveniently constructed with the aid of symmetry groups.[46] I will briefly present few examples below.

With Bob, we constructed a four-fold cover of the Rössler system using the V_4 group based on three rotational matrices

$$\begin{array}{ccc} R_X(\pi) & R_Y(\pi) & R_Z(\pi) \\ \begin{bmatrix} +1 & 0 & 0 \\ 0 & -1 & 0 \\ 0 & 0 & -1 \end{bmatrix} & \begin{bmatrix} -1 & 0 & 0 \\ 0 & +1 & 0 \\ 0 & 0 & -1 \end{bmatrix} & \begin{bmatrix} -1 & 0 & 0 \\ 0 & -1 & 0 \\ 0 & 0 & +1 \end{bmatrix} \end{array}$$

and the identity $\mathbb{I}$ as introduced by Felix Klein.[47] These generators satisfy the relation $\mathbb{I} = \sigma_1^2 = \sigma_2^2 = (\sigma_1 \sigma_2)^2$ and the product of any two generators is the third. This group can be interpreted as the group of rotations by π-radians about the X, Y, and Z-axes. The map $\phi : \mathbb{R}^3(X, Y, Z) \mapsto \mathbb{R}^3(u, v, w)$ is a $4 \to 1$ local diffeomorphism can be written as

$$\begin{cases} u = \frac{1}{2}(X^2 - Y^2) \\ v = \frac{1}{2}(X^2 + Y^2 - 2Z^2) \\ w = XYZ. \end{cases} \qquad (7)$$

Every point in $\mathbb{R}^3(u, v, w)$ has four pre-images in $\mathbb{R}^3(X, Y, Z)$. They are obtained from

$$\begin{cases} X = \pm\sqrt{Z^2 + v + u} \\ Y = \pm\sqrt{Z^2 + v - u}, \end{cases} \qquad (8)$$

the value of Z being determined from

$$w = \sqrt{Z^2 + v + u}\sqrt{Z^2 + v - u}\, Z.$$

This equation has a unique solution.

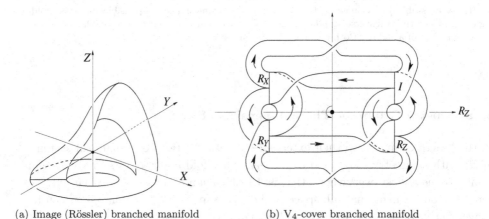

(a) Image (Rössler) branched manifold (b) V_4-cover branched manifold

Fig. 17. Branched manifold of the Rössler attractor — here considered as the image — and the X, Y, Z rotation axes for the symmetry group V_4. The period-one orbit in the orientation-preserving branch 0 links (or not) the three rotation axes. The branched manifold (b) of the resulting V_4 cover is a planar projection of a V_4 symmetric branched manifold.

The rotation axes were located in $\mathbb{R}^3(u,v,w)$ as shown in Fig. 17a. Practically, we started from the modified Rössler system [46]

$$\begin{cases} \dot{x} = -z - y - \mu_z - \mu_y \\ \dot{y} = x + a(y + \mu_y) + \mu_x \\ \dot{z} = b(x + \mu_x) + (z + \mu_z)(x - c + \mu_x) \end{cases} \quad (9)$$

where (a,b,c) are the usual parameter values and (μ_x, μ_y, μ_z) are used to located the rotation axes properly. The V_4-cover is built using the dynamical equation:

$$\begin{cases} \dot{X} = \dfrac{X(Y^2 + 2Z^2)\dot{x} + XY^2\dot{y} + 2YZ\dot{z}}{2\rho^2} \\ \dot{Y} = \dfrac{-Y(X^2 + 2Z^2)\dot{x} + X^2Y\dot{y} + 2XZ\dot{z}}{2\rho^2} \\ \dot{Z} = \dfrac{Z(X^2 - Y^2)\dot{x} - Z(X^2 + Y^2)\dot{y} + 2XY\dot{z}}{2\rho^2} \end{cases} \quad (10)$$

where $\rho^2 = X^2Y^2 + X^2Z^2 + Y^2Z^2$. We varied μ_x and used $(\mu_y, \mu_z) = (-1.0, 1.0)$. We thus obtained three topologically inequivalent V_4-covers of the Rössler system. There is one merging attractor crisis between each of the covers shown in Fig. 18.

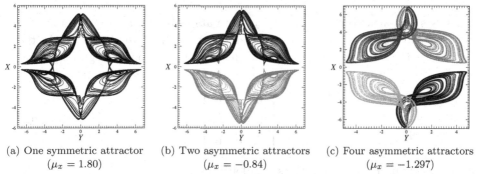

(a) One symmetric attractor $(\mu_x = 1.80)$

(b) Two asymmetric attractors $(\mu_x = -0.84)$

(c) Four asymmetric attractors $(\mu_x = -1.297)$

Fig. 18. V_4-symmetric covers of the Rössler attractor. The topology (connectedness) of the cover depends on the way the different branches link the three rotation axes. Parameter values: $(a,b,c) = (0.42, 2, 4)$.

Another example of a system with V_4 symmetry was proposed by Jinhu Lü and co-workers:[48]

$$\begin{cases} \dot{X} = -\dfrac{ab}{a+b}X - YZ \\ \dot{Y} = aY + XY + \nu \\ \dot{Z} = bZ + XY. \end{cases} \quad (11)$$

This system produces two disconnected Burke and Shaw attractors (Fig. 19b). A first merging attractor crisis reduces the four disconnected attractors (Fig. 19a) into the two attractors shown in Fig. 19b.

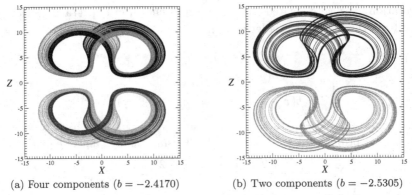

(a) Four components ($b = -2.4170$) (b) Two components ($b = -2.5305$)

Fig. 19. (a) Chaotic attractors solution to the Lü-Chen-Cheng system (11). Parameter values: $a = -10$ and $\nu = 0$.

Among others, another interesting system was proposed by René Thomas:[49]

$$\begin{cases} \dot{x} = -bx + ay - y^3 \\ \dot{y} = -by + az - z^3 \\ \dot{z} = -bz + ax - x^3 . \end{cases} \qquad (12)$$

It possesses a S_6-symmetry which consists of rotations by $\frac{2\pi}{6}$ radians about a symmetry axis, followed by reflection in a plane perpendicular to that axis. In the present case, the rotation axis is chosen in the (1,1,1) direction. The group generator is

$$\Gamma = \begin{bmatrix} 0 & -1 & 0 \\ 0 & 0 & -1 \\ -1 & 0 & 0 \end{bmatrix} \qquad (13)$$

with $\Gamma^3 = -\mathbb{I}$ and $\Gamma^6 = \mathbb{I}$. Equations (12) have the form

$$\dot{x}_i = c_{ij}x_j + c_{ijk}x_jx_k + c_{ijkl}x_jx_kx_l + ... \qquad (14)$$

where $1 \leq i, j, ... \leq 3$. In short, coefficients of the linear and cubic (and odd) terms are constrained, and those of the quadratic and quartic (and even) terms are zero.

In order to have a better representation of the symmetry properties, we used the coordinate transformation

$$\begin{cases} X = \dfrac{\sqrt{3}}{2}(y - x) \\ Y = z - \dfrac{x+y}{2} \\ Z = x + y + z \end{cases} \qquad (15)$$

to map the chaotic attractor (Fig. 20a) in the X-Y plane projection shown in Fig. 20b). These system has 36 fixed points, counting degeneracy, since the fixed point at the origin of the space phase is six-fold degenerated, the two fixed points on the axis $(1,1,1)$ are each three-fold degenerated. The others can be grouped in four sets made of six symmetry related fixed points.

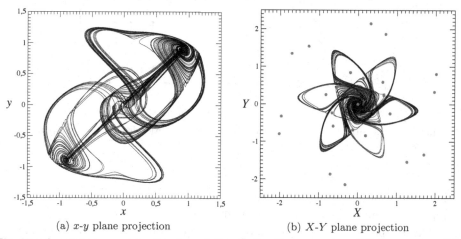

(a) x-y plane projection (b) X-Y plane projection

Fig. 20. A unique symmetric attractor solution to the Thomas system (12) for $a = 1.1$. The 27 fixed points are shown in the X-Y plane projection. The points on the $(1,1,1)$ axis cannot be distinguished from the point at the origin in this representation.

Many more examples can be found in[46] and in the monograph *The Symmetry of Chaos*.[50]

7. Symmetric Toroidal Chaos

Symmetries can also be observed in toroidal structure. A particularly interesting system was proposed by Dequan Li:[51]

$$\begin{cases} \dot{X} = a(Y - X) + dXZ \\ \dot{Y} = kX + fY - XZ \\ \dot{Z} = cZ + XY - eX^2 \end{cases}. \qquad (16)$$

This system can be considered as a Lorenz system to which two nonlinear terms were added. It has a rotation symmetry $\mathcal{R}_z(\pi)$. The toroidal attractor solution to this system is structured around a nontrivial torus as explained below.[52] The attractor (Fig. 21a) is bounded by a torus which can be bounded using quadrilaterals as shown in Fig. 21. Using the Euler-Poincaré formula

$$N_{\text{vertices}} - N_{\text{edges}} + N_{\text{faces}} = 2 - 2g,$$

the genus g of the torus can be determined. With $N_{\text{vertices}} = 32$, $N_{\text{edges}} = 72$, and $N_{\text{faces}} = 36$, we got a genus equal to 3 for this bounding torus.

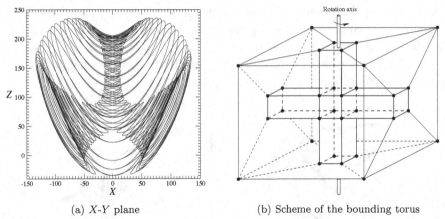

(a) X-Y plane (b) Scheme of the bounding torus

Fig. 21. Limit cycle solution to the Li system (16), and a schematic view of the corresponding bounding torus. Parameter values: $a = 40$, $c = 11/6$, $d = 0.16$, $e = 0.65$, $k = 55$, and $f = 20$.

According to the bounding tori theory developed by Tsankov and Gilmore, a two-component Poincaré section must be used for attractors bounded by a genus-3 torus.[10] The components have to be related to the holes surrounding foci (and not saddles). The two components are thus chosen as shown in Fig. 22a.

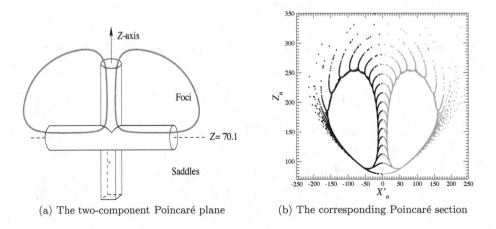

(a) The two-component Poincaré plane (b) The corresponding Poincaré section

Fig. 22. Sketch for choosing the Poincaré plane (a) and the resulting Poincaré section (b) of the toroidal chaotic attractor. Parameter values: $a = 41$, $c = 11/6$, $d = 0.16$, $e = 0.65$, $k = 55$, and $f = 20$.

By modding out the symmetry with the two-to-one coordinate transformation (3), we obtained the corresponding image attractor (Fig. 23a) which obviously intersects the rotation axis (Fig. 23b). As observed in the peeling bifurcation, this implies a nontrivial structure for the two-fold cover (Fig. 23c). A scenario for constructing topologically inequivalent covers of toroidal chaotic image attractors was described with Bob and Timothy Jones.[53]

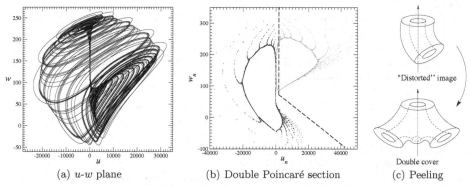

(a) u-w plane (b) Double Poincaré section (c) Peeling

Fig. 23. Image chaotic attractor (a) solution to the Li system (16). The rotation axis intersects the attractor (b), thus inducing a nontrivial two-fold cover.

8. Conclusion and Perspectives

During more than ten years we developed a general approach for injecting symmetry in chaotic systems and, by inverting the procedure, for modding out the symmetry. Most of our applications were performed on three-dimensional systems to take advantage of the beautiful topological analysis whose most representative tool — for dissipative systems — is the branched manifold. Although many examples have been already analyzed, there are yet some "acid" cases for which a more detailed analysis would be welcome. Among others, there is the double scroll attractor solution to the Chua circuit [54] or to the system proposed by Rössler (see Chapter 5). In particular, the link between such branched manifold and Ghrist's template supporting all knots[55] is still to clarify.

Branched manifolds, bounding tori and first-return maps to a Poincaré section are the right tools to imagine a general terminology for designating the attractors by their properties and not by the name of their discoverers. Such a nontrivial task is one of the items Bob and myself have on our to-do-list. No doubt that we have still many years of 'enjoyable fights' to complete our program.

References

1. L. de Broglie, *Ondes et Mouvements*, Gauthier-Villars (1926), reprinted by Jacques Gabay, 1988.
2. L. de Broglie, *La physique nouvelle et les quanta*, Flammarion (1937), 1993.
3. C. Letellier, *Caractérisation topologique et reconstruction d'attracteurs étranges*, Ph'D dissertation, Université de Paris VII, 1994.

4. E. N. Lorenz, Deterministic nonperiodic flow, *Journal of the Atmospheric Sciences*, **20**, 130-141, 1963.
5. O. E. Rössler, An equation for continuous chaos, *Physics Letters A*, **57** (5), 397-398, 1976.
6. G. P. King & I. Stewart, Phase space reconstruction for symmetric dynamical systems, *Physica D*, **58**, 216-228 (1992).
7. C. Letellier & G. Gouesbet, Topological characterization of reconstructed attractors modding out symmetries, *Journal de Physique II*, **6**, 1615-1638 (1996).
8. P. Cvitanović & B. Eckhardt, Symmetry decomposition of chaotic dynamics, *Nonlinearity*, **6**, 277-311 (1993).
9. C. Letellier, P. Dutertre & G. Gouesbet, Characterization of the Lorenz system taking into account the equivariance of the vector field, *Physical Review E*, **49** (4), 3492-3495 (1994).
10. T. D. Tsankov & R. Gilmore, Topological aspects of the structure of chaotic attractors in $\mathbb{R}^3$, *Physical Review E*, **69**, 056206 (2004).
11. O. E. Rössler, Chaotic behavior in simple reaction system, *Zeitschrift für Naturforschung A*, **31**, 259–264 (1976).
12. G. B. Mindlin, X.-J. Hou, H. G. Solari, R. Gilmore & N. B. Tufillaro, Classification of strange attractors by integers, *Physical Review Letters*, **64** (20), 2350-2353 (1990).
13. F. Pappoff, A. Fioretti, E. Arimondo, G. B. Mindlin, H. G. Solari & R. Gilmore, Structure of chaos in the laser with saturable absorber, *Physical Review Letters*, **68** (8), 1128-1131 (1992).
14. G. B. Mindlin, R. López-Ruiz, H. G. Solari & R. Gilmore, Horseshoe implications, *Physical Review E*, 48 (6), 4297-4304 (1993).
15. N. B. Tufillaro, T. Abbott & J. Reilly, *An Experimental Approach to Nonlinear Dynamics and Chaos*, Addison-Wesley, New York (1992).
16. O. E. Rössler, Different types of chaos in two simple differential equations, *Zeitschrift für Naturforschung A*, **31**, 1664-1670 (1976).
17. R. F. Williams, The structure of lorenz attractors, *Lecture Notes in Mathematics*, **615**, 94112 (1977).
18. R. Miranda & E. Stone, The proto-Lorenz system, *Physics Letters A*, **178**, 105-113 (1993).
19. T. Tsankov & R. Gilmore, Strange attractors are classified by bounding tori, *Physical Review Letters*, **91**, 134104 (2003).
20. C. Letellier & G. Gouesbet, Topological characterization of a system with high-order symmetries: the proto-Lorenz system, *Physical Review E*, **52** (5), 4754-4761 (1995).
21. R. Shaw, Strange attractor, chaotic behavior and information flow. *Zeitschrift für Naturforschung A*, **36** (a), 80-112 (1981).
22. C. Letellier, P. Dutertre, J. Reizner & G. Gouesbet, Evolution of multimodal map induced by an equivariant vector field, *Journal of Physics A*, **29**, 5359-5373 (1996).
23. P. Dutertre, *Caractérisation des attracteurs étranges par la population d'orbites périodiques*, Ph'D thesis, Université de Rouen, France (1995).
24. C. Letellier & R. Gilmore, Covering dynamical systems: Two-fold covers, *Physical Review E*, **63**, 16206 (2001).
25. G. E. Hale, On the probable existence of a magnetic filed in sun-spots, *Astrophysical Journal*, **28**, 315 (1908).
26. G. Galileo, *Istoria e Dimostrazioni Intorno Alle Macchie Solari e Loro Accidenti* (History and Demonstrations concerning the Sunspots and their Properties), Rome (1613).
27. C. Scheiner, *Tres epistolae de Maculis Solaribus scriptae ad Marcum Velserium* Augsburg (1612) — *Rosa Ursina*, Bracciano (1630).

28. W. R. Shea, Galileo, Scheiner, and the interpretation of sunspots, *Isis*, **61** (4), 498-519 (1970).
29. A. van Helden, Galileo and Scheiner on sunspots: A case study in the visual language of astronomy, *Proceedings of the American Philosophical Society*, **140** (3), 358-396 (1996).
30. H. Schwabe, Über die Flecken der Sonne, *Astronomische Nachrichten*, **15**, 244-248 (1838).
31. R. Wolf, Sunspot epochs since A.D. 1610: the periodic return of sunspot minima, *Comptes-Rendus de l'Académie des Sciences*, **35**, 704-705 (1852).
32. G. E. Hale, F. Ellerman, S. B. Nicholson & A. H. Joy, The magnetic polarity of sun-spots, *The Astrophysical Journal*, **49**, 153-178 (1919).
33. C. Letellier, J. Maquet, L. A. Aguirre & R. Gilmore, Evidence for low dimensional chaos in the sunspot cycles, *Astronomy & Astrophysics*, **449**, 379-387 (2006).
34. R. N. Bracewell, The sunspot number series, *Nature*, **171**, 649-650 (1953).
35. C. Lainscsek, F. Schrrer & J. B. Kadtke, A general form for global dynamical data models for three-dimensional systems, *International Journal of Bifurcation and Chaos*, **8** (5), 899-914 (1998).
36. P. D. Mininni, D. O. Gomez & G. B. Mindlin, Stochastic relaxation oscillator model for the solar cycle, *Physical Review Letters*, **85**, 5476-5480 (2000).
37. L. A. Aguirre, C. Letellier & J. Maquet, Forecasting the time series of sunspot numbers, *Solar Physics*, **249** (1), 103-120 (2008).
38. S. A. Billings & S. Chen, Extended model set, global data and threshold model identification of severly nonlinear systems, *International Journal of Control*, **50**, 1897-1923 (1989).
39. C. Letellier, T. Tsankov, G. Byrne & R. Gilmore, Large-scale structural reorganization of strange attractors, *Physical Review E*, **72**, 026212 (2005).
40. G. Chen & T. Ueta, Yet another chaotic attractor, *International Journal of Bifurcation and Chaos*, **9**, 1465-1466 (1999).
41. Y. Wang, J. Singer & H. H. Bau, Controlling chaos in a thermal convection loop, *Journal of Fluid Mechanics*, **237**, 479-498 (1992).
42. T. Shimizu & N. Morioka, On the bifurcation of a symmetric limit cycle to an asymmetric one in a simple model, *Physics Letters A*, **76**, 201-204 (1980).
43. A. M. Rucklidge, Chaos in models of double convection, *Journal of Fluid Mechanics*, **237**, 209-229 (1992).
44. J. S. Sprott, Some simple chaotic flows, *Physical Review E*, **50** (2), 647-650 (1994).
45. T. Rikitake, Oscillations of a system of disk dynamos, *Proceedings of the Cambridge Philosophical Society*, **54**, 89-105 (1958).
46. C. Letellier & R. Gilmore, Symmetry groups for 3D dynamical systems, *Journal of Physics A*, **40** (21), 5597-5620 (2007).
47. F. Klein, *Vorlesungen ber das Ikosaeder und die Auflösung der Gleichungen vom fünften Grade*, B. G. Teubner, Leipzig, (1884).
48. J. L, G. Chen & D. Cheng, A new chaotic system and beyond: the generalized Lorenz-like system, *International Journal of Bifurcation & Chaos*, **14**, 1507-1537 (2004).
49. R. Thomas, Deterministic chaos seen in terms of feedback circuits: analysis, synthesis, "labyrinth chaos", *International Journal of Bifurcation & Chaos*, **9**, 1889-1905 (1999).
50. R. Gimore & C. Letellier, *The Symmetry of Chaos*, Wiley, (2007).
51. D. Li, A three-scroll chaotic attractor, *Physics Letters A*, **372**, 387-393 (2008).
52. C. Letellier & R. Gilmore, Poincaré sections for a new three-dimensional toroidal attractor, *Journal of Physics A*, **42**, 015101 (2009).

53. C. Letellier, R. Gilmore & T. Jones Peeling bifurcation of toroidal chaotic attractors, *Physical Review E*, **76**, 066204 (2007).
54. T. Matsumoto, L. O. Chua & M. Komuro, The double scroll, *IEEE Transactions on Circuits & Systems*, **32** (8), 798-818, (1985).
55. R. W. Ghrist, Branched two-manifolds supporting all links, *Topology*, **36** (2), 423-448 (1997).

PART 3
Applications of Chaos Theory

Chapter 11

The shape of ocean color

Nicholas Tufillaro

College of Oceanic and Atmospheric Sciences
Oregon State University
USA

"I want to see gamma rays! I want to hear X-rays! And I want to — I want to smell dark matter! Do you see the absurdity of what I am? I can't even express these things properly because I have to — I have to conceptualize complex ideas in this stupid limiting spoken language! But I know I want to reach out with something other than these prehensile paws! And feel the wind of a supernova flowing over me! I'm a machine! And I can know much more! I can experience so much more. But I'm trapped in this absurd body!"

– Brother Cavil, Battlestar Galactica

Contents

1. Ocean Color Primer 251
2. Shape of Water Spectrum 255
3. Shape of Ocean Spectra 256
4. Derivative Spectroscopy and Embeddings 258
5. Torsion, Writhe, Linking, and All That 261
6. Applications: Edge Detection and Product Indicators 263
References 267

1. Ocean Color Primer

Our ancestors entered the nonlinear places known as manifolds when inquiry moved from configuration to phase space. The manifolds they encountered where sometimes tame. In recent years our wanderings often start with data, not theory. We find ourselves surrounded by wilderness, data generated manifolds in which we are easily lost. The terrain is rough, the directions numerous. The manifolds encountered when making observation of the environment are, more often than not, wild places. In this chapter we begin tramping in the observational landscape known as 'ocean color.'

Plankton fuels life in the ocean, and it creates half the oxygen we breathe. Plankton also drives our fossil fuel possessed culture. The gas in our cars is not really from dinosaurs — its from the 'Great Dying' that occurred 250 million years ago when 96% of all marine species died and dropped to the bottom of the ocean. Then, like now, great fumes of green house gases filled the air and the earth got very warm. The oceans became acidic, and the web of life dissolved. This is the organic carbon reservoir that our machines are eating today. Perhaps this is the shape of things to come. Perhaps the future will be fueled by living algae for organic machines, and their in-organic brethren alike.

Time parameterizes dynamics. Wavelength parameterizes spectra. And fruit-flies like a banana. Dynamics and spectra become one when we look at atoms. Ocean color refers to the spectral study of light as it bounces about the water column in an ocean. Plankton fuels life in the ocean. And photons fuel plankton. So the base of oceanic life, plankton, rise to the light. As the light rains down from the sky, the photons are absorbed and scattered by whatever they hit in the upper ocean. In the open ocean this is water, salt, a few other essential minerals, and plankton. In the coastal ocean the soup is richer, the water brighter. Rivers soak the boundaries of the ocean with sediments, crushed leaves and other 'colored dissolved organic matter' (CDOM). A rich mixture of minerals wells up to the surface at the margins from the deep ocean and feeds large blooms of a dizzying diversity of plankton that have been adapting and evolving for three billion years.

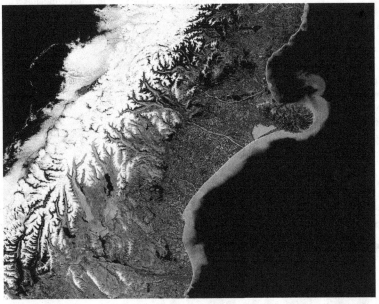

Fig. 1. A RGB image created from ESA MERIS radiance data of the South Island of New Zealand (October 2007). Lake Pukaki is the bright aqua lake at the base of Mt. Cook toward the lower left. Christchurch is in the upper right section north of the circular landmass — Banks Peninsula. Deep clear lakes look almost black in the image since, unlike land, most light is not reflected off the surface but scattered into the water column. Only a few photons are back scattered in an upward direction toward the satellite.

I first really noticed ocean color flying into Dunedin on the South Island of New Zealand.[1] The Tasman glacier at the base of Mt. Cook feeds lake Lake Pukaki with sediments rich in quartz and albite. The glacial flour sets the waters of Lake Pukaki aglow, and the lake drains into the Waitaki river which in turn brightens the coastal waters east of the Canterbury plains. This makes a striking image from 500 miles high, as seen from the orbit of MERIS, the European Space Agencies **ME**dium **R**esolution **I**maging **S**pectrometer (Fig. 1).

MERIS is a multispectral imager, meaning its spectral resolution is limited to fifteen bands in the visible and near infrared (Tab. 1). The bands are chosen to be sensitive to key water or atmospheric properties. MERIS launched in 2002 and its ground sampling distance (GSD), or in other words its pixel size, is either 300 meters at full resolution (FR), or 1200 meters in reduced resolution (RR). Its orbit is periodic, repeating every 35 days [a].

Table 1. MERIS spectral channels.

Band	Wavelength	Width	Applications
1	412.5	10	Detrital Pigments, CDOM
2	442.5	10	Chlorophyll absorption maximum
3	490	10	Chlorophyll and other pigments
4	510	10	Suspended sediment, red tides
5	560	10	Chlorophyll absorption minimum
6	620	10	Suspended sediment
7	665	10	Chlorophyll absorption and fluorescence reference
8	681.25	7.5	Chlorophyll fluorescence peak
9	708.25	10	Fluorescence reference, atmospheric corrections
10	753.75	7.5	Vegetation, cloud
11	760.625	3.75	Oxygen absorption R-branch
12	778.75	15	Atmosphere corrections
13	865	20	Vegetation, water vapour reference
14	885	10	Atmosphere corrections
15	900	10	Water vapor, land

Fig. 2. HICO 16 June 2011.

MERIS is designed to help reveal the web of life in the ocean not immediately apparent to the human eye. MERIS can see the chlorophyll pigments of plankton in

---
[a]http://envisat.esa.int/instruments/meris/

the water column as they drift near the surface of the ocean (Fig. 2). The plankton act as a kind of tracer particle revealing, for example, the chaotic advective mixing off the coast of Christchurch, New Zealand.

HICO^{TM} — the **H**yperspectral **I**mager of the **C**oastal **O**cean — is an experimental spectrometer that became operational on the international space station in 2009.[2] HICO^{TM} was built by the Naval Research Lab, with funding from the Office of Naval Research, as a proof of concept that an ocean imager could be developed and launched quickly using mostly off the shelf parts. HICO^{TM} is aimed at unraveling the complexities of coastal waters. It does this by using a grating spectrometer for viewing the visible and near-IR with a full resolution mode of 1.9 nm, which is binned up by three to provide an operational spectral resolution of 5.7 nm. All spectral channels are measured between 353.4 nm to 1080.9 nm, though only 87 of these channels are normally reported, between 405 nm to 897.6. HICO's ground sampling distance is roughly 90 meters, so its spatial resolution is about three times better than MERIS, and its spectral resolution is about ten times better.

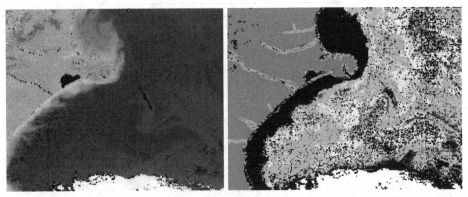

Fig. 3. Left: ESA MERIS RGB image of the east coast of the South Island of New Zealand 28 March 2011. Right: Chlorophyll map computed from ESA MERIS radiance data using ESA's Algal 1 method to estimate chlorophyll concentrations. The image shows a bloom off of Bank's peninsula. The plankton forming the bloom acts as tracer particles that appear to show a chaotic advective mixing near the ocean surface. The red indicates a higher concentration of chlorophyll.

Depending on how opaque the water column is, photos that reach HICO might arrive from up to a depth of 50 meters for very clear water. Still most photons reaching the sensor are from scattering in the atmosphere. Less than one out of ten photons are typically from water scattering. To get any signal at all from water, ocean sensors need a larger dynamic range than land sensors. Both MERIS and HICO typically have a signal-to-noise ratio (SNR) of greater than 200:1. This is also why ocean sensors typically have bigger pixel sizes than land sensors — they get more photons by focusing on a larger section of the earth. The game with HICO is to use its increased resolution both spatially and spectrally to untangle complex waters — that is, to provide estimates of sediment concentration, biological pigments, and even bathymetry for complex water such as the seen on the south east coast of New Zealand (Fig. 3), or the San Francisco Bay (Fig. 4).

Fig. 4. HICO RGB image of San Francisco Bay from 28 September 2011. The color dots indicate sample points for spectra shown in Fig. 5.

2. Shape of Water Spectrum

Much of what we know of ocean microbiological productivity is gleaned from the shape of visible spectra measured from earth orbit. Only visible photons penetrate both the atmosphere and the water column, and interact with biological pigments. This is because those same pigments are busy harvesting energy from those same visible photons. The shape of some typical at-sensor spectra are shown in Fig. 5, for the sample points indicated in Fig. 4. The spectrum marked by a black curve is from relatively clear water off of Pt. Reyes. The spectral shape reflects mostly a composition of the solar irradiance with atmospheric Rayleigh scattering, and falls off roughly as λ^{-4}. This 'clear water' is often called a 'dark pixel' since it does not back scatter photons as much as land or brighter waters say laden with sediments. In particular, there is almost no backscatter from the red end of the spectrum, and this can be used as a quick and dirty method to remove the contribution to the spectrum from atmospheric scattering, namely subtracting off the 'dark pixel' radiance values from the spectra from other sampling points.[3]

In remote sensing lingo, the calibrated at-sensor radiance is called 'L1' (Level 1) data. This radiance data is usually corrected for an atmospheric contribution by modeling procedures that take into account atmospheric scattering and absorption, such as H_2O vapor and O_2 absorption. Atmospherically corrected data results in estimates for above water radiance values, and is called 'L2' data. Lastly, the at-sensor data pixels are usually mapped, or 'warped,' to ground coordinates, which is called 'L3' data.

The dips and bumps in the spectra can be used to identify constituents of the water column. Chlorophyll has an absorption minimum near 660 nm, and a fluorescence peak near 680. However, at very high algal levels this peak tends to shift

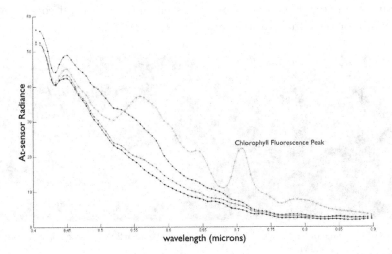

Fig. 5. HICO™ spectra for image of San Francisco Bay from 28 September 2011. The color of the spectra indicates where the sample location is shown in Fig. 4. The green spectra is from the salt ponds in the southern bay and show very high chlorophyll levels as indicated by the prominent chlorophyll fluorescence near 710 nm. The black spectra are from a relatively clear ocean point and show a spectra dominated by Rayleigh atmospheric scattering.

to the red and move as far to the right as 710 nm.[4,5] This can be seen in the green spectrum in Fig. 5 which is sampled from a salt pond in the South Bay.

Other typical spectral features are shown in Fig. 6, a HICO image of the Klamath Lake in Oregon. Spectra from the land vegetation, or thick surface algal mattes, shows a sharp increase around 700 nm. This is called the 'red edge.' We see a muted example of this in shallow lake water shown in the blue spectrum in Fig. 6. The brighter spectrum also highlights any atmospheric absorption features in the spectrum. Most of Klamath Lake is relatively shallow so to get a dark pixel we need to look toward the west to the Lake of the Woods which has clearer waters. The dip in all the spectra near 430 nm is due to a minimum in the solar irradiance.

3. Shape of Ocean Spectra

Scatter and the finite bandwidth of any sensor tends to smooth the data, so even in data that might have a sharp derivative, we can still think of it as outlining a manifold. If we stack the spectral data together along one spatial dimension, we get an outline of a two dimensional manifold, a surface. And tacking on both spatial dimension gives a nonlinear three dimensional manifold.

Fig. 7 shows a line of spectra off of Pt. Reyes in the San Francisco Bay area. To highlight features of the spectra we subtract off the spectrum from a 'dark pixel', at the black dot indicated in the RGB image. The yellow circul is off Bolinas and the mound of spectra between ≈500-620 nm is indicative of higher sediment or CDDOM concentration in the water column. The narrow peak(s) near ≈700 nm are indicative

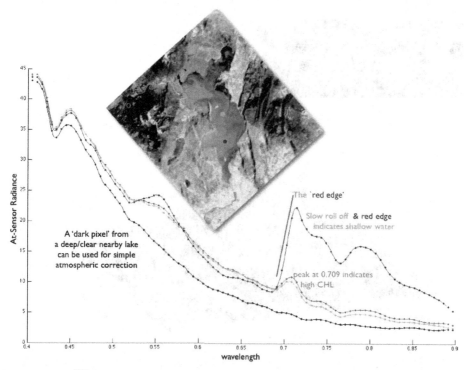

Fig. 6. HICO^{TM} image for Klamath Lake, Oregon, 31 July 2011. The color of the spectra indicates where the sample location is. A 'red edge' is seen in shallow waters. High chlorophyll levels are also indicated by the fluorescence peak. The dip near 430 nm in all the spectra is due to the G Fraunhofer lines in the solar irradiance.

of a surface algal bloom. One occurs north of Bolinas (red dashed circle) and the other south (red circle). Interestingly, a close examination of the RGB image shows that the higher chlorophyll concentrations appear to occupy a strip at the interface between the (fresher) bay water, (salt) sea water tidal interface. Also, the sediment signal also appears to fall off very sharply as we slide across portions of the image that cover fresh water inputs to the ocean. It would be useful to detect 'fronts' in the ocean color data which could be indicative of fresh water, salt water interfaces (perhaps as indicated by sediment concentrations), or biological interfaces (perhaps as indicated by rapid changes in chlorophyll concentrations). In the line spectra shown in Fig. 7, we notice two regions where there looks like an interface between bay and sea water as indicated by the sediment signal. One is the region off of Bolinas, and the second crossing the outlet to San Francisco Bay.

The rich spectral data of HICO should permit the identification of individual pigments in the spectra, and this in turn might allow for species identification. At least three methods are apparent to decompose the hyperspectral data sets as a first step in building product estimators (chlorophyll concentration, sediment concentration, turbidity, ...). First, a global linear decomposition (singular value decomposition

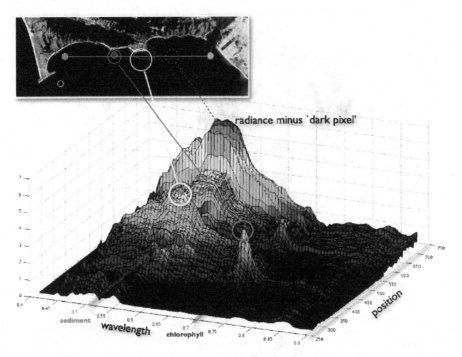

Fig. 7. HICO^{TM} spectral surface for (yellow) line of spectra indicated in RGB image at top, 28 September 2011. The 'dark pixel' (lower left) is subtracted from every spectrum, resulting in a spectra that better highlights extrema like those indicating Chlorophyll ($\approx$700 nm) or sediment ($\approx$500 – 620 nm).

(SVD)). Second, a global nonlinear description, for example, radial basis functions (RBF). And third, a local linear decomposition which is stitched together to estimate the nonlinear manifold. The linear method we are pursing is to create signatures from both *in situ* samples and lab spectra for specific groups of algae based on the principal components of the spectrum and then to match the library of SVD signatures to atmospherically corrected remote sensing data.[6] We are also modeling the data with radial basis functions, a global nonlinear approach.[7] This is very useful for merging data sets, such as HICO and MERIS. Lastly, algorithms developed at the Naval Research Lab stitch together locally linear descriptions of the hyperspectral surface to create an intrinsic coordinate system for the nonlinear manifold. Though computationally intensive, this method provides very good discrimination of similar spectra which are from different water constituents.[8]

4. Derivative Spectroscopy and Embeddings

In laboratory measurements, 'derivative spectroscopy' is commonly used to identify specific elements, molecules, or substances. HICO's high spectral resolution allows for a similar method for remote sensing data. Product algorithms for multispectral sensors like MERIS are not suited to these methods because of their limited spectral sampling. Derivate spectroscopy has previously been used to identify optimal spectral channels for the design of multi-spectral instruments,[9] and to estimate coastal bottom types from aerial hyperspectral data.[10] Here we illustrate the use of methods adapted from derivative spectroscopy for product generation from HICO data.

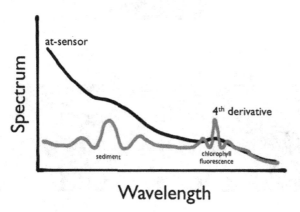

Fig. 8. Sketch showing how derivative spectroscopy can amplify narrow-band features in spectrum. The 4th derivative of the spectrum is marked by red.

Derivate spectroscopy methods are useful in untangling spectral components when the underlying scattering or absorptive features have significantly different half widths. Consider the sketch of a simple at-sensor spectrum shown in Fig. 8. The broadest feature is the Rayleigh scattering which is the monotonically decreasing across the spectrum from blue to red. Taking the first derivative will de-emphasize this broad scattering signature by essentially subtracting a baseline from initial spectrum. Typically either the 2nd or 4th derivatives are examined. The 2nd derivative shows similar extrema locations as the original spectrum, but with the opposite sign. Examining the 4th derivative recovers the sign. To minimize noise in taking the derivative, the original spectrum is smoothed and interpolated before the derivative calculation. This can be accomplished with cubic splines, or more commonly, Savizky-Golay filters are employed.[11] Figure 9 shows where the derivative helps to highlight spectral features sensitive to sediment concentration in recent Columbia River spectra. Spectral signatures identified in this way can create products which are 'regionally tuned,' and built on historical data specific to a coastal area.

Since we are making the effort to compute spectral derivatives, a natural next step is to examine a derivative embedding for each data pixel. Figure 10 shows what we might call a 'spectral braid' for the sampling points around the San Francisco Bay shown in Figs. 4 and 5. The curves are parameterized by wavelength which changes from red (top) to blue (bottom). Each strand is composed of the spectra

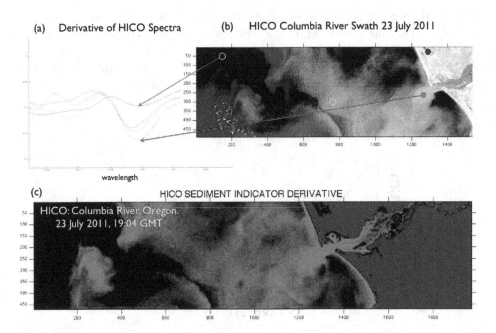

Fig. 9. Analysis of a HICO spectra using derivative spectroscopy. (a) The derivative spectra calculated at points indicated from a HICO image of the Columbia River (b) taken 23 July 2011. (c) A 'regionally tuned' map of sediment concentration (bright red is high sediment) emanating from the Columbia River.

minus a dark pixel, so the black strand is a straight line, and all the other strands are referenced to the dark pixel spectrum.

Derivative embeddings of hyperspectral data can be interesting on several counts. For instance, if derivative data helps to separate close spectra, then a derivative embedding creates a natural metric for measuring this separation. Namely, spectral separation can be measured as a norm in an embedding space, such as a Euclidean norm between between two spectral samples,

$$d(\lambda_n) = \sqrt{(L_b - L_a)^2 + ... + (L_b^{n'} - L_a^{n'})^2} \tag{1}$$

where $L^{n'}$ denotes the nth derivative of radiance w.r.t. λ for two spectral braids a and b), or a functional norm between spectrum by integrating over λ.

In particular, for three dimensional embeddings we can also begin to compare the geometric or topological properties of spectra, and use geometric or topological signatures of an individual spectrum , or collections of spectra, to address questions like identifying rapid changes in spectra (e.g. front identification), or detecting subtle changes in spectra (e.g. product indicators).

Using λ as a parameter for an embedding coordinate is very helpful for computing geometric properties (for instance, linking number can be computed as a single instead of a double integral[13]). However, we can consider other embeddings as well,

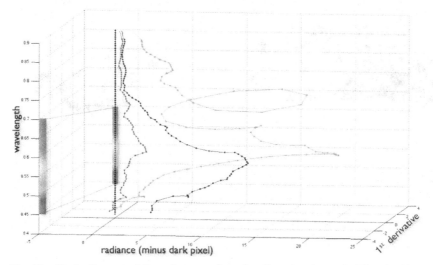

Fig. 10. Derivative embedding of spectra from San Francisco image in Figs. 4 and 5.

such as three dimensional delay embeddings. Figure 11c shows a three dimensional delay embedding for the sample points (red, blue) indicated in Fig. 11a. By connecting the endpoints we can examine the knots and links formed by spectra. A similar view is possible with braids, we simply connect end points there too (Fig. 11d), or, for the more mathematically savvy, map the curve to the unit sphere to form a 'tantrix' curve, which is closed with a spherical geodesic. The writhe of a open curve closed in this way is well-defined (modulo 1).[14]

5. Torsion, Writhe, Linking, and All That

The spectral braid forms a 3-dimensional curve, so we briefly review the intrinsic geometry of 3-D space curves. Instead of just looking at curves, though, we consider 'tubes,' which will enable us to examine the geometry of nearby spectra.[15]

A spectral braid is an open curve in $\mathbb{R}^3$. Surround this 'axis' curve by a circle of almost parallel curves a distance ϵ away. The union of these circular curves forms a tube. The axis curve is parameterized by λ, the wavelength, or the arc length s. Calling this axis curve $\mathbf{x}(s)$, the tangent vector to the axis curve is,

$$\hat{\mathbf{T}}(s) = \frac{d\mathbf{x}}{d\mathbf{s}}. \qquad (2)$$

Define $\hat{\mathbf{V}}(s)$ to be the unit normal to $\hat{\mathbf{T}}(s)$, the normal vector points in the direction $\mathbf{y}(s) = \mathbf{x}(s) + \epsilon \hat{\mathbf{V}}(t)$. The tube inherits a local coordinate system from the axis curve, namely (s, ϕ), where the secondary curve $\mathbf{y}(s)$ has coordinates $(s, 0)$. A choice of $\mathbf{y}$ fixes the coordinate system and is called a 'framing'.[16] If we set $\hat{\mathbf{W}} = \hat{\mathbf{T}} + \hat{\mathbf{V}}$ then the tube surface is traced out by

$$\mathbf{y}(s, \phi) = \mathbf{x}(s) + \epsilon(\cos \phi \hat{\mathbf{V}}(s) + \sin \phi \hat{\mathbf{W}}(s)).$$

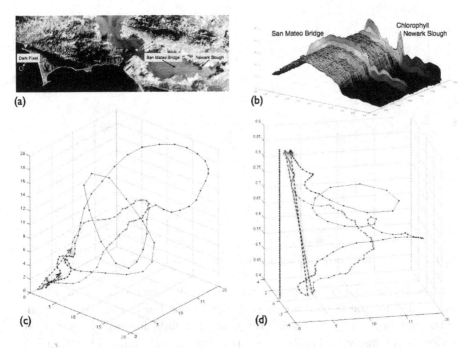

Fig. 11. Embedded spectra for the HICO scene (a) of the San Francisco Bay (28 September 2011). The sample points are indicated by red and blue dots in (a), and the line of spectra between the two end points is shown in (b). The sample points pass over the San Mateo bridge which is easy to see in (b), and the Newark Slough in the South Bay with high chlorophyll levels. A delay embedding of the spectrum for the sample points in (a) is shown in (c). The embedding curve can be 'closed' by attaching the end points which is indicated by a dotted line with an arrow. The associated braid spectra (d) can also be closed by attaching endpoints. The linking between the spectra is $\mathcal{L} = 1$ in both embeddings (c) and (d).

The 'Frenet frame' describes the local geometry of the axis curve. Define the 'curvature,' κ,

$$\kappa = \left| \frac{d\hat{\mathbf{T}}(s)}{ds} \right|. \tag{3}$$

The 'normal' and 'binormal' vectors are ($\kappa \neq 0$),

$$\hat{\mathbf{N}} = \frac{1}{\kappa} \frac{d\hat{\mathbf{T}}(s)}{ds}, \quad \hat{\mathbf{B}} = \hat{\mathbf{T}} \times \hat{\mathbf{N}}. \tag{4}$$

Taken together, the vectors $\{\hat{\mathbf{T}}, \hat{\mathbf{N}}, \hat{\mathbf{B}}\}$ form a orthonormal right handed basis with 'torsion,' τ, described by the Frenet-Serret equations

$$\frac{d\hat{\mathbf{T}}(s)}{ds} = \kappa \hat{\mathbf{T}}, \quad \frac{d\hat{\mathbf{N}}(s)}{ds} = \tau \hat{\mathbf{B}} - \kappa \hat{\mathbf{T}} \quad \frac{d\hat{\mathbf{B}}(s)}{ds} = -\tau \hat{\mathbf{T}}. \tag{5}$$

For the framing choose $\hat{\mathbf{V}} = \hat{\mathbf{N}}$.

The Gauss linking number is typically defined in terms of a double integral.[12] Here we use the the fact the spectral braid is open and parameterized by λ to compute the linking number in terms of a sum of single integral.[13] Consider the rotation, or winding, about the vertical direction, z, along the wavelength axis in Fig. 10. The linking number can be expressed as a 'winding angle' of curves as we trace them out in the z (wavelength) direction. Consider two spectral braids $\mathbf{x}_i$ and $\mathbf{y}_i$, and their difference vector at a height z, $\mathbf{r}_{ij}(z) = \mathbf{x}_j(z) - \mathbf{x}_i(z)$. The rotation rate of $\mathbf{r}_{ij}$ is given by,

$$\frac{d\Theta_{ij}}{dz} = \frac{\hat{\mathbf{z}} \cdot \mathbf{r}_{ij} \times \mathbf{r}'_{ij}(z)}{|\mathbf{r}_{ij}(z)|^2}, \tag{6}$$

and the net winding number is

$$\Delta\Theta_{ij} = \int_{z_1}^{z_2} \frac{d\Theta_{ij}}{dz}. \tag{7}$$

Divide the curve into pieces separated by the turning points $dz/ds = 0$, indexed by z_i to z_j and define

$$\sigma_i(z) = \begin{cases} -1, & z \in (z_i, z_{i+1}) \quad \text{and} \quad dz/ds < 0 \\ +1, & z \in (z_i, z_{i+1}) \quad \text{and} \quad dz/ds > 0 \\ 0, & z \notin (z_i, z_{i+1}). \end{cases} \tag{8}$$

Then the linking number between the two curves can be shown to be[13]

$$\mathcal{L}_k = \sum_{i=1}^{n} \sum_{j=1}^{m} \frac{\sigma_i \sigma_j}{2\pi} \Delta\Theta_{ij}. \tag{9}$$

For our product indicators, it will turn out that the net winding number between two curves, $\tilde{\Lambda} = \int_{z_1}^{z_2} \frac{d\tilde{\Lambda}}{dz} dz$, is initially more useful than the linking number.

The Călugăreanu theorem states that the linking is equal to the twist plus the writhe,

$$\mathcal{L}_k = \mathcal{T}_w + \mathcal{W}_r. \tag{10}$$

The writhe only depends on the axis curve. The derivative of the twist, $d\mathcal{T}_w/ds$ measures the rotation rate of the secondary curve about the axis curve. Using Călugăreanu theorem the writhe can be calculated by first computing the linking and twist. Twist is a local quantity, so we can define a twist over a finite interval for an open curve. A similar quantity for writhe can be computed that is sometimes called the polar writhe.

6. Applications: Edge Detection and Product Indicators

Edge detection is important in ocean remote sensing to help define the biogeochemical boundaries of a region, and to look for correlations between, say, biological productivity and the geochemical substrate. Methods for edge detection in hyperspectral data often use a (amplitude dependent) thresholding method.[17] To some degree, this amplitude dependence can be reduced by considering (atmospherically corrected) normalized water leaving radiance, or reflectance [b]. Still, it would be useful (say when working with L1 data) to have some measures of changes that are, as much as possible, intrinsic to geometry, the shape, of the spectral data.

Similarly, for a 'product indicator,' we are not so much interested in producing a calibrated measurement of a product like chlorophyll, but rather to create an indicator that is sensitive to a change of state from low algal levels to very high concentrations of algae that could be indicative of a harmful algal bloom (HAB).

In both cases we would like the algorithm to be relatively insensitive to amplitude changes. We can imagine methods that depend only on the geometry of an individual braid spectrum, and look for changes in this geometry as we move across a scene — for instance a jump in the winding number as we move across a front, or in and out of a region with a high product concentration. And we can consider properties of pairs, or collections of orbits, for instance a change in linking number between orbit pairs. We call such methods geometric or topological indicators since, as much as possible, we will try to make them insensitive to the ambient metric structure, or put another way, make them relatively robust to changes in absolute signal levels. Here we will just provide a few examples of algorithms we have tried.

Our construction for a spectral braid provides us with a natural phase function from the winding number, which should be relatively insensitive to the overall amplitude, and provides us with a alternative method to compare spectra.

As a first example we return to the HICO^{TM} scene of the San Francisco Bay (Fig. 4) and focus on the San Pablo Bay to the north. In Fig. 12a we indicate by color dots the points where we've taken three spectral samples across the Bay. Fig. 12b shows the phase difference of the spectral samples computed from the winding number of the spectral braid. The cyan curve (the difference between the blue and green sample points in Fig. 12a) shows the difference between what appears to be visually similar water masses when also compared to the red sample point. Toward the red end of the spectrum (between 0.575 nm to 0.75 nm), the phase difference from the blue and red sample points appears to be clearly separated. Defining a phase difference to be greater than the standard deviation below 0.55 nm of the blue and cyan curves (Fig. 12b) allows us to use the phase difference function above 0.575 nm to produce a boundary (edge detection) between these two water masses, which is shown in Fig. 12c. The edge (yellow pixels in Fig. 12c) identifies the San Pablo Strait, a channel in the San Pablo Bay used for shipping. Though not the

---

[b]It is helpful to convert radiance to the apparent reflectance to get a more uniform comparison. The measured at-sensor radiance can be expressed as $L_{obs} = L_a + L_{sun}\tau\rho$ where L_a is the path radiance, ρ is the surface reflectance, L_{sun} is the solar radiance above the atmosphere, and τ is the 2-way transmittance for the Sun-surface-sensor path. Define the satellite apparent reflectance as $\rho^*_{obs} = \frac{\pi L_{obs}}{\mu_0 E_0}$, and $\rho^*_{obs} = T_g[\rho_a + \frac{\tau\rho}{1-\rho s}]$. By inverting the last equation we get $\rho = \frac{\rho^*_{obs}}{T_g - \rho^*_a}/[\tau + s(\rho *_{obs}/T_g - \rho^*_a)]$.

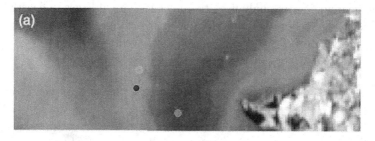

(a) 28 September 2011

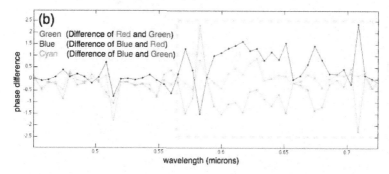

(b) The phase difference computed from the 'winding number' of the spectra

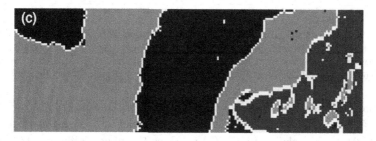

(c) Edge detection is computed from phase difference between $0.575\ nm < \lambda < 0.75\ nm$

Fig. 12. HICO image of San Pablo Bay off of Pinole Point Park. The phase difference (b) is computed from the 'winding number' of the spectra indicated by the colored dots in (a). The edge detection (c) identifies the San Pablo Strait.

only way to identify the channel from remote sensing data, the use of embedding to produce a hyperspectral spectral braid does appear to validate the method, and provides an example of how the intrinsic geometry of spectra — the shape of ocean color — can be used to process hyperspectral remote sensing data.

As a second example we move to the South Bay of the same HICO™ image. Fig. 13 shows two spectral samples from the South Bay, with one sample (green dot) from a salt pond, the Newark Slough, which apparently is an algal mat. The jump in the phase difference of the spectral braid at both 697 and 709 nanometers shows that these channels provide an excellent indicator of algal mats in this image.

The 709 nm channel is commonly used to detect high algal levels in coastal remote sensing.[4] The evidence here suggests that, at least for these salt ponds, the 697 nm channel provides a good classifier for algal mats as well.

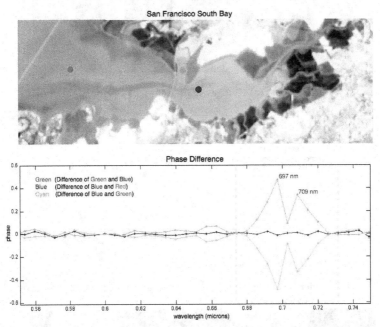

Fig. 13. HICO image of the South San Francisco Bay 28 September 2011 (top). The green dot indicates spectra from the Newark Slough which is an algae mat. The lower plot shows the phase difference between the spectra from the Newark Slough and the surrounding bay water. The extrema at 0.697 and 0.709 can be used to create an indicator function for the algae mat.

As a final example we take a look at the mouth of the San Francisco Bay with HICO™. As we just mentioned, the 709 nm channel is used for coastal remote sensing of high algal levels. In Fig. 14(a) we show the phase difference function for water masses that appear inside and outside of the edge of what appears to be bay water and sea water in Fig. 14(b). Highlighting in red the pixels that contain a jump in the phase difference at 709 nm, reveals three patches of presumably high algal concentrations that exist at the interface of the bay and sea waters. The patches are mostly likely a coastal convergence zone containing a mix of materials from the bay and possibly foam [c].

In this chapter we described methods to process hyperspectral ocean remote sensing data which attempt to use geometrical and topological constructions, copying where possible — "good artist copy, great artist steal" — from the work in topological dynamics of Bob Gilmore. In going from dynamics to spectra we replace the parameter t with λ, and then shamelessly steal from Bob's work in dynamics to begin to unravel the complexities of ocean color using the spirit and language of

---

[c]C. O. Davis, Personal communication (2011).

Bob's program for nonlinear dynamics, and in the process begin to see the shape of ocean color.

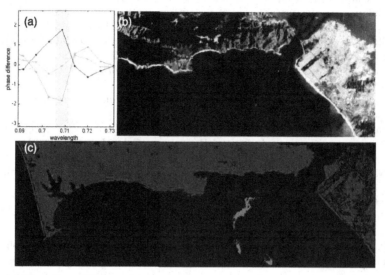

Fig. 14. (a) The phase difference function for spectra at the mouth of the San Francisco Bay showing that the 709 nm HICO channel can be used to indicate chlorophyl rich water. (b) HICO image of mouth of San Francisco Bay, 28 September 2011. (c) Indicator function for high chlorophyll levels which appear to show a high concentration of chlorophyll at the interface of bay water and sea water.

References

1. M. Twain, *Following the Equator*, American Publishing Co., Hartford. (1897). See Chapter 15 for Twain's visit to Dunedin.
2. M. R. Corson & C. O. Davis, A new view of the coastal oceans from the space station, EOS, *Transactions of the American Geophysical Union*, **92** (19), 161 (2011).
3. N. Tufillaro, C. O. Davis & K. B. Jones, Indicators of plumes from HICO^{TM}, *Proceedings of Ocean Optics XX*, Anchorage, AK, (27 September –1st October, 2011).
4. J. Gower, S. King, G. Borstad & L. Brown, Use of the 709 nm band of MERIS to detect intense plankton bloom and other conditions in coastal waters, *Proceedings of the 2004 Envisat & DRS Symposium*, Salzburg, Austria, (6–10 September 2004).
5. A. Gilerson, A. Gitelson, J. Zhou, D. Gurlin, W. Moses, I. Ioannou & S. Ahmed, Algorithms for remote estimation of chlorophyll-a in coastal and inland waters using read and near infrared bands, *Optical Express*, **18** (23), 24109-25 (2010).
6. J. D. Ortiz, D. L. Witter, K. A. Ali, N. Fela, M. Duff & L. Mills, Evaluating multiple color producing agents in Case II waters. *International Journal in Remote Sensing*, in press.
7. H. Wendland, *Scattered Data Approximation*, Cambridge University Press (2010).
8. C. M. Bahmann, T. L. Ainsworth & R. A. Fusina, Exploiting manifold geometry in hyperspectral imagery, *IEEE Transactions in Geoscience and Remote Sensing*, **43** (3), 441-454 (2005).

9. Z-P. Lee, K. R. Carder, R. Arnone & M-X He, Determination of primary spectral bands for remote sensing of aquatic environments, *Sensors*, **7**, 3428-3441 (2007).
10. E. M. Louchard, R. P. Reid, C. F. Stephens, C. O. Davis, R. A. leathers, T. V. Downes & R. Maffione, Derivative analysis of absorption features in hyperspectral remote sensing data of carbonate sediments, *Optics Express*, **10** (26), 1573-1584 (2002).
11. A. Savitzky & M. J. E. Golay, Smoothing and differentiation of data by simplified least squares procedures, *Analytical Chemistry*, **36** (8), 16271639 (1964).
12. R. Gilmore & M. Lefranc *The topology of chaos: Alice in stretch and squeezeland*, Wiley-VHC, 2nd ed., Weinheim, Germany (2011).
13. M. A. Berger & C. Prior, The writhe of open and closed curves, *Journal of Physics A*, **39**, 8321-8348 (2006).
14. E. L. Starostin In: *Physical and numerical models in knot theory including applications to the life sciences. Series on Knots and Everything.* (Calvo J, K. Millett, E. Rawdon, & A. Stasiak eds). World Scientific Publishing, Singapore, pp. 525-545 (2005).
15. G. H. M. van der Heijden & J. M. T. Thompson, Helical and localised buckling in twisted rods: a unified analysis of the symmetric case, *Nonlinear Dynamics*, **21** (1), 71-99 (2000).
16. K. H. Ko & L. Smolinsky The framed braid group and 3-manifolds, *Proceedings of the American Society*, **115** (2), 541-551 (1992).
17. A. G. P. Shaw & R. Vennel, Measurements of an oceanic front using a front-following algorithm for AVHRR SST imagery, *Remote Sensors Environment.* **75**, 47-82 (2001).

Chapter 12

Low dimensional dynamics in biological motor patterns

Gabriel B. Mindlin

Universidad de Buenos Aires
Departamento de Fisica
FCEN-UBA Pabellon I
Ciudad Universitaria
1428 Buenos Aires, Argentina

Bob Gilmore leads a beautiful program of research that aims at providing a classification of chaotic dynamics present in natural systems. That program allows to build confidence, or refute models proposed to account for observed complex dynamics. This strategy has been particularly useful for systems whose behavior can be explained in terms of low dimensional dynamical systems. Years after the time I spent with Bob and his coworkers collaborating in that project, I found experimental evidence of low dimensional, non trivial dynamics in the motor patterns used by some songbirds during their vocalizations. This is intriguing, since motor patterns emerge out of the interaction of an enourmous number of individual dynamical units. Under which conditions do large arrays of out of equilibrium units present low dimensional average activity? Is synchronization necessary to achieve these states? Are there other mechanisms leading to low dimensional average dynamics? In this chapter, I review recent studies on the macroscopic behavior displayed by a large set of coupled excitable units which are periodically forced. I show that low dimensional, yet non trivial average dynamics emerges in a regime presenting a time dependent degree of synchrony.

Contents

1. Average Dynamics of a Large Set of Driven Excitable Units 273
2. Solutions 276
3. Final Remarks 278
References 279

Physics and Biology share a common history of fruitful interactions and yet, their approaches when it comes to understanding nature are very different. In physics one is invited to identify the basic mechanisms behind a phenomenon, and simplifications are part of the daily routine. In fact, many physicists work in conditions which are hard (if not impossible) to find outside their labs. Lasers, as conducted, for example, by Jorge Tredicce, are devices originally conceived to show stimulated emission in all its splendor. In such conditions, the number of variables that are involved in a dynamical process might be reduced. On the other hand, biologists are very cautions when it comes to simplifying. In fact, there is a very good reason: each overlooked detail might be the difference between life and death. There is a minimum complexity beyond which life is simply impossible and therefore, biology is used to dealing with complexity. Physicists frequently get frustrated when interacting with biologists, since the latter are not inclined to build operational, mathematical models that might lead to quantified predictions. Biologists, on the other hand, react strongly and skeptically (in the best case) or with indifference (most likely) when they see a biological system being described by a few equations. Building mathematical models, describing dynamical phenomena by low order differential equations are considered privileges of those dealing with simple (understood as "non living") systems.

Neuroscience, in this regard, might be considered a particularly challenging field. It is arguably said once and again that the brain is among the most complex dynamical devices found in nature. On the other hand, an important part of behavior consists of this highly complicated nervous system generating motor outputs that drive peripheral biomechanical devices: outputs that in many cases consist of reasonably low dimensional signals. One beautiful example is animal gaiting. Animal locomotion usually uses distinct periodic patterns of leg movements, known as gaits. It has been observed that most gaits possess a degree of symmetry. Martin Golubitsky and coworkers have shown a similarity between the generalities of coupled nonlinear oscillator arrangements and the observed symmetries of gaits. That leads them to speculate on the constraints on the general structure of the neural circuits, i.e. central pattern generators, that control locomotion.[1] Another example is birdsong. It has been shown that just a few simple motor gestures responsible for the respiratory patterns and the tension of muscles controlling the avian vocal organ were enough to drive a phonatory model into generating realistic birdsong[2,3,4]. In Figure 1 we show the respiratory patterns involved in birdsong production.

Interesting enough, direct measurements of the respiratory patterns during song production in the case of the domestic canary[5] showed that the behind the diversity of syllables there was an common theme: the respiratory patterns could be approximated by different subharmonic solutions of a driven nonlinear system[5,6,7] (see Figure 2). These examples are illustrative of a large class of problems where complex neural architectures generate patterns of activity that, in average, result in low dimensional motor gestures.

These observations are quite motivating, since in principle there are many ways in which a large ensemble of out of equilibrium units might end up displaying a dynamics which, in average, is low dimensional. Probably the simplest of all is complete synchronization.[8] If all the units end up synchronizing, the average dynamics will present the dimensionality necessary to represent the dynamics of one unit. Needless to say, complete synchronization is only one special solution that an ensemble of units can display, and work in this area has a rich history.

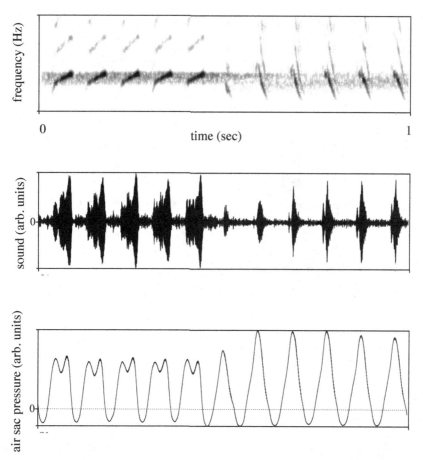

Fig. 1. In order to vocalize, a bird has to pressurize its air sacs. For uttering each syllable, a large pressure fluctuation is necessary. When the pressure exceeds some threshold, the labia in its vocal organ starts to modulate the airflow, and sound is produced. When a canary sings different syllables, different pressure patterns are used. A canary typically repeats a syllable several times before switching to a different syllable type. In the first panel we show the frequency representation of a song fragment consisting of eleven syllables. The actual sound recording is illustrated in the middle panel. The bottom inset displays the air sac pressure.

From a dynamical perspective, the problem of many interacting nonlinear units is a difficult one. Yoshiki Kuramoto[9] studied a simplified model to inspect the dynamics of a population of coupled oscillators. He described each one in terms of its phase, and assumed a simple functional form to describe the coupling between the interacting units. Under these assumptions, he was able to prove that whenever the coupling coefficient exceeds a threshold, mutual synchronization among some units takes place. Kuramoto studied how an order parameter describing the size of the

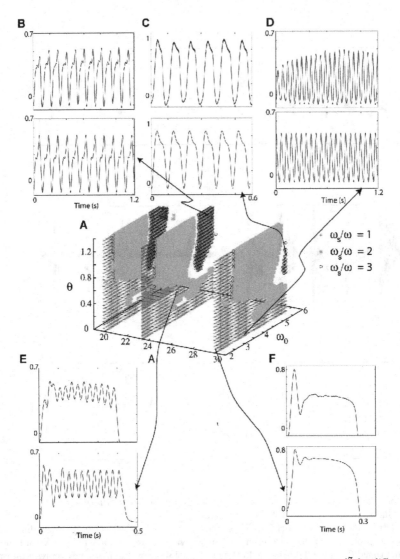

Fig. 2. A periodically forced 2d nonlinear differential equation was integrated[7] for different values of its parameters (A, w, θ) (A). The solutions were very similar to experimentally measured air sac pressure patterns in singing canaries. For (B) to (F), the top panels correspond to experimental recordings, and the bottom ones to the numerical solutions of the low dimensional model. The air sac pressure is a physiological gesture that emerges out of the average over the activities of a large set of neurons driving respiratory muscles.

synchronized subpopulation depends on the coupling, and showed that the stronger the coupling, the larger the degree on synchrony. Moreover, the non synchronic units would not contribute to the average activity in the limit of infinite number of interacting units. In this way, a second mechanism can be described such that low dimensional dynamics in average is obtained: a core of synchronized units coexisting with a set of unlocked units that do not contribute to the average.

In this chapter I will describe a different dynamical regime[10,11] in which low dimensional and yet complex dynamics can be found in the average activity of an array of out of equilibrium units. Interesting enough, in this regime there is a precise time dependent degree of synchrony. The model used to present this regime is inspired in a frequently found neural architecture: it consists of a large set of driven, globally coupled excitable units. Excitability is a dynamical state ubiquitously found in the nervous system. It might be present when a population of excitatory neurons interacts with one of inhibitory neurons (those two populations together constitute a dynamical unit known as "neural oscillator", even if its dynamics might be excitable).

In the spirit of Kuramoto, Winfree[12] and other researchers interested in the dynamics of large nonlinear arrays, I will work with units whose dynamics can be represented by phase equations. The behavior displayed by our array illustrates a different mechanism in which complex, yet low dimensional average behavior can be obtained in a large set of out of equilibrium units.

1. Average Dynamics of a Large Set of Driven Excitable Units

The study of the generic features presented by systems consisting of large number of coupled oscillators has a long history.[8] In part, this is due to the wide range of areas where this problem emerges. Kuramoto made a seminal contribution to the field by introducing a mathematical model which allowed some analytic treatment. The dynamics of each oscillator was described in terms of its phase, and the coupling between the different oscillators was assumed to be sinusoidal, what simplified the analysis of the model. By introducing a mean field function, he obtained an indicator of the coherence of the units. He found that the oscillators are forced by the mean field, and depending on their parameters, they can synchronize with it. Recently, Edward Ott and Thomas Antonsen[13] found that the Kuramoto model presents an invariant manifold, that is, a set of states for which the macroscopic dynamics becomes low dimensional. This strategy was later applied to solve closely related problems, as the dynamics of a set of periodically forced coupled oscillators.[14]

In this chapter I will describe the dynamics of a set of forced coupled excitable systems. As mentioned in the introduction, it is common to find neural architectures consisting of coupled neural oscillators, i.e. interconnected subpopulations of excitatory and inhibitory neurons. For a wide region of their parameter space, neural oscillators present excitable dynamics: unless they are perturbed, they rest on their quiescent state; after being perturbed, the return to equilibrium can follow qualitatively different paths. In neural oscillators, excitability is lost in a saddle-node bifurcation producing a limit cycle, giving rise to oscillations born with non zero amplitude and zero frequency.[15] A simple phase equation can account for this dynamics:

$$\dot{\theta} = \omega - \gamma \sin(\theta). \tag{1}$$

If $\frac{\omega}{\gamma} < 1$, the system has two fixed points, one stable and the other unstable. The separation between these fixed points depends on the $\frac{\omega}{\gamma}$ ratio. The qualitative behavior of this system is understood comparing the response of the system to different initial conditions. If these are close to the quiescent state, the system rapidly decays to the stable fixed point, while for initial conditions above some threshold, the response of the system is a large excursion in the phase space.

In this study, the forcing units will be represented mathematically by phase oscillators. I assume that there is global coupling between all driven excitable units, global coupling between all forcing units, and a directed coupling from the driving units to the network of excitable, driven units.

In this way there are two sets of phase oscillators. Set 1 refers to the driving set, while set 2 refers to the driven set. Greek letters are used to indicate population number and Latin letters are used to index the elements within a population. Using this notation:

$$\dot{\theta}_i^\sigma = \omega_i^\sigma - \gamma_i^\sigma \sin(\theta_i^\sigma) + \sum_{\sigma'=1}^{2} \frac{K_{\sigma\sigma'}}{N^{\sigma'}} \sum_{j=1}^{N^{\sigma'}} \sin(\theta_j^{\sigma'} - \theta_i^\sigma), \tag{2}$$

where for each population the natural frequency of the ith oscillator is denoted by ω_i^σ. The parameter ruling the excitable nature of the oscillator is γ_i^σ, the number of oscillators in the population is given by N^σ and $K_{\sigma\sigma'}$ stands for the connectivity matrix.

In the limit $N^\sigma \to \infty$, these populations can be described in terms of density probability functions $f^\sigma(\theta, \omega, t)$, with $\sigma = 1, 2$. The evolution of f^σ is given by the continuity equation,

$$\frac{\partial f^\sigma}{\partial t} + \frac{\partial}{\partial \theta}(f^\sigma v^\sigma) = 0, \tag{3}$$

where the velocity v is given by:

$$v^\sigma(\theta^\sigma, \omega, t) = \omega^\sigma - \gamma^\sigma \sin(\theta^\sigma) + \sum_{\sigma'=1}^{2} K_{\sigma\sigma'} \int_{-\infty}^{\infty} \int_0^{2\pi} \sin(\theta' - \theta^{\sigma'}) f^{\sigma'}(\theta', \omega, t) d\theta' d\omega. \tag{4}$$

In this way, the state of the system is described by the density functions f^σ. I have defined f in such a way that the fraction of oscillators with phases between θ and $\theta + d\theta$ and natural frequencies between ω and $\omega + d\omega$ is given by $f(\theta, \omega, t)d\theta d\omega$. Therefore, in order to close the system one needs that the following equations are satisfied:

$$\begin{cases} \int_{-\infty}^{\infty} \int_0^{2\pi} f^\sigma(\theta, \omega, t) \, d\theta d\omega = 1 \\ \int_0^{2\pi} f^\sigma(\theta, \omega, t) \, d\theta \quad = g^\sigma(\omega). \end{cases}$$

As suggested by Kuramoto, a system complex order parameter can be defined:

$$r^\sigma(t) = \sum_{\sigma'=1}^{2} K_{\sigma\sigma'} z^{\sigma'}, \qquad (5)$$

where z^σ is the complex average of the oscillators in σ-th population given by the following equation:

$$z^\sigma = \int_{-\infty}^{\infty} \int_0^{2\pi} e^{i\theta} f^\sigma(\theta,\omega,t) d\theta d\omega. \qquad (6)$$

With these definitions, the velocity (4) simplifies to:

$$v^\sigma(\theta^\sigma,\omega,t) = \omega^\sigma + \frac{\gamma^\sigma}{2i}(e^{i\theta^\sigma} - e^{-i\theta^\sigma}) + \frac{1}{2i}(e^{-i\theta^\sigma} r^\sigma - e^{i\theta^\sigma} r^{\sigma*}), \qquad (7)$$

where the super index * denotes complex conjugation. Let us assume that it is sufficient to characterize the forcing population by a harmonic fluctuation. It is possible to represent this by making $\gamma^{\sigma=1} = 0$ and $K_{12} = 0$. In this way one reduces this population to one behaving as in the case described by Kuramoto. It is known that above a certain threshold in the coupling strength K_{11}, the oscillators synchronize. This is the way to obtain the desired coherent fluctuations needed to represent a harmonic driving. Also by changing this parameter, the amplitude of the fluctuations can be controlled. It is possible to simplify the notation and leave $\gamma^{\sigma=2} = \gamma$ as a parameter of the model.

It is a conventional strategy to address this problem by expanding f^σ in a Fourier series in θ,

$$f^\sigma(\theta^\sigma,\omega,t) = \frac{g(\omega)}{2\pi}\left[1 + \sum_{n=1}^{\infty} f_n^\sigma(\theta^\sigma,t)e^{in\theta^\sigma} + c.c.\right], \qquad (8)$$

with c.c. denoting complex conjugation. Replacing (8) and (7) into (3) one obtains, in principle, an infinite dimensional system of equations for f_n^σ.

An important breakthrough in the analysis of the problem was introduced by Antonsen and Ott, who[13] noticed that the following ansatz $f_n^\sigma(\omega,t) = (\alpha_\sigma(\omega,t))^n$ would satisfy all the amplitude equations as long as certain equations would be satisfied by α_σ, namely:

$$\begin{cases} \dot{\alpha}_1 = -i\omega\alpha_1 + \dfrac{K_{11}}{2}(\alpha_1 - |\alpha_1|^2 \alpha_1) \\ \dot{\alpha}_2 = -i\omega\alpha_2 + \dfrac{\gamma}{2}(1 - \alpha_2^2) + \dfrac{K_{22}}{2}(\alpha_2 - |\alpha_2|^2 \alpha_2) + \dfrac{K_{21}}{2}(\alpha_1 - \alpha_1^*\alpha_2^2). \end{cases} \qquad (9)$$

By further assuming that $g^\sigma(\omega)$ is a Lorentzian,

$$g^\sigma(\omega) = \frac{\Delta^\sigma}{\pi[(\omega - \omega_0^\sigma)^2 + (\Delta^\sigma)^2]}, \qquad (10)$$

and that $\alpha_\sigma(\omega,t)$ satisfies certain analyticity conditions in the complex ω-plane, Ott and Antonsen[13] evaluated eq. (9) by contour integration. Multiplying both sides of (9) by $g^\sigma(\omega)$ and using the residue theorem the following equations for the evolution of $\alpha_\sigma(\omega,t)$ is obtained:

$$\begin{cases} \dot{\alpha}_1(\omega_0^1 - i\Delta^1, t) = -i(\omega_0^1 - i\Delta^1)\alpha_1 + \dfrac{K_{11}}{2}(\alpha_1 - |\alpha_1|^2 \alpha_1) \\ \dot{\alpha}_2(\omega_0^2 - i\Delta^2, t) = -i(\omega_0^2 - i\Delta^2)\alpha_2 + \dfrac{\gamma}{2}(1 - \alpha_2^2) + \dfrac{K_{22}}{2}(\alpha_2 - |\alpha_2|^2 \alpha_2) \\ \phantom{\dot{\alpha}_2} + \dfrac{K_{21}}{2}(\alpha_1 - \alpha_1^* \alpha_2^2). \end{cases} \quad (11)$$

The exact distribution function of the oscillators can be obtained by performing the summation in Eq. (8) as discussed in.[13] The order parameter of each subpopulation is given by (5). By using the Fourier expansion (8) of the distribution function and the ansatz in (6), the following relation between the distribution function and the order parameters for each set of units is obtained:

$$z^\sigma(t) = \int_{-\infty}^{\infty} \int_0^{2\pi} e^{i\theta} f^\sigma(\theta,\omega,t)\, d\theta d\omega = \int_{-\infty}^{\infty} \alpha_\sigma^* g(\omega)\, d\omega = \alpha_\sigma^*(\omega_0^\sigma - i\Delta^\sigma, t). \quad (12)$$

We can write the order parameter in its Euler form as, $z^\sigma(t) = \rho^\sigma(t) e^{i\theta^\sigma(t)}$. The average position of the oscillators is given by $\theta^\sigma(t)$ while the modulus $\rho^\sigma(t)$ measures how peaked the distribution is. Remarkably, there is no need to study an infinite set of equations for all the amplitudes of the phase distributions: the success of the ansatz indicates that the dynamics can be restricted to a low dimensional manifold. Notice that in accordance with the results by Yoshiki Kuramoto, the first population can be described in terms of a global oscillation, whose amplitude depends on the coupling among its units. The second population will be described by the complex order parameter α_2, obeying a forced nonlinear differential equation.

2. Solutions

In order to inspect the dynamics of this system, we can perform a simulation for the set of oscillators, and compute the average phases of the forcing and driven sets of oscillators. According to the computations described in the previous section, for an infinite set of oscillators in each population, the results should be equivalent to inspecting the behavior of α_1^* and α_2^*.

Figure 3 displays the average activity of 100 driven oscillators,[10] for parameter values such that the averaged system presents chaotic dynamics. Figure 3 (a) shows the average activity of the driven population of excitable cells: ($< cos(\theta_i) >$, $< sin(\theta_i) >$). The complexity of the projected time series, the delicate structure of forbidden bands (signature of coexisting unstable periodic orbits), are indications that the dynamics might be chaotic. If that is the case, we expect a single trajectory to coexist with infinitely many unstable periodic orbits. The method of close returns[16] allows to find pieces of the data set that resemble periodic orbits. In fact, those periodic orbits can be thought as a backbone of the strange attractor, and their description the is key to classify strange attractors.

We look for a time interval for which the system is close to an unstable periodic orbit (b). In fact, the displayed trajectory is a good approximation of a period 2 orbit. Since we have the phases for the individual oscillators, it is possible to inspect how the extended system achieved this average dynamics. We compute a histogram counting how many oscillators present a phase within $(\theta, \theta + 2\pi/100)$ at each time t (c). Notice that when the order parameter approaches zero, the oscillators disperse: different oscillators are found in different phases. The system behaves with a high degree of synchrony for some time, desynchronizes for a while, recovers some synchrony... in such a way that the average activity of the population displays very well structured chaotic dynamics. In other words, low dimensional dynamics could emerge from an extended system as the result of a core of units locked sub harmonically to a forcing, coexisting with asynchronous units whose average dynamics do not contribute to the average activity of the system. That is the case, for example, in the seminal work by Kuramoto. This is not the case for our system, which requires time dependent levels of synchrony. Notice that the dynamics of the driven population, in average, is described by a periodically forced equation for the complex variable α_2. Therefore, periodic or even quasiperiodic solutions could be obtained with a constant order parameter $\|\alpha_2\|$. Chaotic solutions, on the other hand, can not be constrained to live in a torus and therefore, $\|\alpha_2\| = \|\alpha_2(t)\|$: the existence of chaos in the average dynamics requires a complex dynamics of the amount of synchrony in the system.

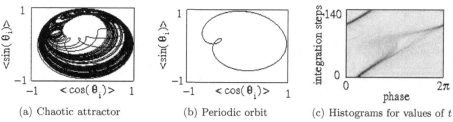

(a) Chaotic attractor   (b) Periodic orbit   (c) Histograms for values of t

Fig. 3. For the parameter values such that the averaged system displays chaotic dynamics, a simulation with one hundred oscillators. In (a), $< \cos(\theta_i) >$ and $< \sin(\theta_i) >$ are displayed. A close return analysis allows to reconstruct unstable periodic orbits coexisting with the attractor. One of such orbits is displayed in (b). The order parameter approaches zero when this orbit is visited. In (c), we display the histograms for different values of t, where the number of excitable driven units with phase between $(\theta, \theta + 2\pi/100)$ is coded with different grey tones. For the periodic orbit to achieve its topological shape, the system has to temporarily lose synchrony. In these simulations, $K_{11} = 5.5$, $\gamma = 6.83$, $\omega_0^1 = 10$, $\omega_0^2 = 5.8$, $K_{22} = 10$, $K_{21} = 5$.

The calculations described in the previous section assumed a infinite number of units. Yet, we chose to show a simulation with one hundred phase oscillators in each population. In what sense can we state that a simulation of the extended system is equivalent to a simulation of our order parameters? The program of research that Bob Gilmore has been pursuing in the last years[17] provides tools to build confidence or refute competing models proposed for a given chaotic data set. An experimental time series corresponding to a chaotic system might never be quite reproduced, even if we have a great model, due to the high sensitivity to the initial conditions. Yet, the skeleton of unstable periodic orbits reconstructed from the data must be consistent with that of the proposed model. The topological description of a dynamical system also provides a beautiful, deep picture of the mechanisms that take place in the phase space so that the observed complex dynamics is obtained. For the system of equations that we presented in the previous section, in Figure 4 we show six panels displaying the time evolution of a close, one dimensional set of initial conditions for the variables describing the average dynamics of the driven population. Between the first panel and the last one, the time is equal to one period of the driving population. Notice that a circle map might be a pertinent tool to inspect the time evolution of an initial condition. The croissant shaped curve shown in the first inset (top, left) is first stretched, then rotated approximately π radians, and finally bended into a shape similar to the initial one. In the process, the tips of the croissant folded over the back of the croissant. If a one dimensional map is built to describe how the original shape is mapped into itself, the foldings will be responsible for the negative slope of the map for some regions of the domain. This opens the door for complex temporal dynamics.

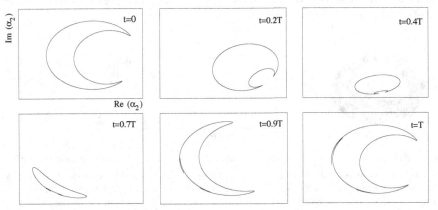

Fig. 4. The time evolution of a close, one dimensional set of initial conditions for the variables describing the average dynamics of the driven population. The parameters used in this simulation: $w_1 = 9$, $w_2 = 5.8$, $\gamma = 6.830$, $k_{21} = 5$, $k_{22} = 10$, $k_{11} = 5.5$, $\delta_1 = 0.01$, $\delta_2 = 1$. From top, left, clockwise, the time evolutions of a croissant shaped curve of initial conditions.

3. Final Remarks

In the last few decades, many researchers from different disciplines began to accept that simple, deterministic rules could be compatible with complex temporal dynamics. Now, once a natural phenomenon displays such behavior: how do we extract the rule? If we attempt to build a model: how do we build confidence/refute it? The research program that Bob Gilmore has been leading over the last years addresses these important issues. It does it by means of deeply inspiring geometrical flow descriptions and precise quantities as topological invariants of periodic solutions, which are robust and informative. It can be argued that many of these tools only apply to very low dimensional systems, and therefore the study of larger sets of interacting agents might not profit from this program. The purpose of this chapter is to show that recently, important advances in the study of large arrays of interacting nonlinear units allow us to explore the average behavior of these out of equilibrium systems. Those studies show that equations for the order parameters describing the collective dynamics of these systems can be obtained, and interesting enough, in some families of problems, they are stated in surprisingly low dimensional manifolds. Is every detail important in a neural architecture? Is there room for random fluctuations in the connectivities of a neural network? Motivated by the need to compute the average activity of a set of driven excitable units we found that the order parameters describing the collective behavior of the network obeyed equations capable of displaying a variety of solutions such as subharmonic responses and chaos. There is a deep relationship between the dynamics of the average and the behavior of the network: the system could not present average chaotic dynamics without time dependent levels of synchrony. It will not be surprising to see in the future, a fruitful interaction between the topological study of low dimensional dynamical systems and the statistical analysis of large arrays of nonlinear units. Let us sit in a table topology and thermodynamics: Bob will show up.

References

1. M. Golubitsky, I. Stewart, P. L. Buono, & J. J. Collins, Symmetry in locomotor central pattern generators and animal gaits, *Nature*, **401**, (6754), 693-695 (1999)
2. T. Gardner, G. Cecchi, M. Magnasco, R. Laje, & G. B. Mindlin, Simple motor gestures for birdsongs, *Physical Review Letters*, **87** (20), 208101 (2001)
3. G. B. Mindlin, T. J. Gardner, F. Goller, & R. Suthers, Experimental support for a model of birdsong production, *Physical Review E*, **68** (4), 041908 (2003)
4. J. D. Sitt, E. M. Arneodo, F. Goller, & G. B. Mindlin, Physiologically driven avian vocal synthesizer, *Physical Review E*, **81** (3), 031927 (2010)
5. M. A. Trevisan, G. B. Mindlin, & F. Goller, Nonlinear model predicts diverse respiratory patterns of birdsong, *Physical Review Letters*, **96** (5), 58103 (2006)
6. L. M. Alonso, J. A. Alliende, F. Goller, G. B. Mindlin, Low-dimensional dynamical model for the diversity of pressure patterns used in canary song, *Physical Review E*, **79** (4), 041929 (2009)
7. J. A. Alliende, J. M. Méndez, F. Goller, G. B. Mindlin, Hormonal acceleration of song development illuminates motor control mechanism in canaries, *Developmental Neurobiology*, **70** (14), 943-960 (2010)
8. A. Pikovsky, M. Rosenblum, & J. Kurths, *Synchronization: A universal concept in nonlinear sciences*, Cambridge University Press (2003)

9. Y. Kuramoto, Self-entrainment of a population of coupled non-linear oscillators, in *International symposium on mathematical problems in theoretical physics*, 420-422, Springer (1975)
10. L. M. Alonso, & G. B. Mindlin, Average dynamics of a driven set of globally coupled excitable units, *Chaos*, **21**, 023102 (2011)
11. L. M. Alonso, J. A. Alliende, G. B. Mindlin, Dynamical origin of complex motor patterns, *The European Physical Journal D*, **60**, 361-367 (2010)
12. A. T. Winfree, *The geometry of biological time*, Springer Verlag (2001)
13. E. Ott, T. M. Antonsen, Low dimensional behavior of large systems of globally coupled oscillators, *Chaos*, **18**, 037113 (2008)
14. L. M. Childs, & S. H. Strogatz, Stability diagram for the forced Kuramoto model, *Chaos*, **18**, 043128 (2008)
15. F. C. Hoppensteadt, & E. M. Izhikevich, *Weakly connected neural networks*, Springer Verlag (1997)
16. G. B. Mindlin, & R. Gilmore, Topological analysis and synthesis of chaotic time series, *Physica D*, **58** (1-4), 229-242 (1992)
17. R. Gilmore, & M. Lefranc, *The topology of chaos*, Wiley (2002)

Chapter 13

Minimal smooth chaotic flows

Jean-Marc Malasoma

Laboratoire DGCB CNRS 3237, Université de Lyon
Ecole Nationale des travaux Publics de l'Etat
Rue Maurice Audin, F-69518 Vaulx-en-Velin Cedex
France

Since the landmark paper of Lorenz in 1963, substantial progresses have been accomplished in the last decades toward understanding the appearance of deterministic chaos in dynamical systems. However, identifying the sufficient conditions so that a differential system exhibits a chaotic behaviour remains an open problem. This paper deals with minimal smooth chaotic flows: considering the right-hand side of an autonomous differential system, what are the minimal functional structures needed to reach chaotic behaviour?

Contents

1. Introduction 281
2. A First Class of Minimal Chaotic Flows 283
   2.1. A first example 285
   2.2. A second example 287
   2.3. A third example 288
   2.4. A fourth example 289
   2.5. Equivalence of the six first minimal examples 290
3. A Second Class of Minimal Chaotic Flows 290
   3.1. A fifth example 290
   3.2. A sixth example 293
   3.3. A seventh and last example 294
   3.4. Equivalence of the three last minimal examples 295
4. What are the Minimal C_2-equivariant Chaotic Flows? 296
5. Conclusion 299
References 300

1. Introduction

The necessary conditions for achieving chaos in systems of smooth differential equations are well known. The Poincaré-Bendixson[1,2] theorem implies first that a three-dimensional phase space is at least required. Besides, linear differential systems are obviously nonchaotic, so at least one nonlinearity is also needed. Since the simplest smooth nonlinear functions are quadratic polynomials, we focus here on three-dimensional autonomous quadratic differential systems.

Two ubiquitous systems are widely studied in the literature on deterministic chaos. These are the celebrated Lorenz[3] system:

$$\begin{cases} \dot{x} = \sigma(y - x) \\ \dot{y} = rx - y - xz \\ \dot{z} = -bx + xy \end{cases} \quad (1)$$

which right-hand side has seven terms including two quadratic nonlinearities, and the so-called Rössler-76 system:[4]

$$\begin{cases} \dot{x} = -y - z \\ \dot{y} = x + ay \\ \dot{z} = b - cz + xz \end{cases} \quad (2)$$

with seven terms as well but only one of them is nonlinear. Hence, the Rössler-76 system is algebraically simpler than the Lorenz one and was even considered for several decades as the simplest chaotic dissipative system.

In February 1979, Otto Rössler published[5] four new prototype equations for continuous chaos. The simplest one of them is the so-called Rössler-79 system:

$$\begin{cases} \dot{x} = -y - z \\ \dot{y} = x \\ \dot{z} = ay - bz - ay^2 \end{cases} \quad (3)$$

It displays a chaotic behaviour for appropriately chosen parameter values, for example $a = 0.386$ and $b = 0.2$.

Independently, Pierre Coullet, Charles Tresser and Alain Arnéodo proposed[6] in July 1979 a forced oscillator for which the state variable x satisfies $\ddot{x} + \beta \dot{x} + x = \eta$ and $\dot{\eta} = \mu x(1 - x)$ where μ is a real parameter and the dissipation β is positive. Therefore a compact form for these two equations is the scalar jerk equation:

$$\dddot{x} + \beta \ddot{x} + \dot{x} - \mu x + \mu x^2 = 0 \quad (4)$$

which can be recast in the form:

$$\begin{cases} \dot{x} = y \\ \dot{y} = z \\ \dot{z} = \mu x - y - \beta z - \mu x^2 \end{cases} \quad (5)$$

and which is also chaotic for well chosen parameter values like $\beta = 0.2$ and $\mu = 0.58$.

Systems (3) and (5) are characterized by six terms including a single quadratic nonlinearity on each of their right-hand sides. Their common algebraic structure is thus simpler than the structure of the Rössler-76 system. Both of these systems have retained this remarkable feature during 18 years but unfortunately, this did not prevent them from being virtually forgotten during this period and even still today.

So much so that, referring to the Rössler-76 system, Lorenz[7] wrote in 1993 : "By this time, a number of systems of differential equations with chaotic solutions had been discovered, but I felt I still had the distinction of having found the simplest. Rössler changed things by coming along with an even simpler one. His record still stands".

The following year, unaware of these simpler systems found much earlier, Clinton Sprott undertook to test Lorenz's claim that the Rössler-76 system is the simplest chaotic flow. His numerical brute-force method was to examine whether there are three-dimensional autonomous ordinary differential equations with six or fewer terms including a single nonlinearity or with five or fewer terms including two nonlinearities and whose solutions are chaotic.

The method employed has uncovered[8] nineteen apparently distinct simple new examples of chaotic quadratic flows. Fourteen of them have six terms including a single nonlinearity so they have the same complexity as the Rössler-79 system, while the remaining five systems are simpler with five terms including two nonlinearities.

No chaotic system was found with either only three or four terms on their right-hand side. Although about 10^8 systems were randomly selected for examination, Sprott's method cannot guarantee that all the chaotic systems with six or fewer terms including a single nonlinearity or with five or fewer terms including two nonlinearities have been discovered. However, these partial results raise the natural question: what is the behaviour of three-dimensional quadratic flows with less than five terms on their right-hand side?

The answer to this question is not fully known at this time. Zhang Fu and Jack Heidel[9,10] have proved that quadratic dissipative systems with a nonzero but constant divergence and with fewer than five terms cannot exhibit chaos. Although dissipative systems with constant divergence appear quite often in the literature, this is a very restrictive assumption. For example, the Rössler-76 system has not a constant divergence.

Later, the same authors have attempted to extend this result to conservative[11] cases with a zero divergence. However, the solution is far from complete even with this restrictive assumption. These authors have failed to provide a rigorous proof for the system $\dot{x} = y^2 - x^2$, $\dot{y} = x$, and $\dot{z} = y$. Moreover, some difficult cases have clearly been forgotten.

Regarding the minimal structure of chaotic flows with nonconstant divergence, Malasoma[12] proved very recently the following:

Theorem: *Jerk equations $\dddot{x} = J(x, \dot{x}, \ddot{x})$, where the jerk function J is a quadratic polynomial with only one or two terms and for which $\partial J(x, y, z)/\partial z$ is not a constant function of time, do not exhibit chaos.*

2. A First Class of Minimal Chaotic Flows

Failing to handle the general case, we restrict our attention to dissipative systems with a constant nonzero divergence. The theoretical results show that the minimal allowable algebraic structure consists in three-dimensional quadratic systems with five terms including a single nonlinearity. Once more, Sprott embarked in a numerical search for chaos but he focused on the special case of jerk equations $\dddot{x}= J(x,\dot{x},\ddot{x})$ where the jerk function J is restricted to a quadratic polynomial.

Sprott[13,14] found two examples, each of them defined by a jerk function constituted by three monomials including a single quadratic nonlinearity:

$$\dddot{x} + \alpha\ddot{x} + x - \dot{x}^2 = 0 \tag{6}$$

$$\dddot{x} + \alpha\ddot{x} + x - x\dot{x} = 0 \tag{7}$$

In their representations as dynamical systems, they are constituted by five monomials on their right-hand sides:

$$\begin{cases} \dot{x} = y \\ \dot{y} = z \\ \dot{z} = -\alpha z - x + y^2 \end{cases} \tag{8}$$

and

$$\begin{cases} \dot{x} = y \\ \dot{y} = z \\ \dot{z} = -\alpha z - x + xy \end{cases} \tag{9}$$

Equations (6) and (7) are the simplest quadratic jerk equations that exhibit chaotic dynamics and the associated chaotic flows (8) and (9) are algebraically simpler than any previously reported three-dimensional quadratic flows and we cannot find a simpler one with constant divergence. However, it is easy to show by a linear change of variables that these two examples are equivalent.

Sprott's examples raise two new questions. How many dissipative chaotic systems with five terms including one quadratic nonlinearity can exist? Are they all equivalent?

In order to bring new elements to answer these questions, Malasoma[15] developed in 2002 a specific strategy. By possibly exchanging the names of variables, without loss of generality we may assume that the nonlinearity belongs to the third equation. We can then notice that (8) and (9) exhibit this structure. The four remaining terms are then necessarily linear or constant.

First, we list all the possible systems of the appropriate type, by using all the possible combinations of variables, in a systematic way. Since we only considered dissipative flows with constant dissipation, at least one linear term gives a contribution to the divergence of the flow which must be negative. We obtain 65 systems which are neither obviously asymptotic to a two-dimensional surface, nor easily reducible to two-dimensional autonomous systems.

Then, about 10^6 dissipative systems of each type were randomly chosen with the five coefficients in the range -5 to 5. For each of these systems, we analyzed numerical solutions for 10^6 randomly chosen initial conditions with $|x| < 50$, $|$

$y \mid < 50$, and $\mid z \mid < 50$. We stopped the numerical calculations of a given solution whenever we detected the condition $\mid x \mid + \mid y \mid + \mid z \mid > 10^4$ or $t > 10^6$. Finally, when bounded stable solutions were found, we numerically computed the spectrum of the Lyapuonv exponents.

When this search was terminated, nine chaotic systems[15] were identified, including seven previously unknown minimal chaotic flows and both Sprott's minimal examples.

2.1. A first example

The method described above has highlighted a first example with the following structure:

$$\begin{cases} \dot{x} = ay \\ \dot{y} = by + cz \\ \dot{z} = dx + exy \end{cases} \quad (10)$$

where $b < 0$ so that the system is dissipative with a constant divergence and $acde \neq 0$. For this system we have found many choices of values for its parameters, allowing the existence of both regular and chaotic stable numerical solutions.

We can show without difficulty that the three differential equations (10) can be combined into the following:

$$\frac{d}{dt}(6acxz - 3a^2y^2 - 2cex^3) = 6acdx^2 - 6a^2by^2 \quad (11)$$

If $acd > 0$, the time derivative of the function $M(x, y, z) = 6acxz - 3a^2y^2 - 2cex^3$ has a positive sign, therefore M is a monotonic increasing function of the time which has a limit L as t tends to infinity. If L is finite, then any attractor for the equation lies on the two-dimensional surface $M(x, y, z) = L$ and is not chaotic thanks to the Poincaré-Bendixson theorem. If $L = +\infty$ then at least one of the three variables is not bounded and cannot be chaotic.

Given this result, we assume $b < 0$ and $acd < 0$, and we define the four rescaling transformations for the three variables and the time, which do not reverse the time:

$$\begin{cases} x = -ade^{-1}(-acd)^{-\frac{1}{3}}\tilde{x} \\ y = -de^{-1}\tilde{y} \\ z = -c^{-1}de^{-1}(-acd)^{\frac{1}{3}}\tilde{z} \\ t = (-acd)^{-\frac{1}{3}}\tilde{t} \end{cases} \quad (12)$$

After removing the tilde over the new variables, we obtain the reduced one-parameter system with $\alpha > 0$:

$$\begin{cases} \dot{x} = y \\ \dot{y} = -\alpha y + z \\ \dot{z} = -x + xy \end{cases} \quad (13)$$

Bounded stable numerical solutions for (13) were found in the quite small range $2.0168 \lesssim \alpha \lesssim 2.3043$. Outside this interval, all stable solutions seem to be unbounded. For $\alpha \lesssim 2.3043$, depending on initial conditions, solutions are either unbounded or settled down on a stable limit cycle created by a saddle-node bifurcation.

When the parameter α is further decreased, this limit cycle loses its stability and a complete subharmonic cascade leads to chaotic attractors. We used the Poincaré section Σ defined by:

$$\Sigma = \{(y, z) \in \mathbb{R}^2 \mid x = 0, \dot{z} > 0\} \tag{14}$$

The successive coordinates y_n of the points (y_n, z_n) on Σ are plotted versus α as illustrated by the bifurcation diagram shown in Fig.1. Outside the classical periodic windows, the system is chaotic in the tiny range $2.0168\ldots < \alpha < 2.0577\ldots$ of its control parameter.

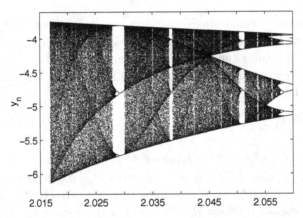

Fig. 1. Bifurcation diagram of the variable y_n plotted versus α for (13) with the Poincaré section $\Sigma = \{(y, z) \in \mathbb{R}^2 \mid x = 0, \dot{z} > 0\}$. The end of the subharmonic cascade leading to chaos is clearly visible.

Figure 2 shows a phase space portrait of the chaotic attractor for (13) with $\alpha = 2.016838$ which is very close to the parameter value for which a boundary crisis occurs and the attractor is destroyed together with its basin of attraction.

By using y_n we can define the first return map displayed in Fig. 3 for the same value $\alpha = 2.016838$ of the control parameter. A unimodal map constituted by two monotonic branches separated by a critical point is clearly observed. Thanks to the unimodal nature of this first return map, the unstable periodic orbits within the chaotic attractor can be encoded by using only two symbols. Moreover, the bifurcation diagram can be predicted by the kneading theory because there is a universal order for the creation or the destruction of periodic orbits when α is varied.

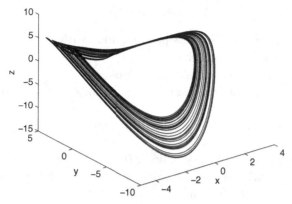

Fig. 2. Chaotic attractor for (13) just before the boundary crisis at $\alpha = 2.0168$.

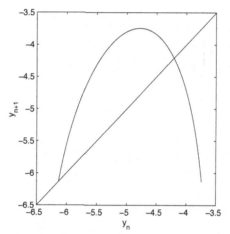

Fig. 3. First return map corresponding to y_n on the Poincaré section $\Sigma = \{(y, z) \in \mathbb{R}^2 \mid x = 0, \dot{z} > 0\}$ for (13) with $\alpha = 2.016838$.

As often observed, the boundary crisis occurs[16,17] when the increasing monotonic branch intersects the bisecting line. The symbolic dynamics is therefore complete, i.e. all periodic orbits which may be encoded using two symbols are embedded within the chaotic attractor.

2.2. *A second example*

The second example found has a general structure similar to the previous one but the nonlinearity xy is replaced by y^2:

$$\begin{cases} \dot{x} = ay \\ \dot{y} = by + cz \\ \dot{z} = dx + ey^2 \end{cases} \quad (15)$$

with again $b < 0$ in order to have a dissipative system with a constant divergence and $acde \neq 0$. Here again, various choices of values for the five parameters allow both regular and chaotic stable numerical solutions.

After some easy calculations the differential equations (15) imply the following:

$$\frac{d}{dt}\left(3b^2y^2 + 6bcyz - 6cdxy + 3c^2z^2 - 2cey^3\right) = 6b(by + cz)^2 - 6acdy^2 \quad (16)$$

If $acd > 0$, the time derivative of the function $M(x, y, z) = 3b^2y^2 + 6bcyz - 6cdxy + 3c^2z^2 - 2cey^3$ has a negative sign. Then M is a monotonic decreasing function of the time. We come up with the same conclusion: the system cannot be chaotic. Therefore we assume $b < 0$ and $acd < 0$ and we use the transformations which do not reverse the time:

$$\begin{cases} x = -a^2e^{-1}(-acd)^{-\frac{2}{3}}\tilde{x} \\ y = -ade^{-1}(-acd)^{-\frac{1}{3}}\tilde{y} \\ z = -ac^{-1}de^{-1}\tilde{z} \\ t = (-acd)^{-\frac{1}{3}}\tilde{t} \end{cases} \quad (17)$$

Again, after removing the tilde over the new variables, the reduced one-parameter system with $\alpha > 0$ is:

$$\begin{cases} \dot{x} = y \\ \dot{y} = -\alpha y + z \\ \dot{z} = -x + y^2 \end{cases} \quad (18)$$

2.3. A third example

The third example of structure allowing chaotic numerical solutions is:

$$\begin{cases} \dot{x} = a + by \\ \dot{y} = cy + dz \\ \dot{z} = exy \end{cases} \quad (19)$$

where $c < 0$ and $abde \neq 0$. From (19) we find after some manipulations:

$$\frac{d}{dt}\left(6acx + 6aby - 6bdxz + 3b^2y^2 + 2dex^3\right) = 6adex^2 + 6c(a + by)^2 \quad (20)$$

If $ade < 0$, the time derivative of the function $M(x, y, z) = 6acx + 6aby - 6bdxz + 3b^2y^2 + 2dex^3$ has a negative sign, therefore M is a monotonic decreasing function of the time. Therefore, the system cannot be chaotic by using the same argument of monotony. Consequently we assume $c < 0$ and $ade > 0$ and we define the following transformations which once again do not reverse the time:

$$\begin{cases} x = a(ade)^{-\frac{1}{3}}\tilde{x} \\ y = ab^{-1}\tilde{y} \\ z = ab^{-1}d^{-1}(ade)^{\frac{1}{3}}\tilde{z} \\ t = (ade)^{-\frac{1}{3}}\tilde{t} \end{cases} \quad (21)$$

and removing the tilde over the new variables gives once again a one parameter system:

$$\begin{cases} \dot{x} = y + 1 \\ \dot{y} = -\alpha y + z \\ \dot{z} = xy \end{cases} \quad (22)$$

with $\alpha > 0$.

2.4. A fourth example

The structure of the fourth example identified is the following:

$$\begin{cases} \dot{x} = az \\ \dot{y} = b + cx \\ \dot{z} = dz + exy \end{cases} \quad (23)$$

where $d < 0$ and $abce \neq 0$. From Eq. (23) we have:

$$\frac{d}{dt}(6bcx + 6bdy + 3c^2x^2 + 6cdxy - 6acyz + 2aey^3) = 6abey^2 + 6d(b + cx)^2 \quad (24)$$

If $abe < 0$, the time derivative of the function $M(x, y, z) = 6bcx + 6bdy + 3c^2x^2 + 6cdxy - 6acyz + 2aey^3$ has always a negative sign, therefore M is here again a monotonic decreasing function of the time and again the system cannot be chaotic. Thus we assume $d < 0$ and $abe > 0$ and we define the following rescaling transformations which preserve the orientation of the time:

$$\begin{cases} x = bc^{-1}\tilde{x} \\ y = b(abe)^{-\frac{1}{3}}\tilde{y} \\ z = a^{-1}bc^{-1}(abe)^{\frac{1}{3}}\tilde{z} \\ t = (abe)^{-\frac{1}{3}}\tilde{t} \end{cases} \quad (25)$$

After removing the tilde over the variables, it is straightforward to obtain the one-parameter system with $\alpha > 0$:

$$\begin{cases} \dot{x} = z \\ \dot{y} = 1 + x \\ \dot{z} = -\alpha z + xy \end{cases} \quad (26)$$

2.5. Equivalence of the six first minimal examples

The six chaotic systems (8), (9), (13), (18), (22), and (26) can be grouped in the same classe by using C^k-equivalence[18] of vector fields.

Let $\dot{x} = f(x)$ and $\dot{y} = g(y)$ be two C^r ($r \geq 1$) vector fields defined on $\mathbb{R}^n$. They are said to be C^k-equivalent ($k \leq r$) if there exists a C^k-diffeomorphism D which maps orbits $\phi_t(x)$ of the first system onto orbits $\psi_t(x)$ of the second system, preserving orientation but not necessarily parametrization by time. Let $\tau(x,t)$ be an increasing function of t along orbits (it must be increasing in order to preserve orientation of orbits), then the systems are C^k-equivalent if $D \circ \phi_t(x) = \psi_{\tau(x,t)}(D(x))$. If D does preserve parametrization by time, then the systems are said to be C^k-conjugate. In this case $\tau(x,t) \equiv t$ and $D \circ \phi_t(x) = \psi_t(D(x))$.

Table 1 shows that systems (8), (9), (18), (22), and (26) are C^∞-conjugate to system (13). Therefore, since equivalence is a transitive relationship, the six members of this class are C^∞-conjugate and therefore C^∞-equivalent. It is worth noting that in each case the C^∞-diffeomorphism D and its inverse D^{-1} are polynomials.

These six flows have a single equilibrium point and eigenvalues of this equilibrium satisfy the same characteristic equation $P(\lambda) = \lambda^3 + \alpha\lambda^2 + 1 = 0$. An elementary study of the roots of this polynomial shows that this equilibrium point has one negative real eigenvalue and two complex conjugate eigenvalues with positive real parts. It is thus a hyperbolic equilibrium point and more precisely a saddle-focus with an instability index of 2.

As it is the case for the system (13), the route to chaos for all of the six examples is the standard subharmonic cascade leading to chaotic attractors in the tiny range $2.0168\ldots < \alpha < 2.0577\ldots$ of their common control parameter.

3. A Second Class of Minimal Chaotic Flows

In this section we will conclude this analysis by determining the reduced forms of the last three systems exhibited by Malasoma. We will show that they belong to a second class of minimal chaotic flows distinct from the first one.

3.1. A fifth example

The structure of the fifth example identified is similar to that of the system (10). The only difference is that in the first equation y is replaced by z:

$$\begin{cases} \dot{x} = az \\ \dot{y} = by + cz \\ \dot{z} = dx + exy \end{cases} \quad (27)$$

where $b < 0$ and $acde \neq 0$. One easily obtains:

$$\frac{d}{dt}(3adx^2 - 3abxz - 3a^2z^2 + 3aex^2y - ecx^3) = -3b(adx^2 + a^2z^2) \quad (28)$$

Table 1. Minimal C^∞-conjugate chaotic systems of the first class. The C^∞-diffeomorphisms D are listed in the second column, and their inverses D^{-1} in the third column.

Systems	D	D^{-1}
$\dot{x} = y$ $\dot{y} = -\alpha y + z$ $\dot{z} = -x + xy$		
$\dot{u} = v + 1$ $\dot{v} = -\alpha v + w$ $\dot{w} = uv$	$u = x$ $v = y - 1$ $w = z - \alpha$	$x = u$ $y = v + 1$ $z = w + \alpha$
$\dot{u} = w$ $\dot{v} = u + 1$ $\dot{w} = -\alpha w + uv$	$u = y - 1$ $v = x$ $w = z - \alpha y$	$x = v$ $y = u + 1$ $z = w + \alpha u + \alpha$
$\dot{u} = v$ $\dot{v} = w$ $\dot{w} = -\alpha w - u + uv$	$u = x$ $v = y$ $w = -\alpha y + z$	$x = u$ $y = v$ $z = \alpha v + w$
$\dot{u} = v$ $\dot{v} = -\alpha v + w$ $\dot{w} = -u + v^2$	$u = x^2/4 - z/2$ $v = x/2$ $w = \alpha x/2 + y/2$	$x = 2v$ $y = -2\alpha v + 2w$ $z = -2u + 2v^2$
$\dot{u} = v$ $\dot{v} = w$ $\dot{w} = -\alpha w - u + v^2$	$u = x^2/4 - z/2$ $v = x/2$ $w = y/2$	$x = 2v$ $y = 2w$ $z = -2u + 2v^2$

If $ad > 0$, the time derivative of the function $M(x, yz) = 3adx^2 - 3abxz - 3a^2z^2 + 3aex^2y - ecx^3$ has a positive sign, therefore M is a monotonic increasing function of the time and the system cannot be chaotic. Consequently, we can assume $b < 0$ and $ad < 0$ while using the rescaling transformations:

$$\begin{cases} x = -ade^{-1}c^{-1}\tilde{x} \\ y = -de^{-1}\tilde{y} \\ z = -de^{-1}c^{-1}(-ad)^{\frac{1}{2}}\tilde{z} \\ t = (-ad)^{-\frac{1}{2}}\tilde{t} \end{cases} \qquad (29)$$

After removing the tilde over the variables, we obtain the one-parameter system with $\alpha > 0$:

$$\begin{cases} \dot{x} = z \\ \dot{y} = -\alpha y + z \\ \dot{z} = -x + xy \end{cases} \qquad (30)$$

Bounded stable numerical solutions for Eq. (30) were found in the small interval $10.284 \lesssim \alpha \lesssim 14.620$ and outside this range all stable solutions seem to be unbounded. For $\alpha \lesssim 14.620$, depending on initial conditions, solutions are either unbounded or settled down on a stable limit cycle created by a saddle-node bifurcation. Its basin of attraction appears to be relatively small, so that initial conditions must be chosen carefully.

When α is decreased the limit cycle loses its stability and a complete subharmonic cascade produces chaotic attractors as illustrated by the bifurcation diagram shown in Fig.4. This system is chaotic over most of the quite narrow

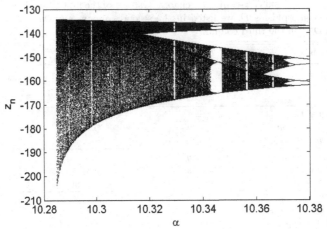

Fig. 4. Bifurcation diagram of the variable z_n on the Poincaré section $\Sigma = \{(y, z) \in \mathbb{R}^2 \mid x = 0, \dot{z} > 0\}$, plotted versus control parameter α for (30).

range: $10.2849\ldots < \alpha < 10.3716\ldots$ This interval is twice as large as the interval $2.0168\ldots < \alpha < 2.0577\ldots$ in which systems belonging to the first class of minimal chaotic flows exhibit chaotic attractors.

Figure 5 displays a phase space portrait of the chaotic attractor for (30) with $\alpha = 10.284910$ which is very close to the critical value for which a boundary crisis occurs. One is struck by the fact that the chaotic attractor is extremely thin in the vicinity of $y = 1$ and $z = \alpha$ which are the equations of two planes. It is worth noting that the intersection of these planes is a line which is the phase portrait of the particular solution of (30) defined for any real number x_0 by:

$$\begin{cases} x(t) = x_0 + \alpha t \\ y(t) = 1 \\ z(t) = \alpha \end{cases} \tag{31}$$

By using for example z_n we can define the first return map displayed in Fig. 6 for the same value $\alpha = 10.284910$ of the control parameter. Once again, a unimodal map constituted by two monotonic branches separated by a critical point can be clearly observed. Therefore, the unstable periodic orbits within the chaotic attractor can be encoded by using two symbols and the bifurcation diagram can be predicted by the kneading theory. Here again, the boundary crisis occurs when the increasing monotonic branch intersects the bisecting line and the symbolic dynamics is complete.

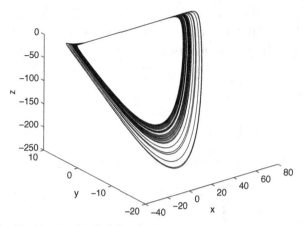

Fig. 5. Chaotic attractor for (30) just before the boundary crisis at $\alpha = 10.284910$.

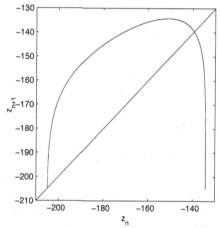

Fig. 6. First return map corresponding to z_n on the Poincaré section $\Sigma = \{(y, z) \in \mathbb{R}^2 \mid x = 0, \dot{z} > 0\}$ for (30) with $\alpha = 10.284910$.

3.2. A sixth example

The sixth example found has the following structure:

$$\begin{cases} \dot{x} = az \\ \dot{y} = by + cz + d \\ \dot{z} = exy \end{cases} \quad (32)$$

where $b < 0$ and $acde \neq 0$. This leads to:

$$\frac{d}{dt}(3abxz + 3a^2z^2 - 3aex^2y + cex^3) = 3a^2bz^2 - 3adex^2 \quad (33)$$

If $ade > 0$, the time derivative of the function $M(x,y,z) = 3abxz + 3a^2z^2 - 3aex^2y + cex^3$ has a negative sign, so M is a monotonic decreasing function of the time and the system cannot be chaotic. Therefore we assume $b < 0$ and $ade < 0$ and use the transformationand use the transformations:

$$\begin{cases} x = -adc^{-1}(-ade)^{-\frac{1}{3}}\tilde{x} \\ y = -d(-ade)^{-\frac{1}{3}}\tilde{y} \\ z = -dc^{-1}\tilde{z} \\ t = (-ade)^{-\frac{1}{3}}\tilde{t} \end{cases} \qquad (34)$$

After removing the tilde over the variables, we obtain the one-parameter system with $\alpha > 0$:

$$\begin{cases} \dot{x} = z \\ \dot{y} = -\beta y + z - 1 \\ \dot{z} = xy \end{cases} \qquad (35)$$

3.3. A seventh and last example

The seventh and last example found is:

$$\begin{cases} \dot{x} = a + bz \\ \dot{y} = cy + dz \\ \dot{z} = exy \end{cases} \qquad (36)$$

where $c < 0$ and $abde \neq 0$. One obtains:

$$\frac{d}{dt}(3acx + 6abz + 3bcxz + 3b^2z^2 + edx^3 - 3ebx^2y) = 3c(a+bz)^2 + 3adex^2 \qquad (37)$$

If $ade < 0$, the time derivative of the function $M(x,y,z) = 3acx + 6abz + 3bcxz + 3b^2z^2 + edx^3 - 3ebx^2y$ has a negative sign, so M is monotonic decreasing and the system cannot be chaotic. Therefore we assume $c < 0$ and $ade > 0$ and use the transformations:

$$\begin{cases} x = a(ade)^{-\frac{1}{3}}\tilde{x} \\ y = ab^{-1}d(ade)^{-\frac{1}{3}}\tilde{y} \\ z = ab^{-1}\tilde{z} \\ t = (ade)^{-\frac{1}{3}}\tilde{t} \end{cases} \qquad (38)$$

After removing the tilde over the variables, we obtain the one-parameter system with $\alpha > 0$:

$$\begin{cases} \dot{x} = z + 1 \\ \dot{y} = -\beta y + z \\ \dot{z} = xy \end{cases} \qquad (39)$$

3.4. Equivalence of the three last minimal examples

It appears that (39) is obtained by a simple translation of the variable z in (35). So systems (35) and (39) are C^∞-conjugate and therefore C^∞-equivalent.

By using the increasing transformation of the time $\tau(t) = \alpha^{1/3}t$ and the polynomial C^∞-diffeomorphism D defined in table 2, the system (30) can be transformed in system (39) with $\beta = \alpha^{2/3}$. Therefore (30) and (39) are C^∞-equivalent and as a consequence (30), (35), and (39) are equivalent and belong to the same class. Besides, since (30) is chaotic in the range $10.2849\ldots < \alpha < 10.3716\ldots$ we can conclude that (35) and (39) are also chaotic in the range $4.7293\ldots < \beta < 4.7558\ldots$

Moreover (30), (35), and (39) have a single equilibrium point and eigenvalues of this equilibrium satisfy the same characteristic equation $Q(\lambda) = (\lambda+\alpha)(\lambda^2+1) = 0$. This is thus a non-hyperbolic point and more precisely a focus-center.

If we consider now two minimal chaotic flows among the nine considered throughout this paper and if the first one belongs to the first class and the other belongs to the second class, we can certify so they can not be C^k-equivalent for $k \geq 1$. Indeed their single equilibria are not of the same type: respectively a saddle-focus versus a focus-center.

Table 2. Minimal C^∞-equivalent chaotic systems of the second class. The increasing transformation of the time is $\tau(t) = \alpha^{1/3}t$ and the C^∞-diffeomorphisms D are listed in the second column, and their inverses D^{-1} in the third column.

Systems	D	D^{-1}
$\dot{x} = z$ $\dot{y} = -\alpha y + z$ $\dot{z} = -x + xy$		
$\dot{u} = w$ $\dot{v} = -\alpha^{2/3}v + w - 1$ $\dot{w} = uv$	$u(\tau) = \alpha^{-2/3}x(\alpha^{-1/3}\tau)$ $v(\tau) = \alpha^{-2/3}(y(\alpha^{-1/3}\tau) - 1)$ $w(\tau) = \alpha^{-1}z(\alpha^{-1/3}\tau)$	$x(t) = \alpha^{2/3}u(\alpha^{1/3}t)$ $y(t) = \alpha^{2/3}v(\alpha^{1/3}t) + 1$ $z(t) = \alpha w(\alpha^{1/3}t)$
$\dot{u} = w + 1$ $\dot{v} = -\alpha^{2/3}v + w$ $\dot{w} = uv$	$u(\tau) = \alpha^{-2/3}x(\alpha^{-1/3}\tau)$ $v(\tau) = \alpha^{-2/3}(y(\alpha^{-1/3}\tau) - 1)$ $w(\tau) = \alpha^{-1}z(\alpha^{-1/3}\tau) - 1$	$x(t) = \alpha^{2/3}u(\alpha^{1/3}t)$ $y(t) = \alpha^{2/3}v(\alpha^{1/3}t) + 1$ $z(t) = \alpha(w(\alpha^{1/3}t) + 1)$

Therefore, we have identified at least two different classes of minimal chaotic flows even if the problem of identifying all the classes of minimal chaotic flows remains still open. Nevertheless, it is noteworthy that the only difference between (13) which is a representative of the first class and (30) which is a representative of the second one lies in the single linear term of the first equation. There is no other solution because $\dot{x} = a$ or $\dot{x} = ax$ would prohibit a chaotic behaviour, it is thus unlikely that any new classes exist. However it is difficult to rule out the possibility of finding new equations which would belong to one of these two classes or to rule out the existence of classes of minimal chaotic flows with nonconstant divergence.

4. What are the Minimal C_2-equivariant Chaotic Flows?

All the minimal chaotic flows as described above do not present any symmetry properties. A natural question arises: what are the minimal chaotic systems which are equivariant under a given symmetry? For example, the famous Lorenz system is equivariant under the transformation $(x, y, z) \to (-x, -y, z)$ which corresponds to the rotation by π radians around the z-axis. With the Schoenflies conventions commonly used to describe crystallographic point groups, C_n denotes a rotation axis where the system is mapped onto itself when rotated by $(360/n)$ degree. Then, the Lorenz system is C_2-equivariant.

Let us consider a C_2-equivariant three-dimensional quadratic flow. Without loss of generality we may assume that the axis of the rotation is Oz. The most general form of such a system is the following:

$$\begin{cases} \dot{x} = a_1 x + a_2 y + a_3 xz + a_4 yz \\ \dot{y} = b_1 x + b_2 y + b_3 xz + b_4 yz \\ \dot{z} = c_0 + c_1 z + c_2 x^2 + c_3 xy + c_4 y^2 + c_5 z^2 \end{cases} \quad (40)$$

If there is no nonlinearity in the two first equations of (40), therefore x and y are solutions of the linear differential system:

$$\begin{cases} \dot{x} = a_1 x + a_2 y \\ \dot{y} = b_1 x + b_2 y \end{cases} \quad (41)$$

obviously x and y are not chaotic.

Similarly, if neither x^2 or xy or y^2 are present in the third equation of (40), then z is solution of the scalar differential equation of first order:

$$\dot{z} = c_0 + c_1 z + c_5 z^2 \quad (42)$$

so z is not chaotic.

Consequently, the presence of at least one of the nonlinearities xz or yz in the first or the second equation and the presence of another nonlinearity x^2 or xy or y^2 in the third equation of (40) constitute the necessary conditions for chaos in this system.

Many C_2-equivariant chaotic systems have been studied in the literature, the oldest known with the simplest structure is undoubtedly the so-called Nosé-Hoover[19,20] system which was considered in 1984 in the field of molecular dynamics:

$$\begin{cases} \dot{x} = y \\ \dot{y} = -x + yz \\ \dot{z} = \alpha - y^2 \end{cases} \quad (43)$$

where α is a positive control parameter. In 1994, in his seminal paper[8] Sprott also reported three C_2-equivariant chaotic systems. The first is the so-called Sprott A system which corresponds to a particular case[21] of the Nosé-Hoover's system obtained with $\alpha = 1$. However, since the divergence of this flow is not constant, the theoretical results recalled in the introduction are worthless to decide whether this system is minimal.

The other two, the so-called Sprott B and C systems, after a permutation of the variables x and y, read:

$$\begin{cases} \dot{x} = -x + y \\ \dot{y} = xz \\ \dot{z} = 1 - xy \end{cases} \quad (44)$$

and

$$\begin{cases} \dot{x} = -x + y \\ \dot{y} = xz \\ \dot{z} = 1 - y^2 \end{cases} \quad (45)$$

Both also contain only five terms including two quadratic nonlinearities, and since they have a constant divergence, these chaotic systems are minimal.

In 2000, Gerard van der Schrier and Leo Maas[22] proposed a simplified version of the Lorenz system obtained in the limit of high Rayleigh and Prandtl numbers. This system is the so-called diffusionless Lorenz system:

$$\begin{cases} \dot{x} = -x - y \\ \dot{y} = -xz \\ \dot{z} = R + xy \end{cases} \quad (46)$$

Clearly, the Van der Schrier and Maas system is a one-parameter version of the Sprott B system (by changing y in -y). These systems generate 2-scroll butterfly-shaped chaotic attractors like the Lorenz attractor.

More recently, in 2008 Malasoma[23] found two C^∞-conjugate but previously unknown C_2-equivariant chaotic systems with the same minimal structure:

$$\begin{cases} \dot{x} = -\alpha x + y \\ \dot{y} = xz \\ \dot{z} = 1 - x^2 \end{cases} \quad (47)$$

and

$$\begin{cases} \dot{x} = y \\ \dot{y} = -\alpha y + xz \\ \dot{z} = 1 - x^2 \end{cases} \quad (48)$$

where α is a positive parameter. If we except the classical periodic windows, the chaotic region falls in the very tiny range $0.8633\ldots < \alpha < 0.8678\ldots$

An interesting observation on the dynamics of (47) and (48) is that, although these systems exhibit the Lorenz system's symmetry, they can generate 1-scroll non butterfly shaped chaotic attractors surrounding all their unstable equilibria points. Both Sprott B and C systems generate 2-scroll butterfly-shaped chaotic attractors like the Lorenz attractor. Figure 7 displays a phase space portrait of such a chaotic attractor for (47) with $\alpha = 0.86335$ just before a boundary crisis.

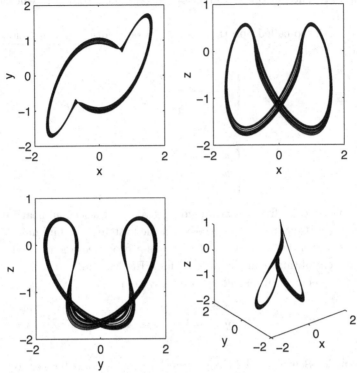

Fig. 7. A non butterfly-shaped chaotic attractor for (47) with $\alpha = 0.86335$ just before the boundary crisis.

The corresponding first return map is shown in Figure 8. At first approximation, this multimodal first return map has four monotonic branches separated by three critical points. Consequently, a topological analysis of this chaotic attractor leads to a template with four branches, and periodic orbits embedded within this chaotic attractor should be encoded using four symbols.

The Malasoma minimal chaotic flow (47) has two equilibria, at $(\pm 1, \pm \alpha, 0)$, and the linearized system around either of these points has eigenvalues satisfying the characteristic equation $P_1(\lambda) = \lambda^3 + \alpha \lambda^2 + 2$. Obviously, the same polynomial is obtained by using the C^∞-conjugate chaotic flow defined (48).

The Sprott B chaotic flow is evidently a particular realization of the more general one-parameter flow:

$$\begin{cases} \dot{x} = -\alpha x + y \\ \dot{y} = xz \\ \dot{z} = 1 - xy \end{cases} \tag{49}$$

which has two fixed points, at $(\pm 1/\sqrt{\alpha}, \pm\sqrt{\alpha}, 0)$, and the linearized system around each of these points has eigenvalues satisfying the characteristic equation $P_2(\lambda) = \lambda^3 + \alpha\lambda^2 + \alpha^{-1}\lambda + 2$.

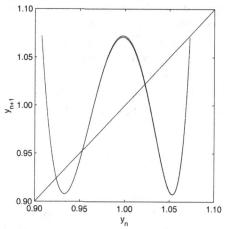

Fig. 8. First return map for (47) with $\alpha = 0.86335$ corresponding to y_n on the Poincaré section $\Sigma = \{(y,z) \in \mathbb{R}^2 \mid x = 0, y > 0\}$.

Similarly, the Sprott C chaotic flow is a particular realization of the more general one-parameter flow:

$$\begin{cases} \dot{x} = -\alpha x + y \\ \dot{y} = xz \\ \dot{z} = 1 - y^2 \end{cases} \tag{50}$$

which has two equilibria, at $(\pm 1/\alpha, \pm 1, 0)$, and the linearized system around each of these points has eigenvalues satisfying the characteristic equation $P_3(\lambda) = \lambda^3 + \alpha \lambda^2 + 2\alpha^{-1}\lambda + 2$.

Consequently, (47), (49), and (50) cannot be C^k-equivalent for $k \geq 1$. They are the representatives of three distinct classes of minimal C_2-equivariant chaotic flows. The first class is constituted by both new, previously unknown chaotic flows defined by Eq. (47) and Eq. (48). The second class contains the generalization of the Sprott B flow defined by Eq. (49), and the third one contains the generalization of the Sprott C flow defined by Eq. (50).

It is noteworthy that the only difference between (47), (49), and (50) lies in the nonlinearity in the third equation and that there is no other possibility of allowed nonlinearity including x or y in this equation. It is thus rather unlikely that new classes of C_2-equivariant minimal chaotic flows with constant divergence do exist, and even if it was the case, their structures should necessarily be different from the previous ones. Moreover it is also difficult to rule out the possibility of finding new systems which would belong to one of the three already known classes.

5. Conclusion

In this paper we have proved the existence of two different classes of minimal chaotic flows. We have also proved the existence of three different classes of C_2-equivariant minimal chaotic flows. Determining all the classes of minimal chaotic flows and all the classes of minimal equivariant chaotic flows under a given symmetry remains interesting still open problems. Throughout this paper we have assumed that the considered systems have a nonzero constant divergence but basic examples such as the Rössler-76 and the Nosé-Hoover systems do not exhibit this particular characteristic. Hence, it becomes essential to find a reliable and conclusive way of resolving our problem without this assumption which is anyway more convenient than essential.

Acknowledgments

The author is grateful to Camille Malasoma and Jean-Marc Zaccardi for a careful reading of the manuscript and for useful suggestions. He also warmly thanks Elisabeth Hanauer for repeated encouragements and stimulating discussions.

References

1. H. Poincaré, Sur les courbes définies par les équations différentielles (III), *Journal de Mathématiques Pures et Appliquées* 4e série, **1**, 167-244 (1885).
2. I. Bendixson, Sur les courbes définies par des équations différentielles, *Acta Mathematica*, **24**, 1-88 (1901).
3. E. N. Lorenz, Deterministic Nonperiodic Flow, *Journal of Atmospheric Sciences*, **20** (2), 130-141 (1963).
4. O. E. Rössler, An equation for continuous chaos, *Physics Letters A*, **57** (5), 397-398 (1976).
5. O. E. Rössler, Continuous chaos - four prototype equations, *Annals of the New York Academy of Sciences*, **316** (1), 376-392 (1979).
6. P. Coullet, C. Tresser & A. Arnéodo, Transition to stochasticity for a class of forced oscillators, *Physics Letters A*, **72** (4-5), 268-270 (1979).
7. E. N. Lorenz, The essence of chaos, University of Washington Press, Seattle (1963).
8. J. C. Sprott, Some simple chaotic flows, *Physical Review E*, **50** (2), R647-R650 (1994).
9. F. Zhang & J. Heidel, Non-chaotic behaviour in three-dimensional quadratic systems, *Nonlinearity*, **10** (5), 1289-1303 (1997).
10. F. Zhang & J. Heidel, Non-chaotic behaviour in three-dimensional quadratic systems: erratum, *Nonlinearity*, **12** (3), 739 (1999).
11. J. Heidel & F. Zhang, Non-chaotic behaviour in three-dimensional quadratic systems II: The conservative case, *Nonlinearity*, **12** (3), 617-633 (1999).
12. J.-M. Malasoma, Non-chaotic behaviour for a class of quadratic jerk equations, *Chaos Solitons and Fractals*, **39** (2), 533-539 (2009).
13. J. C. Sprott, Simplest dissipative chaotic flow, *Physics Letters A*, **228** (4-5), 271-274 (1997).
14. J. C. Sprott, Some simple chaotic jerk functions, *American Journal of Physics*, **65** (6), 537-543 (1997).
15. J.-M. Malasoma, A new class of minimal chaotic flows, *Physics Letters A*, **305** (1-2), 52-58 (2002).

16. C. Letellier, P. Dutertre, J. Reizner, G. Gouesbet, Evolution of a multimodal map induced by an equivariant vector field, *Journal of Physics A*, **29** (17), 5359-5374 (1996).
17. C. Letellier & J.M. Malasoma, Unimodal order in the image of the simplest equivariant chaotic system, *Physical Review E*, **64** (6), 067202 (2001).
18. S. Wiggins, Introduction to applied nonlinear dynamical systems and chaos, Springer-Verlag (1990).
19. S. Nosé, A molecular dynamics method for simulations in the canonical ensemble, *Molecular Physics*, **52** (2), 255-268 (1984).
20. W. G. Hoover, Canonical dynamics: Equilibrium phase-space distributions, *Physical Review A*, **31** (3), 1695-1697 (1985).
21. W. G. Hoover, remark on Some simple chaotic flows, *Physical Review E*, **51** (1), 759-760 (1995).
22. G. Van der Schrier & L.R.M. Maas, The diffusionless Lorenz equations; Shil'nikov bifurcations and reduction to an explicit map *Physica D*, **141** (1-2), 19-36 (2000).
23. J.-M. Malasoma, New Lorenz-like Chaotic Flows with Minimal Algebraic Structure, *Indian Journal of Industrial and Applied Mathematics*, **1** (2), 1-16 (2008).

Chapter 14

The chaotic marriage of physics and financial economics

Claire Gilmore

King's College McGowan School of Business
133 North River Street
Wilkes-Barre, Pennsylvania 18711, USA

By the early 1980s interest in chaos theory was spreading from mathematics and the sciences to other fields, including economics and finance. Initial results, based on the metric approach to testing for chaos in time series data, appeared to lend support to the presence of chaotic behavior in a variety of economic phenomena and in financial markets. Subsequently, a topological approach to the analysis of chaos was developed which led to tests for chaotic behavior more suited to the relatively small, noisy data sets typically available in these fields. This close returns test is demonstrated here and is applied to data from several financial markets. The qualitative topological test does not support evidence of a chaotic generating mechanism in these series. The quantitative form of the close returns test indicates nonchaotic nonlinear behavior that cannot be fully explained by current financial models.

Contents

1. Introduction 303
2. The Close Returns Test: Qualitative Form 305
3. The Close Returns Test: Quantitative Form 306
4. Applications to Empirical Financial Data: Equity Market 308
5. Applications to Empirical Financial Data: Foreign Exchange 311
6. Conclusions 314
References 314

1. Introduction

Work on nonlinear dynamics, particularly chaotic systems, in the natural sciences sparked interest in the fields of economics and finance beginning in the 1980s. The persistent irregularity found in financial markets has generally been attributed to random shocks. However, chaos theory attracted attention as a possible alternative explanation because of its ability to produce complex time series sequences that appear random. A number of theoretical studies began to appear, demonstrating the capacity of economic and financial models to produce chaotic behavior, including in business cycles,[1,2] in international trade,[3] and in equity markets.[4]

Empirical research searching for evidence of chaotic behavior in these areas expanded rapidly, but results were often inconclusive. Early work produced possible evidence of chaotic behavior in such areas as stock returns,[5,6] work stoppage data,[7,8] money aggregates,[9] and Treasury bill returns.[10] A study of business cycle data for Italy, Japan, the UK, and West Germany, however, found no support for a chaotic interpretation.[11] The reliability of these results when applied to relatively small, noisy data sets soon came to be questioned. Applying a procedure to reduce the small sample size bias in dimension calculations,[12,13] Ramsey, Sayers, and Rothman[14] reexamined several of the previously published papers.[6,8,9] They concluded that there was no evidence of chaos in any of these series, with the possible exception of the work stoppage data. These studies have all relied on the metric approach, developed by Peter Grassberger and Itamar Procaccia,[15] to the analysis of data generated from a possibly chaotic model. This approach relies on the calculation of correlation dimensions and Lyapunov exponents; for a detailed discussion of metric methods see Pedrag Cvitanovic.[16] However, there are a number of problems associated with the implementation of metric computations:

(1) The estimate of the correlation dimension converges to the dimension of the chaotic attractor as N, the number of observations, goes to infinity. With small data sets the number of observations may be insufficient for convergence to occur;[17]
(2) Noise in the data may distort dimension calculations;[18]
(3) Estimation of the scaling region to be used in the calculation is subjective;[18]
(4) There is no formal statistical distribution theory for the correlation dimensions, although see the advances by Brock and Baek;[19]
(5) There is substantial bias in correlation dimensions with small data sets;[12,13]
(6) Dimension calculations throw away time-ordering information and cannot lead to information about the dynamics of the chaotic process;[20]
(7) The procedure must be implemented carefully to capture only geometric correlation in the time series and eliminate dynamic correlation. Failure to implement this precaution has rendered many published dimension calculations incorrect.[21]

The topological approach to examination of chaotic processes, which was developed in the physics literature in the early 1990s,[22–24] provided an alternative approach to testing for low-dimensional deterministic chaos in economic and financial data. This method contains a close returns test for detecting chaos which is of particular interest since it works well on small, relatively noisy data sets.[25] Further, the topological approach is able to provide additional information about the underlying system generating chaotic behavior once evidence of chaos is detected. After discussing the close returns method, we will adapt it to test for the presence of deterministic chaos in selected financial time series data.

2. The Close Returns Test: Qualitative Form

Both the metric and the topological approaches to chaos depend on the recurrent nature of a time series flow to test for the existence of a strange attractor. However, while the correlation-integral-based tests try to identify fractal structure, the topological method is based on a more fundamental characteristic of chaotic systems, the recursive behavior of a chaotic time series which over time will nearly, although never exactly, repeat itself. The starting point for implementing the topological algorithm is the time series, x_t, without an embedding. If one of the observations, x_i, occurs near a periodic orbit, then subsequent observations will evolve near that orbit for a while before being repelled away from it. If the observations evolve near the periodic orbit for a sufficiently long time, they will return to the neighborhood of x_i after some interval, T, where T indicates the length of the orbit, measured in units of the sampling rate. This means that $|x_i - x_{i+T}|$ will be small. Further, x_{i+1} will be near x_{i+1+T}, x_{i+2} will be near x_{i+2+T}, and so on. Thus, it makes sense to look for a series of consecutive data elements for which $|x_i - x_{i+T}|$ is small.

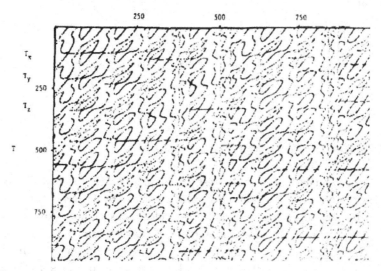

Fig. 1. Close returns plot of chaotic time series generated from Rössler system. First 1960 points of 5000-point observation set.

To detect regions of close returns in a data set a graph can be constructed. Differences less than a small threshold value, ϵ, are coded black, while differences greater than ϵ are coded white. The appropriate size for ϵ will typically be between 0.01 and 0.1 of the maximum difference between any two observations in the data set. The horizontal axis indicates the observation number, i, where $i = (1, 2, \cdots, N)$, and the vertical axis is designated as T, where $T = (1, 2, \cdots, N - i)$. Close returns in the data set are indicated by horizontal line segments appearing in the graph.

However, if the data set is stochastic, a generally uniform array of black dots will appear. A chaotic close returns plot using data generated from the Rössler model[26] appears in Fig. 1, while a pseudorandom graph is presented in Fig. 2.

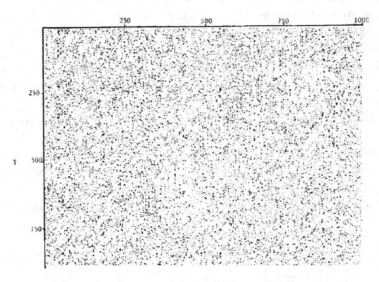

Fig. 2. Close returns plot of pseudorandom time series.

3. The Close Returns Test: Quantitative Form

To develop the qualitative close returns test into a quantitative form we proceed as follows. Using the 500-by-300 observation plot for the chaotic series (Fig. 1), we construct a histogram recording the incidence of close returns hits. For each value of T on the vertical axis of the plot, the number of close returns is summed across the row, with $H_T = \sum \Theta(\epsilon - |x_i - x_{i+T}|)$. Here, Θ is the Heaviside theta function: $\Theta(x) = +1$ if $x \geq 0$, $\Theta(x) = 0$ if $x < 0$. The histogram displays T along the horizontal axis and H_T along the vertical axis. For the chaotic data the histogram will contain a series of peaks, more or less evenly spaced (Fig. 3). In contrast, a histogram for the pseudorandom time series will exhibit a scattering around a uniform distribution (Fig. 4).

A χ^2 test is applied to the histogram in question to determine whether the null hypothesis of *iid* (independent and identically distributed) for the series can be rejected. If the series under study is *iid*, then $H_T = \langle H \rangle$, a constant. That is, the sum of close returns hits will be constant for any value T for which the histogram is constructed. This is tested by calculating the sample χ_c^2. If the data are *iid*, the distribution described in the close returns plot will be binomial, with the probability

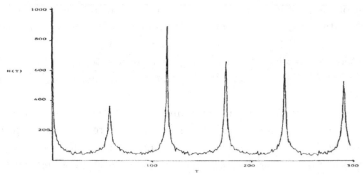

Fig. 3. Close returns histogram of Rössler chaotic series, using the full 5000-observation series over first 300 values of T.

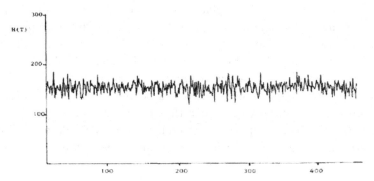

Fig. 4. Close returns histogram of pseudorandom data set.

of a hit, or p, determined by

$$p = \frac{\text{number of close returns}}{\text{total area of plot}} \quad (1)$$

The average value $\langle H \rangle$ is equal to this probability times the number of observations over which the number of close returns hits is counted, n:

$$\langle H \rangle = np \quad (2)$$

The sample χ_c^2 for this binomial distribution is then expressed as

$$\chi_c^2 = \frac{\sum [H_T - \langle H \rangle]^2}{np(1-p)} \quad (3)$$

The calculated value, χ_c^2, is then compared to the test χ_t^2 statistic. If $\chi_c^2 > \chi_t^2$, the null hypothesis that the data are *iid* is rejected at the 95% confidence level. The decision rule can be conveniently expressed as a ratio, χ_c^2/χ_t^2, where a ratio greater than 1.0 causes rejection of the *iid* null. Tests on simulated data have established the reliability of this test.[27] Rejection rates for random series of size 100, 500, and

1,000 observations are all close to the five percent level, and rejection occurs in all cases at all data set sizes for chaotic series produced by the logistic map and by the Hénon map. For the family of autoregressive (AR) series, autoregressive conditional heteroskedasticity (ARCH), and generalized autoregressive conditional heteroskedasticity (GARCH) models the rejection rate depends of the value of the coefficient assigned.[27]

This quantitative test can be further refined. Since the close returns plot is specifically designed to search for chaos by identifying the recurrence property, the projection in the horizontal direction to build up the histogram for the quantitative test emphasizes the presence of the regularly spaced horizontal line segments which result from the recurrence property. If the data have structure which is not chaotic, this particular one-dimensional projection can "wash out" other types of signal. Thus, structure which is not chaotic may be more easily detected by applying the χ^2 test directly to the two-dimensional plot itself to detect departures from *iid*. For computational reasons the simplest way to proceed is to arrange our graph with x_i and x_j (where $i, j = 1, 2, \cdots N$ and $i \neq j$), rather than with x_i and x_{i+T}, as before. The information contained in both forms of the plot is identical. We then divide the new plot into boxes of s observations on a side. The number of close returns is counted up for each box, the probability of a black point and the expected count is computed as previously, and the χ^2 ratio can then be calculated.

The close returns test is next applied to several sets of financial data (a US stock index and several exchange rates) that have been studied using the metric tests to detect chaos.

4. Applications to Empirical Financial Data: Equity Market

For the equity market a value-weighted (with dividends) weekly stock index for US securities was obtained from the Center for Research on Securities Prices (CRSP) for the period July 1962 through December 1989, yielding 1434 weekly observations. This is somewhat longer than samples examined by Scheinkman and LeBaron,[6] Brock et al.,[28] and Hsieh,[29] using metric-based methods. Applying the topological test to the data, we find that the close returns plot reveals no evidence of chaotic behavior (Fig. 5). The quantitative version of the test on the full series produces a χ^2 ratio of 1.49, rejecting the null hypothesis of *iid*. The plot does, however, indicate that the returns are not *iid* and that some type of structure, linear or nonlinear, is present. When the data are examined more closely for stationarity with the Kolmogorov-Smirnov test,[30] it is evident that the first third of the data come from a distribution different from that of the remaining two thirds. Consequently, the data set was divided into two parts: part 1 is the first third and part 2 is the remainder of the observations. Testing each part separately yields χ^2 ratios of 1.0557 and 1.0801, respectively. Again, the *iid* null is rejected in favor of the alternative of some kind of structure, linear or nonlinear. Consequently, a linear filter was applied using autoregressive models[31] of the form

$$x_t = b_0 + \sum b_i x_{t-i} + \theta_t \tag{4}$$

The χ^2 ratios calculated on the standardized residuals from these models again rejected the *iid* null (1.038 for part 1 and 1.0327 for part 2). Further, GARCH models were also applied to provide a better fit of the data series.[31] These models take the following form:

$$\begin{aligned} x_t &= b_0 + \sum b_i x_{t-i} + z_t \\ E|z_t| &= 0 \qquad \qquad \text{Var}|z_t| = h_t \\ h_t &= \beta_0 + \sum \beta_j z_{t-j}^2 \end{aligned} \tag{5}$$

In this case the standardized residuals for each part of the data set did produce χ^2 ratios below 1.0, indicating that these models adequately capture the structure in the equity returns. However, when the χ^2 test was applied directly to the two-dimensional close returns plot, using box sizes of side s ranging from $s = 20$ to $s = 50$ observations, all ratios except one were over 1.00, allowing us to reject the null hypothesis of *iid* for the standardized residuals in favor of some remaining structure (Table 1). The last model which was applied to the two parts of the data set to capture its structure was an exponential GARCH (EGARCH) model,[31] in which the variance equation becomes

$$h_t = \beta_0 + \sum \beta_j z_{t-j}^2 + \sum \delta_k h_{t-k} \tag{6}$$

This type of model within the GARCH family is generally considered to be especially appropriate for financial data. In this case the χ^2 ratio resulting from the histogram of the data was 0.79 for the first part and 1.0078 for the second. However, as in the case with the GARCH models, calculations of the ratio on the two-dimensional close returns plot led to rejection of the *iid* null for most box sizes, as shown in Table 2.

Table 1. χ^2 ratios for box plots of GARCH standardized residuals of CRSP data. The data set length is 478 observations for Part 1 and 956 observations for Part 2.

	Stock Return Series			
Box Size	Part 1		Part 2	
(s)	$\epsilon = 1\%$	$\epsilon = 2\%$	$\epsilon = 1\%$	$\epsilon = 2\%$
20	1.05	1.30	1.14	1.18
25	0.94	1.00	1.08	1.25
30	1.10	1.33	1.36	1.43
35	1.25	1.29	1.26	1.43
40	1.49	1.80	1.46	1.59
45	1.29	1.59	1.50	1.67
50	1.03	1.22	1.33	1.55

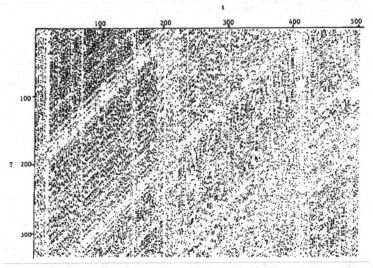

Fig. 5. Close returns plot of weekly stock returns, July 2, 1962, through November 21, 1973.

Table 2. χ^2 ratios for box plots of EGARCH standardized residuals of CRSP data. The data set length is 478 observations for Part 1 and 956 observations for Part 2.

	Stock Return Series			
Box Size	Part 1		Part 2	
(s)	$\epsilon = 1\%$	$\epsilon = 2\%$	$\epsilon = 1\%$	$\epsilon = 2\%$
20	0.94	1.04	1.27	1.42
25	0.97	1.12	1.29	1.68
30	0.99	1.12	1.54	1.91
35	0.99	1.37	1.48	2.08
40	0.93	1.18	1.72	2.26
45	1.58	1.84	1.73	2.65
50	1.03	1.60	1.77	2.58

Compared to the earlier studies, the qualitative close returns test rejects chaos for the CRSP series, while use of the correlation dimension test by Scheinkman and LeBaron[5] led to results that were inconclusive as to the presence of chaos. Brock, Hsieh, and LeBaron[28] used the Brock, Dechert, and Scheinkman (BDS) test[32] derived from the correlation dimension procedure and found that GARCH models adequately modeled the CRSP data for the period March 1974 through December 1985 but not for other periods, differing in part from the application of the quantitative close returns test to the data. Similarly to the quantitative close returns test, the BDS test also applies an *iid* null hypothesis. Finally, the results presented here do agree with Hsieh,[29] who used a range of tests that failed to detect evidence of chaos and also showed that ARCH models did not fully capture the nonlinear structure in the CRSP returns.

5. Applications to Empirical Financial Data: Foreign Exchange

Research using correlation-dimension-based methods with respect to chaotic behavior in foreign-exchange markets also produced inconsistent results. Application of the BDS test by Brock, Hsieh, and LeBaron[28] found no evidence of possible chaotic behavior, while indicating the presence of nonlinear dependence in several exchange-rate series, a necessary although not sufficient condition for chaos. In contrast De Grauwe et al.[33] reported indications of possible chaos on the yen/dollar and pound/dollar markets.

We examined daily exchange rates for the British pound, German mark, and Japanese yen against the US dollar.[34] For the first two foreign currencies the data cover the period January 7, 1976, through December 1, 1995, and for the yen the period is January 7, 1976, through June 17, 1994. The rate of change is calculated as $100 \log\left(\frac{S_t}{S_{t-1}}\right)$, giving 5197 observations for the pound and the mark and 4814 observations for the yen.

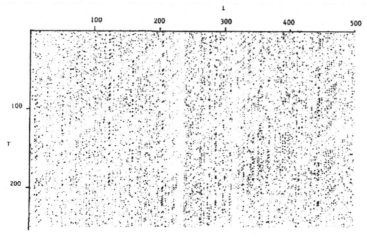

Fig. 6. Close returns plot of German mark — US dollar exchange rate, January 7, 1976, through March 23, 1979.

The qualitative close returns test provides no evidence at any point in the plots for the presence of chaos in any of these series (see Fig. 6 for plot of German mark). These results do not, therefore, support the tentative evidence of chaos in the British pound and for the Japanese yen found by De Grauwe et al.,[33] using the correlation dimension and the BDS tests. Application of the quantitative version of the test to the full series yields χ^2 ratios which consistently exceed 1.0, leading to rejection of the *iid* null and indicating the presence of some type of dependence. Stationarity tests led to dividing each series into two nearly equal parts, with the ratio continuing to reject the *iid* null for the first half in each exchange-rate series but not for the second half (see Table 3). The results for the second half of each series are further analyzed by application of the χ^2 test directly to the two-dimensional plot itself. In this form the χ^2 ratio uniformly exceeds 1.0 (Table 4). Some type of dependence, linear or nonlinear, therefore appears to be present throughout each exchange-rate series.

Table 3. χ_c^2 rations for histograms of exchange-rate series.

Currency	Time period					χ^2
British pound	Jan.	1976	—	Dec.	1995	3.59
	Jan.	1976	—	Dec.	1985	7.24
	Jan.	1986	—	Dec.	1995	0.88
German mark	Jan.	1976	—	Dec.	1995	1.86
	Jan.	1976	—	Dec.	1985	2.34
	Jan.	1986	—	Dec.	1995	0.92
Japanese yen	Jan.	1976	—	June	1994	3.37
	Jan.	1976	—	Mar.	1985	15.68
	Apr.	1985	—	June	1994	0.83

Table 4. Box-plot χ^2 ratios for second time period, from January 1986 to December 1995, for the pound and the mark, from April 1985 to June 1994, for the yen.

Box Size (s)	Exchange-rate Series		
	British pound	German mark	Japanese yen
20	2.93	2.87	3.94
25	4.67	4.08	5.76
30	4.79	4.95	5.11
35	4.76	5.40	7.81
40	6.42	7.29	8.32
45	6.63	7.29	10.08
50	8.64	9.44	9.95

Since the behavior of the exchange rates appears to have changed substantially over the time period under examination, a more detailed analysis was confined to the second half of each time series.[34] To account for possible linear dependence the data were first filtered by autoregressive models:

$$x_t = b_0 + b_1 D_{M,t} + b_2 D_{T,t} + b_3 D_{W,t} + b_4 D_{Th,t} + b_5 D_{h,t} + b_i x_{t-i} + e_t \quad (7)$$

where $D_{M,t}$, etc., are dummy variables for the days of the week and holidays (excluding weekends). When the close returns test was applied to the standardized

Table 5. Box-plot χ^2 ratios on residuals of yen linear model, 4/1985-6/1994.

Box size (s)	χ^2 Ratio
20	3.08
25	3.16
30	4.38
35	5.53
40	4.83
45	7.44
50	4.74

Table 6. χ^2 ratios for box plots of GARCH standardized residuals of exchange-rate series. The data set length is 2598 observations for the mark and 2407 for the yen.

Box Size (s)	Exchange Rate Series	
	German mark	Japanese yen
20	1.70	2.38
25	1.94	2.15
30	2.23	2.72
35	2.96	3.52
40	2.87	3.04
45	3.12	4.21
50	3.28	2.75

residuals of each series, the χ^2 ratio exceeded 1.0 for the pound and the mark but not for the yen. However, when the box-plot χ^2 ratios were calculated on the yen residuals, all ratios exceeded 1.0 (Table 5). Therefore, there are indications of nonlinear dependence for all three series.

Table 7. χ^2 ratios for box plots of EGARCH standardized residuals of exchange-rate series. The data set length is 2598 observations for the pound and the mark and 2407 observations for the yen.

Box Size (s)	Exchange Rate Series		
	British pound	German mark	Japanese yen
20	3.27	1.88	2.75
25	3.91	2.37	3.34
30	5.03	2.65	4.40
35	6.47	3.23	4.69
40	7.59	3.86	5.93
45	8.82	4.10	5.67
50	10.06	5.13	6.90

To model the nonlinear dependence each series was fitted to both GARCH and EGARCH models. For the GARCH standardized residuals the χ^2 test for the pound indicates remaining nonlinear structure, while the χ^2 ratios for the mark and the yen were 0.98 and 0.90 respectively, leading to failure to reject the *iid* null for the standardized residuals. However, the box-plot form of the test detects evidence of nonlinearity in each series which is not accounted for by the GARCH models (Table 6); results for EGARCH models produced comparable results (Table 7).

The close returns tests thus disagree to an extent with other published analyses of exchange rates using the metric approach. Our results do not support the evidence of chaotic behavior reported by De Grauwe[33] using the metric approach. They do agree with Hsieh's conclusion[29] that the popular GARCH-type models used in finance do not adequately model the structure present in exchange rates, but not with Fernandes,[35] who argues that those models do appropriately capture nonlinearity in the British pound.

6. Conclusions

The close returns test is a powerful tool to detect evidence of chaotic behavior or other structure in time series data. The qualitative form of the test uses a fundamental topological characteristic of chaos, the recurrence property, to identify chaotic behavior. Quantifying this test involves the tradeoff of using the *iid* null hypothesis, with the alternative including both linear and nonlinear structure. When applied to data in several financial markets where possible evidence of chaos has been found using the metric approach, the findings of the close returns test disagree and do not support a chaotic interpretation of the behavior of these markets. Instead, this test supports results indicating that there is no evidence of chaotic behavior.[28,29] The box-plot χ^2 ratios for GARCH and EGARCH models of these data indicate that the models have not captured all of the nonlinear dependence.

The methods which have been developed to date to test for chaos can at best provide indications of low-dimensional chaos, that is, behavior resulting from a fairly simple model. Yet, there is no reason to assume that financial behavior is the result of such simple models. Consequently, further theoretical advances are required to establish the basis for tests to detect more complex forms of chaotic behavior.

References

1. J. M. Grandmont, On endogenous competitive business cycles, *Econometrica*, **53**, 995-1045 (1985).
2. A. C.-L. Chian, E. L. Rempel, & C. Rogers, Complex economic dynamics: chaotic saddle, crisis and intermittency, *Chaos, Solitons and Fractals*, **29**, 1194-1218 (2006).
3. J.-W. Lorenz, International trade and the possible occurrence of chaos, *Economics Letters*, **23**, 135-138 (1987).
4. S. Shaffer, Structural shifts and the volatility of chaotic markets, *Journal of Economic Behavior & Organization*, **15**, 201-214 (1990).
5. J. A. Scheinkman & B. LeBaron, Nonlinear dynamics and GNP data, in *Economic Complexity: Chaos, Sunspots, Bubbles, and Nonlinearity*, Proceedings of the Fourth International Symposium in Economic Theory and Econometrics, W. A. Barnett, J. Geweke & K. Shell (Eds), Cambridge University Press (1989).
6. J. A. Scheinkman, & B. LeBaron, Nonlinear dynamics and stock returns, *Journal of Business*, **62**, 311-337 (1989).
7. C. L. Sayers, Work stoppages: Exploring the nonlinear dynamics, *Working Papers*, **27**, Houston - Department of Economics (1988).

8. C. L. Sayers, Diagnostic tests for nonlinearity in time series data: An application to the work stoppage series. *Working Papers*, **12**, Houston - Department of Economics (1988).
9. W. Barnett & P. Chen, The aggregation-theoretic monetary aggregates are chaotic and have strange attractors: An econometric application of mathematical chaos, in *Dynamic Econometric Modeling*, Proceedings of the Third International Symposium in Economic Theory and Econometrics, W. Barnett, E. Berndt & H. White (Eds), Cambridge University Press, pp. 199-246 (1988).
10. W. A. Brock & A. G. Malliaris, *Differential equations, stability and chaos in dynamic economics*, North Holland, Elsevier (1989).
11. M. Z. Frank, R. Gençay, & T. Stengos, International chaos?, *European Economic Review*, **32**, 1569-1584 (1988).
12. J. B. Ramsey & H. J. Yuan, Bias and error bars in dimension calculations and their evaluation in some simple models, *Physical Review Letters*, **A 134**, 287-297 (1989).
13. J. B. Ramsey & H. J. Yuan, The statistical properties of dimension calculations using small data sets, *Nonlinearity*, **3**, 155-175 (1990).
14. J. B. Ramsey, C. L. Sayers & P. Rothman, The statistical properties of dimension calculations using small data sets: Some economic applications, *International Economic Review*, **31** (4), 991-1020 (1990).
15. P. Grassberger & I. Procaccia, Measuring the strangeness of strange attractors, *Physica D*, **9**, 189-208 (1983).
16. P. Cvitanovic, *Universality in chaos*, Adam Helger, Bristol, UK (1984).
17. J.-G. Caputo, Practical remarks on the estimation of dimension and entropy from experimental data, in *Measures of Complexity and Chaos*, N. B. Abraham, A. M. Albano, A. Passamante & P. E. Rapp (Eds), Plenum Press, New York (1989).
18. W. A. Brock, Distinguishing random and deterministic systems: Abridged version, *Journal of Economic Theory*, **40**, 168-195 (1986).
19. W. A. Brock, & E. G. Baek, Some theory of statistical inference for nonlinear science, *Review of Economic Studies*, **58**, 697-716 (1991).
20. G. J. Gunaratne, P. S. Linsay & M. J. Vinson, Chaos beyond onset: A comparison of theory and experiment, *Physical Review Letters*, **A 63**, 1-4 (1989).
21. P. Grassberger, An optimized box-assisted algorithm for fractal dimensions, *Physics Letters A*, **148**, 63-68 (1990).
22. G. B. Mindlin, X.-J. Hou, H. G. Solari, R. Gilmore & N. B. Tufillaro, Classification of strange attractors by integers, *Physics Review Letters*, **64**, 2350-2353 (1990).
23. N. B. Tufillaro, H. G. Solari & R. Gilmore, Relative rotation rates: Fingerprints for strange attractors, *Physical Review A*, **41**, 5717-5720 (1990).
24. G. B. Mindlin, H. G. Solari, M. A. Natiello, R. Gilmore & X.-J. Hou, "Topological analysis of chaotic time series data from the Belousov-Zhabotinskii reaction," *Journal of Nonlinear Science*, **1**, 147-173 (1991).
25. C. G. Gilmore, A new test for chaos, *Journal of Economic Behavior & Organization*, **22** (2), 209-237 (1993).
26. O. E. Rössler, An equation for continuous chaos, *Physics Letters A*, **57**, 397-398 (1976).
27. C. G. Gilmore, Detecting linear and nonlinear dependence in stock returns: New methods derived from chaos theory, *Journal of Business Finance & Accounting*, **23**, 1357-1377 (1996).
28. W. A. Brock, D. A. Hsieh & B. LeBaron, *Nonlinear dynamics, chaos and instability: Statistical theory and economic evidence*, (MIT Press, Cambridge, MA (1991).
29. D. Hsieh, Chaos and nonlinear dynamics: Application to financial markets, *Journal of Finance*, **46** (5), 1839-1877 (1991).

30. M. H. DeGroot, *Probability and Statistics*, Addison-Wesley (1975).
31. J. Y. Campbell, A. W. Lo & A. C. MacKinlay, *The Econometrics of financial markets*, Princeton University Press, Princeton (1997).
32. W. A. Brock, W. D., Dechert, B. LeBaron & J. A. Scheinkman, A Test for independence based on the correlation dimension. *Working Papers*, **9520**, Wisconsin Madison - Social Systems (1995).
33. P. De Grauwe, H. Dewachter & M. Embrechts, *Exchange rate theory: Chaotic models of foreign exchange markets*, Blackwell, Oxford (1993).
34. C. G. Gilmore, An Examination of nonlinear dependence in exchange rates, using recent methods from chaos theory, *Global Finance Journal*, **12**, 139-151 (2001).
35. M. Fernandes, Non-linearity and exchange rates, *Journal of Forecasting*, **17**, 497-512 (1998).

Chapter 15

Introduction of the sphere map with application to spin-torque nano-oscillators

Keith Gilmore

The Molecular Foundry
Lawrence Berkeley National Laboratory
Berkeley, CA 94720, USA

Robert Gilmore

Physics Department
Drexel University
Philadelphia, PA 19104, USA

We generalize the circle map $S^1 \to S^1$ to a nonlinear mapping of the sphere to itself. The sphere map $S^2 \to S^2$ depends on the three parameters that describe the rotation operations in the group SO(3) and one nonlinear parameter that describes the strength of a folding acting on the sphere surface. This new map is used to study the properties of a spin valve that obeys equations of the form $\dot{X}_i = -b_{ij}X_j + c_{ijk}X_j X_k$ and satisfies the conservation condition $\frac{d}{dt}\sum X_i^2 = 0$. With periodic driving, we find regions in the parameter space that describe quasiperiodic, mode-locked, and chaotic behavior.

Contents

1. Introduction 317
2. Background 318
3. Landau-Lifshitz-Slonczewski Equations 319
4. Connection with Lorenz Equations 321
5. Definition of Sphere Map 322
6. Scanning Procedure 324
7. Bifurcation Cuts 325
8. Trajectories 327
9. Summary and Discussion 328
References 329

1. Introduction

In this work we formulate tools to study the spectrum of possible behaviors of a magnetic nano-oscillator that is described by the Landau-Lifshitz-Slonczewski (LLS) equations. These equations have a form introduced by Edward Lorenz in his study of truncations of the Navier-Stokes equations.

Within the LLS equations, the magnetization is a conserved quantity, so is described by a point on the surface of a sphere S^2. Under periodic forcing the phase space is enlarged to $S^2 \times S^1$. We introduce a mapping of the sphere surface to itself, which is a generalization of the circle map, to study the dynamics of this system.

In Sect. 2 we review the circle map, as defined by Kolmogorov and Arnold. We also review the truncated form of the Navier-Stokes equations, as described by Lorenz. Sect. 3 discusses spin-transfer-torque and introduces the Landau-Lifshitz-Slonczewski equations. In Sect. 4 we bring the LLS equations into the form proposed by Lorenz. Section 5 defines the sphere map $S^2 \to S^2$. In Section 6 we introduce three large-scale scans for behavior under rotations about three coordinate axis. Section 7 examines some appropriate bifurcation diagrams, and in Section 8 we investigate how some trajectories behave under variation of the nonlinear coupling strength. Finally, we summarize our findings in Section 9.

2. Background

The circle map

$$\theta_{n+1} = \omega_0 + \theta_n - k\sin(\theta_n) \tag{1}$$

was originally introduced by Andrei Kolmogorov to describe the behavior of periodically driven mechanical oscillators.[1] Here θ represents the phase of the oscillator and θ_n is its value stroboscopically recorded (i.e., at the same phase during each driving period). The term $-k\sin(\theta_n)$ describes the nonlinearity in the response of the oscillator to the driving force. Vladimir Arnold later used this map to study the breakup of invariant tori under perturbations.[2] The circle map exhibits behavior including quasiperiodic orbits (regular but not closed), mode locked regions, Arnold tongues, and period-doubling routes to chaotic behavior.[3]

A standard way to simplify the Navier-Stokes equations involves expanding the various fields present (the velocity, temperature, and pressure fields) in Fourier modes or empirical orthogonal modes, placing these representations in the Navier-Stokes equations, and then integrating out the spatial dependence. Lorenz[4] noted that the resulting set of ordinary differential equations for the time-dependent mode amplitudes assumes the form of a set of ordinary differential equations whose source terms are polynomials of degree no greater than two:

$$\frac{dX_i}{dt} = a_i - b_{ij}X_j + c_{ijk}X_jX_k \tag{2}$$

where the tensor c is symmetric in the last two indices: $c_{ijk} = c_{ikj}$ and we use the summation convention. Lorenz further observed that all initial conditions lead to bounded solutions if b_{ij} is positive definite and the quadratic terms satisfy the condition $c_{ijk}X_iX_jX_k = 0$. These conditions are satisfied by the Lorenz equations.

The Lorenz equations exhibit a behavior that was very surprising for its time: deterministic but nonperiodic flow.[4] The Lorenz strange attractor has been a model for understanding chaotic behavior and a test-case for tools developed to study this behavior.[5]

Spintronics, the control of magnetic systems with spin-currents, holds great technological promise. Some of the interactions of spin-currents with magnetic systems are described by the Landau-Lifshitz-Slonczewski equations. These equations have the form of Eq.(2) with $a_i = 0$, b_{ij} antisymmtric and c_{ijk} with mixed symmetry and $c_{ijk} X_i X_j X_k = 0$. As a result, the phase space for the magnetic moment is the sphere surface S^2. The LLS equations for a periodically driven system involve flows on a three-dimensional phase $S^2 \times S^1$. Since the phase space is three-dimensional, chaotic behavior is possible.[3] A Poincaré section can be chosen as $\phi =$ cst., $\phi \in S^1$. The Poincaré first return map is the map $S^2 \to S^2$. The sphere map is a generalization of the circle map. We report some of the properties of the sphere map in this work.

3. Landau-Lifshitz-Slonczewski Equations

Ferromagnetism is a state of matter in which strong Coulomb exchange interactions align the spins of electrons, producing a net magnetization in the absence of any applied external field. The exchange interaction is short range. On longer length scales the magnetostatic dipole interaction favors antiparallel alignment of spins, leading to the formation of magnetic domains. The electron spins within a given domain are essentially uniformly aligned. Neighboring domains are typically oriented so as to minimize the magnetostatic dipole energy. Between two domains the magnetization rotates smoothly from one domain orientation to the orientation of the neighboring domain producing a magnetic domain wall. The characteristic length scale of domains depends on material parameters and sample geometry, but is large enough for many materials, such as the transition metals, that nanoscale samples can be comprised of a single domain and thought of as having one macroscopic magnetic moment.

It is energetically favorable for the magnetization to align with the effective magnetic field, which consists of the combination of internal and external fields. Because the magnetization carries angular momentum, the effective field applies a torque to the magnetization, which causes the magnetization to precess about the effective field according to

$$\partial_t \hat{m} = -\gamma \hat{m} \times \mathbf{H}^{\text{eff}} \qquad (3)$$

where $\hat{m}$ is a unit vector designating the direction of the magnetization, $\mathbf{H}^{\text{eff}}$ is the effective magnetic field, and γ is the gyromagnetic ratio. If $\mathbf{H}^{\text{eff}}$ is constant in time then $|\gamma \mathbf{H}^{\text{eff}}|$ is the precession frequency.

In reality, some loss mechanism damps the precession of the magnetization, causing it to align in the low energy direction specified by the effective field. This damping is a phenomenological observation that is typically accounted for by the inclusion of another torque, $-(\lambda/|m|)\, \hat{m} \times [\hat{m} \times \mathbf{H}^{\text{eff}}]$, that is orthogonal to both the magnetization direction and the precession direction. Together, these two torques form the Landau-Lifshitz (LL) equation[6]

$$\partial_t \hat{m} = -\gamma \hat{m} \times \mathbf{H}^{\text{eff}} - \frac{\lambda}{|m|} \times \hat{m} \times \mathbf{H}^{\text{eff}}. \qquad (4)$$

The parameter λ is a sample dependent dissipation rate. The damping term is often rewritten in the equivalent Gilbert form[7–9] as $\alpha \hat{m} \times \partial_t \hat{m}$ and the resulting equation referred to as the Landau-Lifshitz-Gilbert equation (LLG equation). In the present work we use the Landau-Lifshitz form.

The magnetization of a material may also be manipulated by an electrical current.[10–13] Due to the inequivalent density of states of each spin type at the Fermi level, a current flowing through a magnetic material will be spin-polarized – it will carry angular momentum in addition to charge. If the magnetization direction of a material is spatially non-uniform along the direction of current flow the spin-polarized current will apply a torque to the magnetization. This torque is known as the spin-transfer torque. The spin-transfer torque shows promise for use in magnetic random access memories[14,15] and telecommunication applications.[16,17]

The spin-transfer torque can have different forms depending on the physical circumstances under consideration.[18] If the magnetization changes very slowly in space, such as in a wide domain wall, then the spin-orientation of the transport electrons will be able to track the local magnetization adiabatically as the electrons flow through the system. In this case, the torque due to the spin current is referred to as the adiabatic spin torque.[10] If the spatial variation of the magnetization is abrupt, the spin orientation of the transport electrons will not be able to respond adiabatically to the local magnetization and the resulting torque is the non-adiabatic spin torque.[13] In this paper we will consider the non-adiabatic spin torque such as occurs in a nano-pillar magnetic multilayer.

The nano-pillar magnetic multilayer system consists of a multilayer nano-pillar connected on both ends to non-magnetic electrodes. The multilayer stack has a physical cross-section of a few tens of nanometers and two (or more) magnetic layers separated by non-magnetic spacer layers. The magnetization of one of the magnetic layers will be fixed (either through exchange biasing or simply by being much thicker than the second magnetic layer); this is referred to as the fixed layer or polarizer. The magnetization of the other magnetic layer will be free to rotate; this layer is referred to as the free layer.

As a current flows through the multilayer stack it becomes spin-polarized by the polarizer layer. The spin-polarized current then impinges upon the free layer, applying a non-adiabatic spin-torque, $\Omega \hat{m} \times [\hat{m} \times \hat{m}_P]$, to the magnetization. The prefactor Ω is a constant with dimensions of frequency that depends on the current density, material parameters, and fundamental constants. The unit vector $\hat{m}_P$ is the direction of magnetization of the fixed layer, which dictates the polarization axis for the spin current. Adding this spin-torque to the LL equation gives gives the Landau-Lifshitz-Slonczewski (LLS) equation

$$\partial_t \hat{m} = -\gamma \hat{m} \times \mathbf{H}^{\text{eff}} - \frac{\lambda}{|m|} \hat{m} \times \hat{m} \times \mathbf{H}^{\text{eff}} + \Omega \hat{m} \times \hat{m} \times \hat{m}_P. \qquad (5)$$

All three torques in this equation are perpendicular to the magnetization and act only to rotate the magnetization, not change its magnitude. Therefore, the trajectory of $\mathbf{m}(t)$ evolves on the surface of a sphere S^2. Since the sphere is two-dimensional the asymptotic behavior can only be stationary or periodic, corresponding to the the only behaviors allowed by the Poincaré-Bendixon theorem.[3]

The dynamic responses allowed by Eq. (5) have been studied both experimentally[19] and theoretically.[20] Three types of behavior are observed. Either the magnetization of the free layer will remain largely unperturbed, will reverse, or will be driven into a stable precession orbit. The particular response depends on the current density, degree of polarization, strength and direction of external and internal fields, and the strength of the damping. For low spin-current driving, the damping torque keeps the magnetization oriented in the low energy direction. Very large spin-currents will cause the direction of the magnetization of the free layer to reverse and align in the oppositely oriented low energy direction. Intermediate spin-current driving can establish stable precession orbits. The frequency of these precession orbits may be tuned by adjusting the current density. This observation suggests an application for the spin-torque multi-layer as a voltage-controlled tunable-frequency oscillator.[16,17] These observations assume a dc-current. In this work we investigate what dynamic responses may be induced by an ac-current.

4. Connection with Lorenz Equations

The Landau-Lifshitz-Slonczewski equations (4) have the form of the Lorenz equations (2) under the identification $m_1 = X_1 = X$, $m_2 = X_2 = Y$, and $m_3 = X_3 = Z$. In this form $a_i = 0$, $b_{ij} = -b_{ji}$, and c_{ijk} is of mixed symmetry with $c_{ijk} X_i X_j X_k = 0$.[4]

If any one of the terms in Eq.(4) is periodically driven, the phase space is enlarged to $S^2 \times S^1$. This occurs in the same way and for the same reasons that the phase space for nonlinear two-dimensional oscillators (e.g., R^2 for the Duffing and van der Pol oscillators) is enlarged to $R^2 \times S^1$ when they are periodically driven.[21,22] In this three-dimensional phase space more exciting behaviors can occur.[3] These include quasiperiodic behavior, periodic behavior due to mode locking, and chaotic behavior. All these types of behavior have been seen in simulations[23-25] and discussed in a recent review article.[26] At the present time there is no systematic understanding of the parameter ranges under which these three types of behavior can be exhibited by the solutions of the periodically driven LLS equations or the nano-oscillators that they model. It would be extremely useful to have a systematic understanding, comparable to that provided by the circle map, for a large class of nonlinear dynamical systems. The purpose of the present contribution is to provided the tools for such a systematic understanding of the parameter space describing the behavior of the solutions of Eq.(5).

Since the phase space is $S^2 \times S^1$, it is useful to study dynamics by studying the first return map on the Poincaré section. The Poincaré section can be chosen as a stroboscopic section $2\pi t/T = $ cst. mod 2π. This is a sphere. We call the first return map $S^2 \to S^2$ the *sphere map* in analogy with maps $S^1 \to S^1$, the widely studied circle maps.[1–3] These maps have been enormously useful for the study of nonlinear oscillators in the plane that have undergone a Hopf bifurcation and are periodically driven (e.g., van der Pol oscillator). We expect sphere maps to be equally useful for studying the properties of nonlinear spherical oscillators such as spin valves.

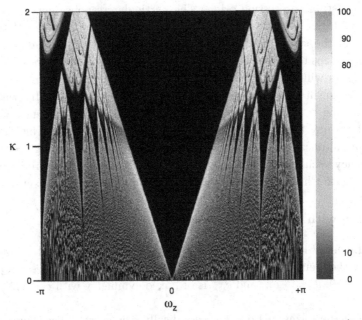

Fig. 1. (color online) Behavior of the sphere map for rotations about the z-axis $(0, 0, 1)$. Color coding: number of distinct points in a long trajectory. Cool colors (blues) indicate low period orbits that exist because of mode locking. Hot colors (reds) indicate either high period orbits or quasiperiodic trajectories. Below $k = 1$ quasiperiodic regions are interrupted by mode-locked regions which generally increase in width as k increases. Evidence of period-doubling cascades can be seen above $k = 1$.

5. Definition of Sphere Map

The circle map is the composition of two simpler maps. The first is a rigid rotation of the circle through the angle ω_0: $\theta \to \theta' = \theta + \omega_0$. This is a linear map. The second map is a nonlinear, sometimes noninvertible deformation of the circle: $\theta \to \theta' = \theta - k\sin(\theta)$. This nonlinear map involves a folding centered on the angle $\theta = 0$. It is noninvertible for $|k| > 1$.

We construct a nonlinear map of the sphere to itself by generalizing these two transformations. The linear step applies a rigid rotation of the sphere onto itself: $(X, Y, Z) \to (X, Y, Z)'$, where the sum of the squares of the coordinates is $+1$. It consists of a rigid rotation through an angle ω about an axis of unit length $\mathbf{n}$: $\omega \mathbf{n} = (\omega_x, \omega_y, \omega_z)$.

The nonlinear step folds the sphere about the half of a great circle through the poles and the point $(1, 0, 0)$ ("Greenwich meridian"). The cut point for this map is the negative X axis, so that the azimuthal angle ϕ satisfies $-\pi \leq \phi = \tan^{-1}(Y/X) \leq +\pi$. This angle is stretched according to $\phi \to \phi' = \phi - k\sin(\phi)$. The image coordinates on the sphere are $Z' = Z, X' = R\cos(\phi'), Y' = R\sin(\phi')$, where $R = \sqrt{X^2 + Y^2}$. This simple two-step map (rigid rotation + fold) is an extension of the circle map from S^1 to S^2. In this case the wrinkling exhibited in the circle map occurs as a folding centered on the half great circle described above.

The sphere map depends on $3 + 1$ parameters. Three identify an element in the rotation (Lie) group SO(3), which acts through a linear representation on the coordinates $(X, Y, Z) \in S^2$. The fourth identifies the strength of a certain type of nonlinearity acting on the sphere surface S^2.

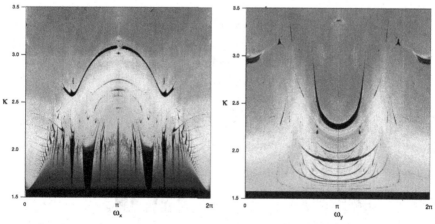

Fig. 2. (Color online) Behavior of the sphere map for rotations about the: x-axis $(1, 0, 0)$ (left) and the y-axis $(0, 1, 0)$ (right). Color coding as in Fig. 1. Mode-locked regions can be seen somewhere for many values of k.

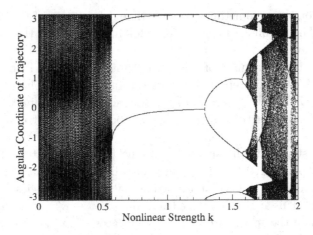

Fig. 3. Bifurcation diagram taken along $\omega = \pi \times 99/100$ in Fig. 1. Quasiperiodic behavior ($0 \leq k < 0.6$) is followed by mode locking ($0.57 \leq k \leq 1.3$), which is followed by period-doubling ($k \leq 1.3 \leq 1.6$) which is followed by chaotic behavior intermingled with periodic windows ($1.6 \leq k \leq 2.0$).

6. Scanning Procedure

To explore the sphere map we choose a rotation axis and describe the trajectories as the rotation angle ω and the nonlinear strength k are varied. The trajectory is characterized by the value of the azimuthal angle ϕ. For a period-p orbit there will be p distinct angular values. Each successive value of this angle is recorded in a

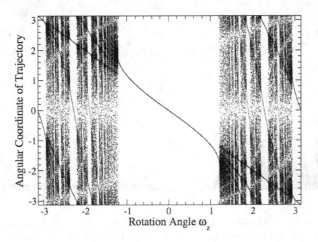

Fig. 4. Bifurcation diagram taken along $k = 1$ in Fig. 1. This bifurcation diagram shows intermingling of quasiperiodic behavior with mode locking. The largest mode-locked region occurs in the range $-1.2\pi < \omega_z < +1.2\pi$. Period-two mode locking occurs around π radians, mode-locking of period three around $\pm\frac{2\pi}{3}$, period-four around $\pm\frac{2\pi}{4}$, etc.

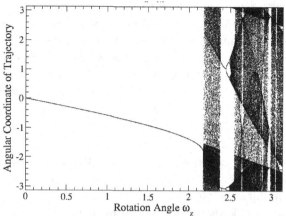

Fig. 5. Bifurcation diagram taken along $k = 1.8$ in Fig. 1 in the restricted range $0 \leq \omega_z \leq \pi$. This bifurcation diagram shows (left to right) how mode-locking gives way to quasiperiodicity which then gives way to period-doubling and then to chaos intermingled with periodic windows. This bifurcation diagram exhibits a two-fold rotational symmetry around the origin.

trajectory and the angular values of these coordinates are binned. The number of bins visited by a trajectory is determined and color coded in Figs. 1 and 2. Mode-locked orbits of period one and two are coded dark and light blue. Orbits of high period are generally indistinguishable from quasiperiodic orbits in this color plot. They can be distinguished by the structure of the bifurcation diagrams presented in Figs. 3 to 8. Further, chaotic behavior cannot occur unless the map is noninvertible, that is, for $|k| > 1$.

The results are shown in Fig. 1 for rotations around the z axis $\mathbf{n} = (0, 0, 1)$. For $k < 1$ the motion is either quasiperiodic or mode-locked, with the mode-locked regions indicated by the dark blue color. These windows (tongues) generally increase in width with increasing k. Above $k = 1$ period doubling begins to occur (shades of lighter blue) leading to chaotic behavior. The chaotic regions are intermixed with mode locked regions, and the chaotic region has diminishing measure as k continues to increase.

For rotations around the x- and y-axes the result is substantially different. In Fig. 2a we show the results for rotations around the x-axis. In this case mode-locking occurs for $k < 1.5$. As k increases above 1.5 locking to higher period orbits occurs, leading eventually to quasiperiodic and chaotic behavior. Figure 2b shows a similar plot for rotations around the y-axis. Just as for rotations around the x-axis, mode-locking occurs for $k < 1.5$ and appears in isolated regions for larger values of k.

7. Bifurcation Cuts

We construct bifurcation diagrams for the results presented in Fig. 1 along various cuts. The first cut (Fig. 3), along k in the range $0 \leq k \leq 2$, at constant rotation angle $\omega = \pi * (99/100)$, shows that the motion is quasiperiodic up to $k = 0.6$. After this point the invariant density increases to the point where a period-two orbit emerges. This continues until $k \simeq 1.3$ where a period-doubling cascade begins. The evolution along this cut terminates in a chaotic attractor for $k \simeq 1.6$. The chaotic regions show windows of mode locking intermingled with the chaotic behavior, typical of the logistic map. The largest window, at $k \simeq 1.7$, has period 2×3. It is the period-three window in the chaotic attractor based on the period-two orbit.

Figure 4 shows a cut at $k = 1$, below the range in which chaos is observed in Fig. 3. This figure shows in detail the behavior indicated in Fig. 1. There is an intermingling of mode-locked and quasiperiodic behavior as ω_z is scanned from $-\pi$ to $+\pi$. Mode locking to a period-one orbit occurs around $\omega_z = 0$. Mode locking to a period-two orbit occurs around $\omega_z = \pi = \frac{2\pi}{2}$. Mode locking to orbits of period $d = 3, 4, \ldots$ occurs around $\omega_z = \frac{2\pi}{d}$.

Another cut, at $k = 1.8$, is shown in Fig. 5. This figure only shows $0 \leq \omega_z \leq \pi$ because of two-fold rotational symmetry about the origin. As ω_z increases from 0 to π mode locking is replaced by quasiperiodicity and then by mode-locking at twice the period. A period-doubling cascade to chaos follows. As is typical, there are periodic windows in the chaotic region.

Figure 6 shows a bifurcation diagram for the scan in Fig. 2 at $k = 2.2$ as a function of the rotation angle ω_x. Only the first half of this bifurcation diagram is shown since the diagram is symmetric under $\omega_x \to 2\pi - \omega_x$. The windows that appear in Fig. 6 generally undergo a period-doubling route to chaos. A region of period-doubling is magnified in Fig. 7.

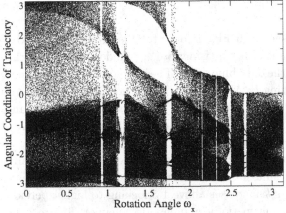

Fig. 6. A cut along $k = 2.2$ in Fig. 2 in the restricted range $0 \leq \omega_x \leq \pi$. This bifurcation diagram shows (left to right) how quasiperiodicity gives way to mode-locking and then to period-doubling to chaos.

A bifurcation diagram taken along the symmetry axis $\omega_y = \pi$ is shown in Fig. 8. A mode-locked period-one orbit undergoes period-doubling to chaos. At $k \simeq 2.85$, the chaotic attractor begins to wind around itself in the sense that its boundaries begin to overlap.

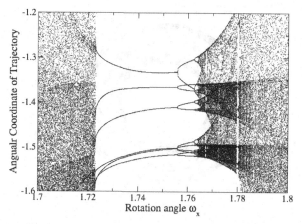

Fig. 7. This blow-up shows clearly how the mode-locked state goes through a period-doubling cascade to chaos, which then exhibits a crisis back to quasiperiodicity.

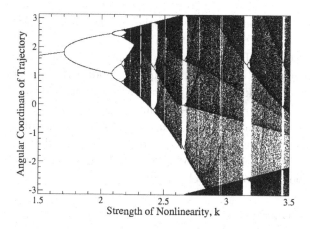

Fig. 8. Bifurcation diagram taken along the scan of Fig. 2 along the symmetry axis. This diagram exhibits the classic period-doubling scenario with a twist: the attractor begins to wind around itself for $k > 2.85$ when the boundaries of the basin of attraction begin to overlap.

8. Trajectories

A trajectory consists of an iterated sequence of points (X_i, Y_i, Z_i) on the sphere surface. The nonlinear transformation maps points from a given latitude (Z value) to the same latitude, as do rigid rotations around the Z-axis. Composition of these two operations maps points into points with the same latitude. Projecting trajectories onto the equatorial plane maps them to a circle with fixed radius. As the strength of the nonlinearity k is varied, the invariant density on this circle varies but its radius does not. These projections provide a simple visual method to classify trajectories. An ensemble of trajectories for k values in the range $0.5 \leq k \leq 2.0$

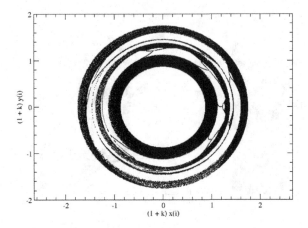

Fig. 9. Scaled projection of a trajectory for $0.5 \leq k \leq 2.0$. Trajectories are scaled by $(1+k)$. Three rings of mode locked behavior are evident. These have period $p = 5, 3, 2$, respectively (from the inside). Each mode locking is followed by a period-doubling cascade to chaos.

is shown in Fig. 9. Since the latitude is invariant under these transformations all trajectories lie on the same circle. For visual clarity we scale the projection by a k-dependent factor $(1 + k)$. The resulting projections are organized into concentric rings. For small values of k (interior region) quasiperiodic orbits predominate. For larger values of k mode locking begins to occur. The first evident mode locking occurs for $p = 5$, followed by a period-doubling route to chaos. The next mode locking occurs for $p = 3$, again followed by a period-doubling cascade. The outside ring shows a period-two mode locking, also followed by a period-doubling route to chaos. For larger values of k a period-one mode locking occurs. This behavior contrasts with the behavior of the circle map.

9. Summary and Discussion

Spin-torque oscillators are nanodevices with important technological promise. Isolated oscillators are described by the Landau-Lifshitz-Slonczewski equations. Since the magnitude of the magnetic moment is assumed constant within the LLS equations, the phase space for a spin valve is the surface of the sphere S^2. If a spin valve is periodically driven the phase space expands to $S^2 \times S^1$, so that chaotic behavior is possible. It is useful to study the dynamical properties of a periodically driven spin valve by investigating the first return map on a Poincaré surface of section — the sphere under stroboscopic measurements at $\phi = \text{cst.}$, $\phi \in S^1$.

To accomplish this, we have generalized the circle map $S^1 \to S^1$ to a sphere map $S^2 \to S^2$. This sphere map is the composition of two maps. The first is a rigid rotation of the sphere to itself which is parametrized by three variables: the Euler angles or a rotation angle ω together with a rotation axis $\mathbf{n}$, or $(\omega_x, \omega_y, \omega_z)$. The second map is a deformation of the density on the sphere surface. We choose this nonlinear map as a fold around the Greenwich meridian, using exactly the same form as occurs in the circle map. The strength of this nonlinearity is defined by a parameter k.

Three scans have been made showing behavior in an $\omega - k$ plane. The first scan is for rigid rotations around the z axis, and the results are similar to those encountered for the circle map. The other two scans involve rotations around the x- and y-axes. In these cases there is a mixture of quasiperiodicity, mode-locking, and chaos. However, the organization of these types of behavior is not yet well-understood.

Several bifurcation plots were made to better understand the information in the initial scans. The bifurcation plots show a systematic organization of the quasiperiodic, mode-locked, and chaotic regions.

Finally, we calculated trajectories under iterates of the sphere map. For rotations around the z axis all iterates lie on a fixed latitude circle. These trajectories have been projected down to the equatorial plane and scaled by a k-dependent factor for visual clarity. These projections (c.f., Fig. 9) emphasize the intermingling of the quasiperiodic (dark) and mode-locked (mostly white) regions.

The sphere map depends on four parameters. Three identify an element in the rotation (Lie) group SO(3). The fourth identifies the strength of a certain type of nonlinearity.

This paper developed out of a long series of discussions during which my father and I attempted to explain our research topics to one another. While we are both physicists, we pursue substantially different subtopics within physics. My father has constructed a career around non-linear dynamics and chaos theory while my initial research direction centered on magnetism in condensed matter systems. Fortunately, magnetic systems display a rich dynamic range, particularly because they can be driven both by magnetic fields and electric currents. This provided a natural opportunity for us to explore the possibility of chaotic dynamics in magnetic systems. We were pleasantly surprised by the wealth of behavior we found.

This work was supported in part by NSF Grant PHY 0754081.

References

1. V. I. Arnold, Small denominators. I. On the mappings of the circumference onto itself, *Transactions of the American Mathematical Society*, II, **46**, 213 (1965).
2. V. I. Arnold, Small denominators and problems of stability of motion in classical and celestial mechanics, *Russian Mathematical Surveys*, **18**(6), 85-191 (1963).
3. E. A. Jackson, *Perspectives on Nonlinear Dynamics*, Cambrdige: University Press (1989).
4. E. N. Lorenz, Deterministic nonperiodic flow, *Journal of the Atmospheric Science*, **20**, 130-141 (1963), c.f., Eq. (4).
5. H. D. I. Abarbanel, R. Brown, J. J. Sidorowich & L. Sh. Tsimring, The analysis of observed chaotic data in physical systems, *Reviews in Modern Physics*, **64** (5), 1331-1393 (1993).

6. L.D. Landau & E.M. Lifshitz, Theory of the dispersion of magnetic permeability in ferromagnetic bodies, *Physikalische zeitschrift der Sowjetunion*, **8**, 153-169 (1935).
7. T. L. Gilbert, A Lagrangian formulation of the gyromagnetic equation of the magnetic field, *Physical Review*, **100**, 1243 (1955).
8. T.L. Gilbert, *Armor Foundation Research Project No. A059*, Supplementary Report, May 1, 1956.
9. T.L. Gilbert, A phenomenological theory of damping in ferromagnetic materials, *IEEE Transactions on Magnetism*, **40** (6), 3443-3449 (2004).
10. L. Berger, Low field magnetoresistance and domain drag in ferromagnets, *Journal of Applied Physics*, **49**, 2156-2161 (1978).
11. L. Berger, Domain drag effect in the presence of variable magnetic field or variable transport current, *Journal of Applied Physics*, **50**, 2137-2139 (1979).
12. L. Berger, Emission of spin waves by a magnetic multilayer traversed by a current, *Physical Review B*, **54**, 9353-9358 (1996).
13. J. C. Slonczewski, Current-driven excitation of magnetic multilayers, *Journal of Magnetism and Magnetic Materials*, **159**, L1-L7 (1996).
14. J. Akerman, Toward a universal memory, *Science*, **308**, 508-510 (2005).
15. S. S. P. Parkin, M. Hayashi & L. Thomas, Magnetic domain-wall racetrack memory, *Science* **320**, 190-194 (2008).
16. W. H. Rippard, M. R. Pufall, S. Kaka, S. E. Russek, T. J. Silva, Direct-current induced dynamics in $Co_{90}Fe_{10}/Ni_{80}Fe_{20}$ point contacts, *Physical Review Letters*, **92**, 027201, (2004).
17. O. Boulle, V. Cros, J. Grollier, L. G. Pereira, C. Deranlot, F. Petroff, G. Faini, J. Barnas, and A. Fert, Shaped angular dependence of the spin-transfer torque and microwave generation without magnetic field, *Nature Physics*, **3**, 492-497 (2007).
18. P. M. Haney, R. A. Duine, A. S. Nunez & A. H. MacDonald, Current-induced torques in magnetic metals : Beyond spin-transfer, *Journal of Magnetism and Magnetic Materials*, **320**, 174412 (2008).
19. S. I. Kiselev, J. C. Sankey, I. N. Krivorotov, N. C. Emley, R. J. Schoelkopf, R. A. Buhrman & D. C. Ralph, Microwave oscillations of a nanomagnet driven by a spin-polarized current, *Nature*, **425**, 380383 (2003).
20. D. C. Ralph & M. D. Stiles, Spin transfer torques, *Journal of Magnetism and Magnetic Materials*, **320**, 1190-1126 (2008).
21. R. Gilmore & M. Lefranc, *The Topology of Chaos*, Wiley, (2002).
22. R. Gilmore & C. Letellier, *The Symmetry of Chaos*, Oxford University Press (2008).
23. Z. Li, Y. C. Li & S. Chang, Dynamic magnetization states of a spin valve in the presence of dc and ac currents: Synchronization, modification, and chaos, *Physical Review B*, **74**, 954417 (2006).
24. Z. Yang, S. Zhang & Y. C. Li, Chaotic dynamics of spin-valve oscillators, *Physical Review Letters*, **99**, 134101 (2007).
25. M. Lakshmanan & K. Nakamura, Landau-Lifshitz equation of ferromagnetism: Exact treatment of the Gilbert damping, *Physical Review Letters*, **53**, 2497-2499 (1984).
26. M. Lakshmanan, The fascinating world of the Landau-Lifshitz-Gilbert equation: An overview, *Philosophical Transactions of the Royal Society A*, **369**, 1280-1300 (2011).

Chapter 16

Robert Gilmore, a portrait

Hernán G Solari

Departamento.Fsica-FCEyN-UBA and IFIBA-CONICET
Universidad de Buenos Aires, C.Universitaria - Pab. I
1428 Buenos Aires
Argentina

To present the personality of Bob Gilmore is a formidable task, as his scientific contributions include group theory, laser physics, non-linear dynamics, catastrophe theory, thermodynamics, dynamical systems, quantum theory and more. But even if we succeed in describing his contributions, much of Gilmore's being would be lost. Bob as advisor, Bob as father, Bob as teacher, Bob as scientific communicator reveal as much of Bob Gilmore as his scientific papers and his books. Very much as in the Group Theory so close to him, there is a Robert Gilmore in abstract as well as representations of Robert Gilmore. We will make an attempt to find the "principle of the rule", the abstract level of Robert Gilmore as well as Robert Gilmore, himself, as a representation of the duality science-humanism.[*]

Contents

1. Some Representations of BG 331
2. From Lasers to Topology: the Context of Discovery 336
3. Epilogue 338
References 339

1. Some Representations of BG

Force, for example, is not a metaphysical conatus of an unknown kind which hides behind its effects (accelerations, deviations, etc.); it is the totality of these effects. Similarly an electric current does not have a secret reverse side; it is nothing but the totality of the physical–chemical actions which manifest it (electrolysis, the

---

[*]Disclaimer: Every understanding implies the connection of the new knowledge with our old knowledge, the links we find depend as much on the new as they depend on the old. Every vision is biased. If the visions of physics are not personally biased it is because for several generations and in all places around the world, physics has been taught following the same books and the same masters producing a common and uniform background. But this is not the case with a personal portrait, it is then unavoidable that my brush will be revealed as much as we reveal Bob. Every statement, every memory, every observation should be preceded by expressions such as "from my point of view", "in my observation", "in my understanding", etc. I will not be repeating such phrases as they would be disruptive, but the reader should keep them in mind. Thanks!

> incandescence of a carbon filament, the displacement of the needle of a galvanometer, etc.). No one of these actions alone is sufficient to reveal it. But no action indicates anything which is *behind itself*; it indicates only itself and the total series.
>
> ...That is why we can equally reject the dualism of appearance and essence. The appearance does not hide the essence, it reveals it; it *is* the essence.
>
> ...But the appearance, reduced to itself and without reference to the series of which it is part, could be only an intuitive and subjective plenitude, the manner in which the subject is affected. If the phenomenon is to reveal itself as *transcendent*, it is necessary that the subject himself transcend the appearance toward the total series of which it is a member.
>
> Jean Paul Sartre in Being and Nothingness.

Very much as in Sartre's explanation of electricity, revealing who is Robert (Bob) Gilmore requires to present and examine the forms in which he transcended his self, the "Bob at work" or as I would like to say it, the representations of Bob Gilmore, calling for the group theory language so close to his heart. I will try to highlight in what comes the particularities of the Bob I worked with and have known for more than 25 years, although most memories come from the period in which I was a postdoctoral fellow and visiting faculty under his advising (1985-1989).

Gilmore and his time

> The axiom that nothing on earth is so important as the highest power, and most varied cultivation of the individual, and that, therefore, the primary law of true morality is, educate yourself, and only the second, influence others by what you are; these axioms are so firmly impressed upon my mind that nothing can change them.
>
> Wilhelm von Humboldt, letter to Georg Forster, 1791.

If "I am I and my circumstances" as José Ortega y Gasset put it, we can begin to understand Robert Gilmore's standpoint in the academic world by considering his circumstances, in particular as compared to those of Wilhem von Humboldt.

Von Humboldt's major contributions, including the creation of Berlin's Free University (1810) came in the wake of major philosophical and political debates by Immanuel Kant. Kant in "The contest of the faculties"[1] established the idea of an university life free from the surveillance of religion, in which faculties dedicated themselves to develop their reason, and with it their freedom. Faculties would help students to reach their goal of becoming illustrated.[2] The main concern of the Philosophy faculty, and its department of Science or Natural Philosophy, was to develop the freedom of faculties and students by developing their reasoning. To understand was the goal and the academic dogma became "knowledge is good in it itself",[3] a dogma adopted by modern universities around the world. In practice, it meant that no longer reason would be subservient of religion. The modern university was born and it would flourish for about one hundred years seeking universal understanding, and about 50 years seeking specialized knowledge, departing from Wilhem von Humboldt's program and limiting its scope.

In contrast, Robert Gilmore's times are signed by successive reforms of the university system that began after the Second World War. Such reforms make for todays mining of science seeking for possibilities of high-added-value marketable goods - the so-called society of knowledge, making place to Robert Nisbet's observations of the early process in America as "the degradation of the Academic dogma"[3] knowledge is not longer good in itself but rather knowledge is "good for". This utilitarian point of view will necessarily make more difficult and finally end the conquering of freedom by the development of reason, this is, it puts an end to the program of the illustration throwing the academic world into a new era in which all matters can be examined from their own point of view as well as from the point of view of the "social interest" (mainly represented by the interest of the capitalists), very much like in the pre-modern times all matters could be considered from their own point of view as well as from the religious point of view as represented by the Church. If the modern times of the universities are coming to an end, what started in 1945 with Vannevar Bush seminal paper[4] is properly termed the postmodern era for the university, a time where interest (in the sense of obtaining a profit) ranks above understanding.

The postmodern era is characterized by the intervention of the state and business partners in the University with a substantial flow of resources (money) for those programs that are concurrent with the interest of the non-academic society coupled with "productivity pressures" (the publish or perish policy can be tracked to up to the late 1950's). Such measures have the secondary effect (or perhaps it is the primary effect) of introducing the interest in the academic world. Von Humboldt warns us "For it is the property of anything which charms us by its own intrinsic worth, to awaken love and esteem, while that which only as a means holds out hopes of ulterior advantage, merely interests us; and the motives of love and esteem tend as directly to ennoble human nature, as those of interest to lower and degrade it."[5]

The introduction of the interest as the main motor of academic life cannot be executed without a re-signification of the roles, structures and values of the academic world. Where once we had open sharing of ideas, we often find concealment of them; where we had cooperation we are told competition should reign supreme; where we had freedom and responsibility we encounter accountability; where self-appraisal was we are imposed external evaluation rules. The forms and values of a society that had not reached the majority of age,[2] supported in domination-submission relations, are imposed upon the university life. Student advisors become managers mainly concerned with keeping production high, getting funds and advertising the product; graduate students and postdoctoral fellows become human resources, politically correct words for labour force.

But, where does Bob stand in this revolution of the academic life?

The advisor A good part of Bob efforts have gone toward advising young people, not just those working under his guidance but rather all graduate students at his Department as well as graduate students requesting his advice coming from other schools. Any one familiar with the idea of the Socratic maieutics would recognize it in Bob's practice. The central point in maieutics is to aid the student to reach himself the answers. Thus, the main concern is not to hand out a "correct solution" but rather to develop the ability of the student to find and recognize correct solutions.

Teaching to be critic A distinctive method of Bob was the open invitation to criticism, all of Bob's students and collaborators have experienced it. Bob pops by his student/collaborator office saying "Hi! I have this crazy idea I want to try on you", the qualification of the idea as "crazy" is indeed and invitation to criticize it. In so doing the student/colleague develops his critical abilities so much dimmed through years of education. And indeed, some ideas were crazy, but other were simply brilliant, and all of them are always beautiful. The voice of Lupercio Leonardo de Argensola could be applied with little changes to really crazy ones:

> The blue sky that we all see
> is neither sky nor blue.The pity
> is so much beauty is untrue!
>
> (Lupercio Leonardo de Argensola, translated by Michael Smith)

But the "crazy idea events" had another outcome, we learned under Bob guidance that there is no dishonor in **Yielding to Reason**, which is indeed, a signature of the old academic world.

Academic freedom Students under Bob's guidance explored and continue to explore a wide range of topics. At the same period of time different students could be exploring nuclear physics, computational physics, group theory, nonlinear dynamics, quantum mechanics or Pythagorean relations in pine-cones. As in the old academic world, students came to Bob seeking advice that will help them to develop according to their own judgment of what was worth studying. Academic freedom was not just the privilege of the professor, much less it was negotiated in exchange for research funds, but rather it was shared with the students. In so doing, Bob created an atmosphere of enthusiasm and cooperation under his protective umbrella.

Funding In the mass society (see the meaning given to mass by Ortega y Gasset) such a strong standing behind freedom and true knowledge is certain to pay a price: requests for economic support of the early program were turned down under the argument that Bob was a generalist, as if the wise man should not be supported at least as much as the specialist. Thus, research was supported by Drexel University as Bob's home institution and a number of other institutions and research projects that supported the education of several students. In the early time students support was coming from Argentina, Canada and Italy. Later, the "French connexion" dominated the scene.

Communication I still remember Bob's comment regarding a young scholar whose articles were below Bob's expectations "He is good, but he is too eager to get tenure". One of the sharpest differences between the old scientific tradition and todays administered-science is the re-signifying of the papers. Once upon a time a paper was a scientific communication. For the administered science, it is the accounting unit for the scientific labour. The consequences of such change in meaning for the papers are vast. One of them is that the optimal strategy is to turn off personal quality controls so that the only controls that apply are the community controls of peer review. Perhaps nothing contrasts more sharply with Bob's requisites. For Bob, the main and more demanding control are his own (and those of his co-authors) controls for beauty and meaning. The sense of beauty needs some explanation. The beauty that we seek is very much the consequence of harmony, the consequence of all the facts and all the logic (or mathematics) connecting facts relating naturally without tensions and without loose strands.

Perhaps because of this sense of beauty, Gilmore is found at pleasure when writing books, a deprecated style of communication indeed. Writing a book requires to put together in harmony large scientific pieces, very much as the final examination forced us in college to put all the subject in perspective. Gilmore has written several such treatises, all of them celebrated, such as: *Lie groups and Lie Algebras and some of their Applications*, *Castrophe Theory for Scientists and Engineers*, *The Topology of Chaos: Alice in Stretch and Squeezeland* (with M. Lefranc) and *The Symmetry of Chaos* (with C. Letellier).

Gilmore believes that the old art of book writing is a cultural-hereditary trade. He encouraged his (advanced) students/friends to write books and/or to participate in his writing. To prove this point we shall mention several books by Nicholas Tufillaro (see for example *Experimental Approach to Nonlinear Dynamics*), books by Gabriel Mindlin (ex. *The Physics of Birdsong (Biological and Medical Physics, Biomedical Engineering)*, co-authored by Rodrigo Laje, scientific-grandson of Gilmore) and the collaboration by Mario Natiello and Hernán Solari (*Nonlinear Dynamics: A Two-Way Trip from Physics to Math*, with BG Mindlin and *The User's Approach to Topological Methods in 3D Dynamical Systems*).

Lie groups, the model program The definite model of knowledge for Bob has been Lie groups and Lie algebras. The deep knowledge, enthusiasm and love for the subjected made evident early in his book *Lie groups and Lie algebras and some* *of their applications*, presides over Bob's research and becomes the model research program.

Group theory, the systematic study of symmetries, is a subject dominated by dualities. The abstract group presents itself in manifold "practical" forms depending on the object displaying the symmetry, and this dualism is at the same time useful and beautiful. The abstract forms permanently remind us about the unity of knowledge, and the unity of knowledge calls for the unity of the human being.

Thus, the title of this section intended to reveal, that as much as electrolysis, the incandescence of a carbon filament, the displacement of the needle of a galvanometer and other phenomena reveal electricity that transcends them. The real Bob, the plenitude of Gilmore, is present and revealed in his actions and we can see in them the particularities of the phenomena as well as the abstract expression, Bob itself.

2. From Lasers to Topology: the Context of Discovery

Let me present the program and developments that are the subject matter of this book not from the standard form of the context of justification, but rather in the unusual presentation of the context of discovery. I believe that new generations can profit from this story.

By 1985 when this program started, nonlinear dynamics has recently irrupted into physics. For several years it had been absent, almost banned, as the classical mechanics textbooks presented a version of rational mechanics supported completely in integrable systems, for which several and rather sophisticated methods were presented. Physicists had not been exposed to non-integrable systems for several generations until the works of Edward Lorenz .[6] The work of Henri Poincaré, David Birkhoff, Stephen Smale and the Russian school of mathematics had been carefully neglected. A relatively modest contribution by Mitchell Feigenbaum[7,8] (the right man at the right moment) unveiled the neglect and ignited the interest of the physics community. It also had the effect of directing the studies towards one dimensional maps of the interval or the circle. Such maps where easy to implement in small computers (the personal computer was emerging in those days) and presented a rich catalogue of unexpected behavior. A good deal of this behavior was already known after the work of Nicholas Metropolis, Paul Stein and Myron Stein[9] (MSS), and Oleksandr Šarkovsky.[10] These later works exploited the particularities of surjective functions in a one dimensional space (unimodal or multimodal maps), the topological restriction was indeed the main element in MSS work. The regularities observed in one-dimensional maps prompted the physics community to develop a program known as the "Roads to chaos"[11] aiming to understand the succession of events that lead to chaotic motion. The period doubling cascade, intermittency and four (three in other versions) successive Hopf bifurcations were the indications that supported the conjecture. Despite the counter-example given by Philip Holmes[12] the program continued for several years and became the standard presentation of chaos in physics textbooks. I shall not restrain myself on this point: this story has a lot to say about the relative power of social-consensus and reason in a community organized along social rules as peer-approval: consensus ignorance reigns supreme.

Back to our main story, we became interested in the laser experiment being run at the lab by the end of the corridor (literal meaning). There, Jorge Tredicce was exploring the dynamics of a CO_2 laser with modulated losses, a laser that produced periodic output as well as chaotic dynamics depending on parameters and/or initial conditions.[13,14] The CO_2 laser falls outside the one-dimensional paradigm and was represented with sufficient realism by a two dimensional forced oscillator. The initial task was to survey the periodic orbits present in the model, understand their relation with parameters and their dynamical relations.[15] The particularities of forced oscillators lead us to study topological (algebraic) properties of vector differences such as

$$(x_1, v_1)(t) - (x_2, v_2)(t) \tag{1}$$

where (x_i, v_i) stand for the position and velocity of the nonlinear oscillator (the logarithm of the intensity and the population inversion in the laser system) and the subscript refers to different periodic orbits, the third coordinate is identical for both orbits as it corresponds to $\phi(0) + \omega t$ in both cases. The relation 1 can only be $(0,0)$ at a bifurcation point involving both orbits, a topological obstruction of the kind of those exploited by MSS in their study of one dimensional maps.

The idea then was rapidly developed producing the Relative Rotation Rates, RRR, that began the program I am evoking. But the results, interesting as they where, were not satisfactory for Bob, and for a short time the program was about to die before it was born.

Everything changed when Robin Carr, then a postdoctoral fellow with Bob mostly interested in pithagorean mathematics came from the library with a gift: a collection of papers he though would be relevant to our research. Among this papers we count Joan Birman and Robert Williams works on the Lorenz map[16,17] and Holmes and Williams work on the Smale construction.[18] It then became evident that by using the idea of a knot-holder or template we could compute the RRR for the Smale's horseshoe, and much to our surprise, the numbers came identical to those computed for the laser system. These new elements lifted Bob's demur.[19]

The comparison of the laser system with the Smale horseshoe represented a paradigm shift. Up to that point Smale's horshoses had been associated with high iterations of maps restricted to tiny regions of phase space in the proximity of transverse homoclinic points but now they were useful to compare the orbit organization of all the orbits in all the phase space available. The shift was possible because of the change of focus between enumerating the orbits present, to the description of their relative organization.

The next important steps in the early development of the research program came after the International Workshop on Quantitative Measures of Dynamical Complexity in Nonlinear Systems organized by and held at Bryn Mawr University

(June, 1989). At the meeting the detection and reconstruction methods for periodic orbits were presented by Daniel P. Lathrop and Eric J. Kostelich [20,21] as well as our results on RRR. It was Nicholas Tufilaro, a Ph.D. student at Bryn Mawr and frequent visitor of Robert Gilmore who made the proposition of putting both ideas at work together. At this point the project suddenly enlarged its scope including autonomous systems and even more important, it incorporated a method for dealing with experimental data, but at the (high) price of having to deal with phase space reconstructions and embedology. The main actor of this effort was Gabriel Mindlin, then a Ph.D. student with Robert Gilmore, and the results can be considered the first works putting together the new paradigm.[22,23]

Much of the difficulties of the program came from the difficulties in reconstructing a three dimensional phase space with three coordinates (as opposed to embedded in a higher dimensional space). While the natural candidates where differential embeddings and time-delay embeddings[24,25] it soon became evident that a naive approach to this matter would lead to problems[26] and that standard results[27] were of no help, the difficulty being that the forcing of other periodic orbits, and then of chaos, rest upon the deformations imposed to the complement of the periodic orbits in a disc (see Natiello and Solari article in this book). As such, all the phase space needs to be faithfully represented and not only the attractor that may have a substantially smaller fractal (box counting) dimension, thus standard results assure that the reconstruction is possible in six dimensions, twice as many as we need. A good part of the program was then devoted to overcome this problem.

The program initiated in 1986 spans now twenty five years of research, it has produced tens of communications including a few books. In contrast, community wide programs have lasted for about three to five years, and quite often they have left very little when completed. This relative longevity is a tribute to Gilmore's research style, to the moderately unrealistic goals set for the program, the presence of internal criticism and debate, and above all, the solid grounding of the program in relation to experimental physics. The continuous relation of the "topological characterization program" with laser physics has been a healthful source of realism for it, as the context of discovery shows, the program is a generalization of the tools we found were needed to analyse a particular laser system.

3. Epilogue

The old scientific ideal, seeking knowledge for itself, seeking the unity of all knowledge as well as a consequence and requisite for the unity of the human being corresponds to a duality such as those that we seek in groups and their representations. Identifying Bob Gilmore in itself, requires to recognize the Gilmore at different works. The same Gilmore that was found as teacher, advisor, researcher, mathematician or physicist can be recognized as well in the father, friend and colleague.

The uniqueness of Robert Gilmore in our times arises not only in the trivial sense, that makes us equal as being different, but emerges as a scientist that has carried the lighted torch of illustration through particularly hostile times, constantly paying the costs for his humanist and scientific stance. Thus, R Gilmore can be thought as himself a representation of the old human being fostered by Kant and W. von Humboldt among others.

His torch is lightening the candles of his students, family and colleagues. As for every scientist, it might come a time at which his scientific paths are shown to be not the most appropriated and synthesized with its refutations give birth to a new scientific adventure. This is not more than the cycle of science, unavoidably analogous to the cycle of life, yet the light of reason Gilmore nurtured will continuously shine.

THANKS BOB

References

1. I. Kant, The conflict of the faculties (Der Streit der Fakultäten, 1798), in *Kant: Political Writings*, H. Reiss (Ed), Cambridge University Press, 2nd Ed (1991).
2. I. Kant. *What enlightement was?* Berlinischen Monatsschrift, (December 1783) — See J. Schmidt, What enlightenment was: How Moses Mendelssohn and Immanuel Kant answered the Berlinische Monatsschrift, *Journal of the History of Philosophy*, **30** (1), 77-101 (1992).
3. R. A. Nisbet, *The Degradation of the Academic Dogma : the University in America, 1945-1970*, Heinemann Educational Publishers (1971).
4. V. Bush, *Science the endless frontier*, United States Government Printing Office, Washington (1945), http://www.nsf.gov/od/lpa/nsf50/vbush1945.htm
5. W. von Humboldt, *The Sphere and Duties of Government (the Limits of State Action)* (1854), Translated by J. Coulthard, John Chapman, London (1792), http://oll.libertyfund.org/EBooks/Humboldt_0053.pdf
6. E. Lorenz, Deterministic non-periodic flow, *Journal of Atmospheric Science*, **20**, 130-141 (1963).
7. M. J. Feigenbaum, Quantitative universality for a class of nonlinear transformations, *Journal of Statistical Physics*, **19** (1), 25-52 (1978).
8. M. J. Feigenbaum, The universal metric properties of nonlinear transformations, *Journal of Statistical Physics*, **21** (6), 669-706 (1979).
9. N. Metropolis, M. L. Stein & P. R. Stein, On finite limit sets for transformations on the unit interval, *Journal of Combinatorial Theory A*, **15** (1), 25-44 (1973).
10. A. N. Šarkovskii, Coexistence of cycles of a continuous map of a line into itself, *Ukrainian Mathematical Journal*, **16**, 61-71, (1964).
11. L. P. Kadanoff, Roads to chaos, *Physics Today*, **36**, 46-63 (1983).
12. P. Holmes & D. Whitley, Bifurcations of one- and two-dimensional maps, *Philosophical Transactions of the Royal Society A*, **311**, 43-102 (1984).
13. F. T. Arecchi, R. Meucci, G. Puccioni & J. Tredicce, Experimental evidence of subharmonic bifurcations, multistability, and turbulence in a q-switched gas laser, *Physical Review Letters*, **49**, 1217-1220 (1982).
14. J. R. Tredicce, F. T. Arecchi, G. P. Puccioni, A. Poggi & W. Gadomski, Dynamic behavior and onset of low dimensional chaos in a modulated homogeneously broadened single mode laser: experiments and theory, *Physical Review A*, **34**, 2073-2081 (1986).

15. H. G. Solari, E. Eschenazi, R. Gilmore & J. R. Tredicce, Influence of coexisting attractors on the dynamics of a laser system, *Optics Communications*, **64**, 49-53 (1987).
16. J. S. Birman & R. F. Williams, Knotted periodic orbits in dynamical systems I: Lorenz equations, *Topology*, **22** (1), 47-82 (1983).
17. J. S. Birman & R. F. Williams, Knotted periodic orbits in dynamical systems ii: knot holders for fibered knots, *Contempory Mathematics*, **20**, 1-60 (1983).
18. P. Holmes & R. F. Williams, Knotted periodic orbits in suspensions of Smale's horseshoe: torus knots and bifurcation sequences, *Archives for Rational Mechanics and Analysis*, **90**, 115-194 (1985).
19. H. G. Solari & R. Gilmore, Relative rotation rates for driven dynamical systems, *Physical Review A*, **37**, 3096-3109 (1988).
20. D. P. Lathrop & E. J. Kostelich, Analyzing periodic saddles in experimental strange attractors, In *Measures of Complexity and Chaos*, N. B. Abraham, A. M. Albano, A. Passamante & P. E. Rapp (Eds) p. 147. Plenum, New York (1989).
21. D. P. Lathrop & E. J. Kostelich, Characterization of an experimental strange attractor by periodic orbits, *Physical Reviews A*, **40**, 4028-4031, (1989).
22. G. B. Mindlin, X.-J. Hou, R. Gilmore, H. G. Solari & N. B. Tufillaro, Classification of strange attractors by integers, *Physical Review Letters*, **64**, 2350-2353 (1990).
23. G. B. Mindlin, H. G. Solari, M. Natiello, R. Gilmore & X.-J. Hou, Topological analysis of chaotic time series data from the Belusov-Zhabotinskii reaction, *Journal of Nonlinear Sciences*, **1** 147-173 (1991).
24. Th. Buzug & G. Pfister, Optimal delay time and embedding dimension for delay-time coordinates by analysis of the global static and local dynamical behavior of strange attractors, *Physical Review A*, **45** (10), 7073-7084 (1992).
25. M. B. Kennel, R. Brown, & H. D. I. Abarbanel, Determining embedding dimension for phase-space reconstruction using a geometrical construction, *Physical Review A*, **45**, 3403-3411 (1992).
26. G. B. Mindlin & H. G. Solari, Topologically inequivalent embeddings, *Physical Review E*, **52**, 1497-1502 (1995).
27. T. Sauer, J. A. Yorke & M. Casdagli, Embedology, *Journal of Statistical Physics*, **65** (3-4), 579-616 (1991).

Author Index

Abdenur, Flavio, 83
Abraham, Ralph, 23
Aguirre, Luis A., 235
Alaoui, Aziz, 190
Andronov, Aleksandr, 101, 103, 104, 111, 113
Anosov, Dmitri, 83, 150
Antonsen, Thomas, 273, 275
Arimondo, Ennio, 182
Arnéodo, Alain, 282
Arnold, Vladimir, 318

Bartlett, John, 53
Berezovsky, S. A., 95
Bergé, Pierre, 228
Bestvina, Mladen, 150
Bielawski, Serge, 213
Billings, Stephen, 235
Birkhoff, David, 26, 27, 53, 56, 107, 336
Birman, Joan, 107, 127, 128, 136, 142, 147, 164, 175, 200, 208, 337
Bonhoeffer, Karl, 102
Bowen, Rufus, 83, 88
Boyland, Philip, 220
Bracewell, Ronald, 235
Bush, Vannevar, 333
Byrne, Greg, 237, 238

Carr Robin, 337
Cartwright, Mary-Lucy, 2
Chen, Sheng, 235

Chirikov, Boris, 53
Chua, Leon, 93
Coullet, Pierre, 282
Coven, Ethan, 83
Croquette, Vincent, 228
Cross, Daniel, 199
Crutchfield, James, 2, 197
Culler, Marc, 144
Cvitanovic, Pedrag, 228, 304

de Argensola, Lupercio Leonardo, 334
de Broglie, Louis, 228
Dehornoy, Pierre, 143
Diamond, Patrick, 74
Díaz, Lorenzo, 83
Doria, Francisco, 120
Dubois, Monique, 228
Dunfield, Nathan, 144
Dutertre, Pascal, 231

Eckhard, Bruno, 229
Eilenberg, Samuel, 26
Einstein, Albert, 27
Elhadj, Zeraoulia, 67
Euler, Leonhard, 4, 188

Farmer, Doyne, 2, 197
Fatou, Pierre, 41–43
Feigenbaum, Mitchell, 70, 169, 336
Fiol, Clarisse, 85
Ford, Joe, 53
Forster, Georg, 332
Fourier, Joseph, 56

Franks, John, 142, 150, 155, 159, 162

Galias, Zbigniew, 90
Galileo, Galilei, 234
Gauss, Carl-Friedrich, 4, 128, 200
Ghrist, Robert, 147
Ghys, Etienne, 135, 143, 144, 146
Gilmore, Claire, 303
Gilmore, Keith, 317
Gilmore, Robert, 1, 4, 117, 121, 169, 208, 211, 225, 227, 229, 231, 239, 244, 266, 269, 277, 317, 332, 338
Giraud, Alain, 44, 50–52
Glass, Leon, 3
Golubitsky, Martin, 270
Gouesbet, Gérard, 228
Grévy, Auguste, 42
Grassberger, Peter, 304
Grebogi Celso, 2
Guckenheimer, John, 2, 48, 87, 104, 109, 111, 116, 135
Gumowski, Igor, 39, 40, 43, 47, 52, 53, 55

Hadamard, Jacques, 42
Hale, George, 235
Handel, Michael, 150
Harriott, Thomas, 234
Hartmann, Georg, 121
Hayashi, Chihiro, 40
Heidel, Jack, 283
Hénon, Michel, 53, 77, 81, 85, 87
Hoffmann, Dietrich, 102
Holmes, Philip, 336
Hoppensteadt, Frank, 104
Hyman, James, 104

Ipaktchi, A., 104, 109, 111

Jablon, Slavik, 142
Jones, Timothy, 196
Jones, Vaughan, 128

Jonker, Leo, 48
Julia, Gaston, 41–43

Kan, Ittai, 83
Kant, Immanuel, 332, 338
Kawakami, Horishi, 40, 55
Khaikin, Sëmen, 101, 102, 104
King, Greg, 228
Kiriki, Shin, 84, 90
Klein, Felix, 240
Koenigs, Dénes, 42
Kofman, Ilya, 139
Kolmogorov, Andrei, 318
Kostelich, Eric, 179, 337
Kovalevsky, Sophie, 29, 30
Kuczma, Marek, 53
Kuhn, Hans, 100
Kuramoto, Yoshiki, 271, 273, 275, 276

Lémeray, Ernest, 42
Laje, Rodrigo, 335
Lanford, Oscar, 70, 72
LaSalle, Joseph, 40
Lathrop, Daniel, 179, 337
Lattès, Samuel, 41, 42
Leau, L., 42
Lefranc, Marc, 205, 208, 211, 213, 335
Lefschetz, Solomon, 27, 28, 40
Leibnitz, Gottfried, 4
Letellier, Christophe, 1, 35, 91, 99, 192, 227, 335
Letterer Erich, 100
Levinson, Norman, 2
Li, Dequan, 243
Li, Tien-Yien, 48, 104, 109, 111
Littlewood, John, 2
Lopez-Ruiz, Ricardo, 183, 230
Lorenz, Edward, 2, 6, 63–65, 79, 85, 104, 106, 111, 133, 134, 170, 175, 229, 230, 281, 317, 318, 336
Lorenz, Konrad, 100
Lozi, René, 3, 63, 80, 85, 110

Lü, Jinhu, 241
Lyapunov, Alexsandr, 30

Maas, Leo, 297
Malasoma, Jean-Marc, 281, 290, 297
Martin, Juan Carlos, 186
Matsumoto, Takeshi, 93, 94
Maxwell, James, 4
May, Robert, 39, 54, 69, 70, 104, 109, 111
Mayer-Kress, Gottfried, 112
Melnikov, V., 53
Messager, Valérie, 91
Metropolis, Nicholas, 336
Mindlin, Gabriel, 3, 4, 174, 183, 209, 216, 230, 269, 335, 338
Mininni, Pablo, 235
Minorsky, Nicolai, 27
Mira, Christian, 25, 39
Miranda, Rick, 230–232
Misiurewicz, Michal, 81, 150, 155, 159, 162
Mittag-Leffler, Gösta, 30
Mittelstädt, Horst, 100
Montel, Paul, 43
Morse, Marston, 56
Myrberg, Pekka, 48, 49

Natiello, Mario, 3, 149, 335, 338
Nëımark, Ju, 46
Nisbet, Robert, 333

Oden, Jérémy, 212, 213
Ortega y Gasset, José, 332, 334
Ortoleva, Peter, 104, 119
Oster, George, 104, 109, 111
Ott, Edward, 2, 273, 275

Packard, Norman, 2, 197
Palmer, Ken, 83
Papoff, Francesco, 182, 183

Peixoto, Mauricio, 27
Pelikan, Steve, 2
Phragmén, Edvard, 25
Pilyugin, Serguei, 83
Pinsky, Tali, 146, 147
Poincaré, Henri, 3, 7, 24–27, 30, 53, 56, 107, 121, 171, 188, 336
Pomeau, Yves, 79
Procaccia, Itamar, 304
Pulkin, C., 48

Rand, David, 48
Rashevsky, Nicholas, 101
Rosen, Robert, 100
Rössler, Otto E., 2, 6, 9, 10, 91, 92, 99, 114, 171, 175, 229, 230, 282
Ruelle, David, 1, 77, 88, 107, 109, 111

Sakurai, Akio, 84
Šarkovsky, Oleksandr, 336
Sartre, Jean-Paul, 332
Schäfer, Fritz-Peter, 100
Scheiner, Christoph, 234
Schwabe, Heinrich, 234
Sciamarella, Denisse, 216
Seelig, Friedrich, 92, 100
Sharkovskji, Aleksandr, 48, 95
Shaw, Chris, 26
Shaw, Robert, 2, 198, 231
Shea, William, 234
Sinai, Yakov, 88
Smale, Stephen, 28, 55, 111, 336
Smith, Paul, 26, 27
Solari, Hernán, 3, 149, 171, 174, 209, 230, 331, 335, 338
Soma, Teruhiko, 84, 90
Sprott, Clinton, 67, 282, 283
Stein, Myron, 336
Stein, Paul, 336
Stewart, Ian, 228
Stone, Emily, 230–232
Swinney, Harry, 2, 179

Tait, Peter Guthrie, 4, 128
Takens, Floris, 1, 2, 77, 109, 111, 197
Thom, René, 28
Thurston, William, 143, 144, 150
Tredicce, Jorge, 170, 171, 269, 336
Tresser, Charles, 282
Tsankov, Tsvetelin, 229, 231, 237, 238, 244
Tucker, Warwick, 86, 88, 92, 135
Tufillaro, Nicholas, 230, 251, 335, 337
Turing, Alan, 101

Ueda Yoshisuke, 40
Ulam, Stan, 70
Urabe, M., 53
Used, Javier, 186

Valentin, Luc, 228
van der Schrier, Gerard, 297
van Helden, Albert, 234
Vandermonde, Alexandre-Téophile, 4
Verhulst Pierre-François, 69
Vitt, Aleksandr, 101, 104

von Humboldt, Wilhelm, 332, 333, 338
von Neumann, John, 70
von Weizäcker, Carl-Friedrich, 100

Waible, Reimara, 100
Weierstrass, Karl, 29
Whitney, Hassler, 198
Wiener, Norbert, 27
Williams, Robert, 6, 87, 107, 116, 135, 136, 138, 142, 147, 164, 175, 200, 208, 230, 337
Winfree, Art, 91, 103, 109, 273
Wolf, Rudolf, 234

Yamaguti, Masaya, 95
Yorke, James, 2, 48, 75, 83, 104, 109, 111, 121
Young, Lai-Sang, 84
Yuan, Guo-cheng, 75, 83

Zeeman, Erich, 105
Zhang, Fu, 283
Zygliczyński, Piotr, 90, 92

Subject Index

adjacency matrix, 220
adjoined cycles, 55
alphabet, 161
analysis situs, 111
animal gaiting, 270
Arnold tongue, 318
attractor, 134
    Belousov-Zhabotinskii, 172
    Burke and Shaw, 232, 238, 239, 241
    chaotic, 45, 181
    double scroll, 121
    Duffing, 177, 190
    flare, 120
    hyperchaotic, 112, 210
    image, 193, 234, 244
    Kremliovsky, 234
    Lorenz, 86, 134, 176, 177, 182, 183, 192, 193, 230, 318
    Lorenz-like, 237
    Milnor, 120
    Rössler, 176, 177, 190, 192
    Shimizu-Morioka, 182, 183
    strange, 45, 206
    superfat, 100
    toroidal chaotic, 244
    van der Pol, 177, 190, 196
autoregressive model, 312

basin boundary, 43
Belousov-Zhabotinskii reaction, 179
bifurcation
    diagram, 286, 324–326
    Hopf, 336

    Nĕımark, 46
    peeling, 234, 235, 244
    period-doubling, 3, 318, 325–328
    saddle-node, 273
birdsong, 270
Birman-Williams projection, 178
blender, 117, 229
Bogus transition, 155
boîtes-emboîtées, 48
boundary crisis, 49
bounding tori, 188, 189, 191, 244, 245
bounding torus, 8
Bracewell trick, 235
braid, 152, 184, 208, 215
    generator, 218
    group, 150
    index, 141, 142
    Lorenz, 136
    permutation, 139
    spectral, 259, 261
    T, 140
    type, 152
brain, 270
branch line, 176
branched
    manifold, 107, 108, 116, 175, 176, 180, 191, 208, 245
    surface, 135
business cycle, 303
butterfly, 210
buzz saw, 187

canaries, 272

Cantor set, 2, 77
cardiac arrhythmia, 103
catfish, 186
caustic, 214
CDOM, 252
chaos, 109, 321, 326
   sandwich, 100, 119
   screw, 100
   spiral, 100
   symmetric toroidal, 243
   unimodal cut, 104
chemical waves, 103
circadian rhythm, 103
circle diagram, 152
circuit
   universal, 102, 104, 106, 112, 120
close returns, 5, 215
   events, 207
   method, 304
   plot, 305, 306, 308–311
   technique, 9
   test, 305, 306, 308, 312
coherent fluctuations, 275
collapsing, 74, 156
   effect, 67
   step, 156, 159
communication cells, 55
complexity, 157
confidence bands, 236
connectedness, 241
connection chains, 55
connectivity matrix, 274
correlation dimension, 304, 311
cover, 230
   n-fold, 239
   two-fold, 232
   V_4 symmetric, 241
covering group, 192
critical
   curve, 43
   point, 43
   set, 43

crossing, 155, 162
   number, 142
curvature, 262
cusp, 210, 212, 225
   singularity, 210–212, 224

derivative spectroscopy, 258, 259
dogfish, 186
dynamical
   automata, 103
   estimates, 181
   measure, 170
   regime, 273
dynamics, 214

Eccles-Jordan trigger, 101, 103
economics, 303
Edelstein switch, 92, 112
edge detection, 263
embedded boxes, 48, 55
embedding, 179, 197
   derivative, 259
   differential, 198, 229, 338
   dimension, 235
   integral-differential, 179
   time delay, 198, 206, 235, 338
energy integral, 196
entropy, 219
equity market, 304
equivariant, 192
error estimates, 170
Euler-Poincaré formula, 243
exchange rate, 311
eye-candy, 196

fat
   representative, 150, 155, 167
   vertex, 155
feedback, 49
Feigenbaum's constant, 70
ferromagnetism, 319
field-effect transistor, 103

finance, 303
financial
   market, 314
   time series, 304
fingerprints, 174, 208
flaring, 120
flatten-and-fold, 120
flip-flop, 101
fold, 155, 186, 210
   double, 193
   line, 212
folded pancake, 116
folding, 120, 186, 187, 239
forcing, 152, 185
foreign exchange, 311
fractal, 43, 44, 187
   dimension, 170, 181
fractalization, 44
frame
   Frenet, 262
framing, 261
Frenet-Serret equations, 262
front identification, 260
full shift, 162, 183
fundamental domain, 228, 230

Galois theory, 70
gâteau roulé, 185
Gauss
   integral, 172
   invariant, 130
   linking number, 4, 172, 200, 263
Geiger counter, 181
genus, 142, 188, 189, 243
geodesic, 144
geometric model, 87
geometrization conjecture, 143
germ, 43
gigaperiodic, 77
global modeling, 228, 235
global Poincaré surface, 189
global torsion, 152, 174, 199

great dying, 252
group representations, 199
growth rate, 220

harmonic driving, 275
HICO, 254
homoclinic
   point, 337
   tangle, 25
homotopy group, 189
horseshoe dynamics, 15, 185, 337
hyperbolic, 177
hyperspectral, 265, 266

intermittency, 49
international trade, 303
intinerary space, 218
invariant
   measure, 70
   torus, 318
   triangulation, 222
irregular reversals, 235
isopleths, 107, 108, 230
isotopy, 10, 130, 175

jelly roll scroll, 185
jerk equation, 282, 283
joint density function, 214
Jones polynomial, 128, 131, 132
Julia set, 43, 55

kinematics, 214
Klein bottle, 209
knot, 214
   carrier, 198
   diagram, 129
   fibered, 144
   generalized index, 192
   harmonic, 198
   holder, 107, 150, 208, 337
   Lorenz, 133, 135, 139
   oriented type, 191

pretzel, 138
prime, 129
theory, 128, 207
torus, 131, 135
trefoil, 138
type, 199
Kolmogorov-Smirnov test, 308

L symbol, 162
Landau-Lifshitz-Slonczewski
  equations, 318, 319, 321, 328
laser, 269
  fiber optic, 185, 186
  modulated CO_2, 206
  with modulated losses, 170, 337
  with saturable obsorbers, 182
leave, 162
level of structure, 191
Li-Yorke box, 115
Lie
  algebra, 197
  group, 197
lift, 192
limit cycle, 25
line diagram, 152–154, 167
link, 214
  invariant, 130, 208
linking number, 130, 179, 260
linking numbers, 10
longitude, 189, 191
Lorenz
  braid template, 136
  mechanism, 186
Lyapunov exponent, 170, 175, 304

Möbius loop, 118
magnetic
  field, 234, 235
  field lines, 178
magnetization, 318
Malthusian growth, 69
map

Arnold's, 111
cap-shaped, 111
circle, 278, 318, 319, 321, 322, 328
coupled logistic, 211
first-return, 107, 108, 232, 245, 322, 328
folded-towel, 100
Hénon, 64, 76, 77, 81, 82, 151, 308
horseshoe, 55, 111, 116, 151, 174
induced return, 222
logistic, 69, 71, 72, 308
Lorenz, 229, 337
Lozi, 64, 80–82, 84
modified Hénon, 67
monodromy, 144
multimodal, 239, 336
Poincaré, 66, 89, 111, 115, 150
pseudo-Anosov, 152
return, 179
sandwich, 118
sphere, 319, 322
stroboscopic, 213
tent, 73
unimodal, 3, 9, 118, 222, 286, 336
walking-stick, 100
Markov
  matrix, 154
  partition, 152, 159
mechanism, 186
megaperiodic, 77
merging attractor crisis, 241
meridian, 189, 191
MERIS, 253
method
  analytical, 40
  Cigala's, 46
  Krylov-Bogoliubov-Mitropolski, 41
  qualitative, 40
metric measure, 170
Misiurewicz domain, 82
mixing transformation, 117
mod out, 195

mode locking, 321, 322, 324, 325, 328
modular surface, 145
motion
   fast, 105
   slow, 105
multichannel analyzer, 179
multiply connected, 44
multiply connected basin, 44
multivibrator, 99, 104
   chemical, 101

NARMAX, 235
   model, 237
Navier-Stokes equations, 133, 317
neuron
   excitatory, 273
   inhibitory, 273
nonconnected, 44
normal loop, 117

obstruction, 199
ocean color, 251
orbit
   basis set, 184, 191
   forcing, 151, 160
   image, 192
   periodic, 171
   pruning, 151, 178
   trimming, 151, 162
order parameter, 275
oscillator
   Bonhoeffer, 102
   coupled, 271
   Duffing, 170, 185, 195, 321
   forced, 337
   neural, 273
   nonstationary, 221
   phase, 274
   spin-torque, 328
   van der Pol, 170, 195, 321

paddlefish, 186

paper sheet model, 229
parallel trajectories, 207
parity, 199
partition
   generating, 222
perestroika, 187
periodic orbits, 9
periodic window, 286
phase portrait, 24, 228, 229
phase space, 106, 251, 321, 328
plankton, 251
playdough, 120
Poincaré period, 152
Poincaré section, 3, 8, 108, 150, 173, 209, 229, 244, 286, 319, 322
   multi-component, 229
   stroboscopic, 213, 214
point group, 194
power spectrum, 228, 229
pseudo-Anosov, 152, 167
   diffeormorphism, 150
pseudo-periodic, 50

quadratic box, 115
quartic cover, 231
quasiperiodic, 321, 322, 324, 326–328

Rademacher function, 145
radial basis function, 258
rainbow, 214
recurrence property, 314
recurrent behavior, 207
red edge, 257
rejection
   criterion, 181
   step, 181
relative rotation rate, 173, 337
repère mobile, 198
representation
   inequivalent, 199
   label, 199
repulsive cycles, 43

residual ripple, 49
resonator
    coupled diode, 211, 225
    diode, 213
Rössler mechanism, 186
rotation sequence, 48

S-shaped
    structure, 186
    surface, 112
saddle-focus, 290
Savizky-Golay filter, 259
scaling ratios, 169
scenario
    van der Pol, 174
    Duffing oscillator, 174
    Smale horseshoe, 174
Schoenflies conventions, 296
school
    Andronov, 41
    Hayashi, 40
Schrödinger equation, 100
sector, 155
self-relative rotation rate, 174
semiflow, 116, 135, 209
sensitivity to initial conditions, 107, 134, 146
shadowing, 76, 83
    orbit-shifted, 84
    parameter-shifted, 83, 89
    property, 83
Sharkovskij's cycles ordering, 48
simplicial space, 223
simply connected, 44
singular value decomposition, 258
singularity, 210
    line, 212
    structure, 211
skeleton, 171
small parameter, 40
solar
    11 year cycle, 195
    22 year cycle, 195
space group, 195
splitting point, 176
squeezing, 186, 208, 220, 223
SRB measure, 88
stationarity test, 312
stochastic instability, 47
strange repeller, 45
stretching, 120, 139, 186, 187, 208, 223
stroboscopic measurement, 328
strongly attracting, 178
structural stability, 27
subharmonic cascade, 286, 290, 291
substitution operator, 218
sunspot, 193–195, 234
    group, 234
surrogate, 171
suspension, 112, 174
swallowtail, 210
switch curve, 51
symbolic
    dynamics, 138
    name, 223
    phase plane, 55
symmetry
    group V_4, 240
    inversion, 233
    rotation, 231, 325
    S_6, 242
synchronization, 270
system
    aperiodic, 318
    Chua, 63, 64, 85, 94, 245
    Duffing, 175
    equivariant, 296, 299
    four-dimensional, 225
    hybrid, 52
    hyperchaotic, 213, 225
    image, 239
    Kremliovsky, 233
    Li, 244, 245

Lorenz, 63, 64, 79, 89, 95, 138, 175, 228, 282, 296, 297, 321
Malasoma minimal, 298
minimal, 283, 290
modified Rössler, 241
Nosé-Hoover, 296
proto-Lorenz, 230, 231
Rössler, 63, 64, 91, 92, 175, 228, 233, 282, 305
Sprott A, 297
Sprott B, 297
Sprott C, 297
Sprott D, 8, 12
Thomas, 243
van der Pol, 175, 322

tearing, 186, 239
template, 107, 134, 162, 208, 209, 214, 230–232, 237, 337
   horseshoe, 208
   Lorenz, 135, 136
   maximal, 167
   universal, 147
theorem
   Birman-Williams, 175, 178
   Călugăreanu, 263
   Cartan's, 192, 196
   Li-Yorke, 3
   Liouville's, 215, 216
   Poincaré-Bendixson, 281, 285, 321
   shadowing, 72
   Thurston's, 152
   uniqueness, 206
theory
   group, 196
   kneading, 181, 292
   representation, 199
thyristor, 49
topological
   analysis, 111, 224
   analysis program, 181
   approach, 305

entropy, 150, 152, 174, 184, 194, 221, 225
   index, 197
   indicator, 264
   invariant, 172, 208
   organization, 214
torsion, 176, 262
   integral, 196
torus, 189
Toulouse, 39
train track algorithm, 221, 223
transverse stability, 238, 239
tree, 154
   fat, 154
triangulation, 225
trimming, 152, 155, 162
trinion, 189, 190
Turing
   cell, 112
   oscillator, 92
twist, 263
twisting, 120

umbilic, 210
unimodal order, 232
universal image, 197
universality, 169
unstable centers, 47
unstable dimension variability, 210
unstable manifold, 208, 209, 222

vocalize, 271

wave-corpuscule duality, 228
web of life, 252
winding
   angle, 263
   number, 264
Wolf index, 234
writhe, 263
   polar, 263